085137

THE LIBRARY
ST. MARY'S COLLEGE OF MARYLAND
ST. MARY'S CITY, MARYLAND 20686

Benchmark Papers in Animal Behavior

Series Editor: Martin W. Schein
West Virginia University

PUBLISHED VOLUMES

HORMONES AND SEXUAL BEHAVIOR / Carol Sue Carter
TERRITORY / Allen W. Stokes
SOCIAL HIERARCHY AND DOMINANCE / Martin W. Schein
EXTERNAL CONSTRUCTION BY ANIMALS / Nicholas E. Collias and Elsie C. Collias
PSYCHOPHYSIOLOGY / Stephen W. Porges and M. G. H. Coles
SOUND RECEPTION IN FISHES / William N. Tavolga
SOUND PRODUCTION IN FISHES / William N. Tavolga
IMPRINTING / E. H. Hess and Slobodan B. Petrovich
VERTEBRATE SOCIAL ORGANIZATION / Edwin M. Banks

RELATED TITLES IN OTHER BENCHMARK SERIES

BEHAVIOR AS AN ECOLOGICAL FACTOR / D. E. Davis
CYCLES OF ESSENTIAL ELEMENTS / L. R. Pomeroy
NICHE: Theory and Application / R. H. Whittaker and S. A. Levin
ECOLOGICAL ENERGETICS / R. G. Wiegert
ECOLOGICAL SUCCESSION / Frank B. Golley
UNDERWATER SOUND / V. M. Albers
ACOUSTICS: Historical and Philosophical Development / R. B. Lindsay
PHYSICAL ACOUSTICS / R. B. Lindsay

Benchmark Papers
in Animal Behavior 9

A BENCHMARK® Books Series

SOUND PRODUCTION IN FISHES

Edited by

WILLIAM N. TAVOLGA
City University of New York
and
American Museum of Natural History

Dowden, Hutchinson & Ross, Inc.
STROUDSBURG, PENNSYLVANIA

Copyright © 1977 by **Dowden, Hutchinson & Ross, Inc.**
Benchmark Papers in Animal Behavior, Volume 9
Library of Congress Catalog Card Number: 76-28352
ISBN: 0-87933-261-1

All rights reserved. No part of this book covered by the copyrights hereon may be reproduced or transmitted in any form or by any means—graphic, electronic, or mechanical, including photocopying, recording, taping, or information storage and retrieval systems—without written permission of the publisher.

79 78 77 1 2 3 4 5
Manufactured in the United States of America.

Library of Congress Cataloging in Publication Data

Main entry under title:
Sound production in fishes.
   (Benchmark papers in animal behavior 9)
   Includes index.
   1. Fishes—Behavior—Addresses. essays, lectures.
2. Sound production by animals—Addresses, essays, lectures. 3. Air-bladder (in fishes)—Addresses, essays, lectures. I. Tavolga, William N., 1922–
QL639.3.S68   597'.05'908   76-28352
ISBN 0-87933-261-1

AUG 21 1978

Exclusive Distributor: **Halsted Press**
A Division of John Wiley & Sons, Inc.
ISBN: 0-470-99084-8

SERIES EDITOR's FOREWORD

Not many years ago virtually all research publications dealing with animal behavior could be housed within the covers of a few hardbound volumes that were easily accessible to the few workers in the field. Times have changed. Present-day students of animal behavior have all they can do to keep abreast of developments within their own area of special interest, let alone in the field as a whole.

It was even fewer years ago that those who taught animal behavior courses could easily choose a suitable textbook from among the few available; all "covered" the field, according to the bias of the author. Students working on a special project used *the* text and *the* journal as reference sources, and for the most part successfully covered their assigned topics. Times have indeed changed. Today's teacher of animal behavior is confronted with a bewildering array of books to choose among, some purporting to be all-encompassing, others confessing to strictly delimited coverage, and still others professing to be collections of recent and important writings.

In response to the problem of the steadily increasing and overwhelming volume of information in the area, the Benchmark Papers in Animal Behavior was launched as a series of single topic volumes designed to be some things to some people. Each volume contains a collection of what an expert considers to be the significant research papers in a given topic area. Each volume serves several purposes. For teachers, a Benchmark volume serves as a supplement to other written materials assigned to students; it permits in-depth consideration of a particular topic while confronting students (often for the first time) with original research papers of outstanding quality. For researchers, a Benchmark volume saves countless hours of digging through the various journals to find the basic articles in their area of interest; often the journals are not easily available. For students, a Benchmark volume provides a readily accessible set of original papers on the topic in question, a set that forms the core of the more extensive bibliography that they are likely to compile; it also permits them to see at first hand what an "expert" thinks is important in the area and to react accordingly. Finally, for librarians, a Benchmark volume represents a collection of important papers from many diverse sources that makes readily available materials that

Series Editor's Foreword

might otherwise not be economically possible to obtain or physically possible to keep in stock.

The choice of topics to be covered in this series is no small matter. Each of us could come up with a long list of possible topics and then search for potential volume editors. Alternatively, we could draw up long lists of recognized and prominent scholars and try to persuade them to do a volume on a topic of their choice. For the most part, I have followed a mix of both approaches: match a distinguished researcher with a desired topic, and the results should be outstanding. And so it is with the present volume.

Dr. Tavolga was one of the early workers in the area of acoustics and behavior in fishes and is still actively engaged in such studies. His extensive research and publications over the years have earned for him an international reputation in the field of behavior. Few persons would have been as qualified to undertake a Benchmark project on acoustics in fishes. He has wisely elected to cover this topic in two volumes: the present one on sound production and a companion volume on sound reception in fish. Taken singly or together, these volumes admirably reflect the basic philosophy underlying the Benchmark Series, that of tracing the development of ideas through a confrontation with the original literature.

MARTIN W. SCHEIN

PREFACE

The study of sound production in fishes, a substantial segment of the broader field of marine bioacoustics, can serve as an excellent example of the link between human affairs and basic scientific research. This study has served as a "bridge between 'useless' research and applied information,"* and in making this keynote statement, Dr. S. R. Galler noted how the study of underwater sound brought together such diverse specialists as marine biologists, acoustic physicists, electronic engineers, and even psychologists. In another aspect of this link, the naval battles of World War II were directly responsible for the technological developments that made much of marine biology possible today, especially the part dealing with underwater sound.

Accordingly, the history of this field shows only slow gains and few reports up to about 1939, but an exponential rise in productivity after 1946. The selection of papers for this volume reflects this abrupt change. Part I consists of a review of the state of the art. Part II contains a few examples of the major researches on sonic fishes prior to 1940. Part III demonstrates the quantum jump in the field after World War II, following the release of wartime technological advances. Part IV exemplifies some of the significant work on sonic mechanisms in fishes, and Part V shows some of the early attempts to correlate sonic output with behavior.

I am indebted to Dr. Martin W. Schein for initially suggesting such a compilation of benchmark papers, and to my colleagues for their advice on what to include. The facilities of the Library and the Photography department at the American Museum of Natural History were essential to this project.

This volume is the second in the series on acoustics and behavior in fishes, the first being devoted to the capacities of fishes to detect sound. (*Sound Reception in Fishes,* William N. Tavolga, ed., Dowden, Hutchinson & Ross, Inc., Stroudsburg, 1976.)

WILLIAM N. TAVOLGA

*From *Marine Bio-Acoustics,* vol. 2, Pergamon Press: Oxford.

CONTENTS

Series Editor's Foreword	v
Preface	vii
Contents by Author	xiii

PART I: REVIEW AND RECENT ADVANCES

Editor's Comments on Papers 1 and 2   2

1   TAVOLGA, W. N.: Sound Production and Detection   3
  Fish Physiology, Vol. 5 (W. S. Hoar and D. J. Randall, eds.), Academic Press, 1971, pp. 135–162, 183–191, 192–205

2   TAVOLGA, W. N.: Recent Advances in the Study of Sound Production in Fishes   47
  Original article prepared especially for this volume

PART II: HISTORY AND EARLY DESCRIPTIONS

Editor's Comments on Papers 3 Through 7   56

3   SÖRENSEN, W.: Are the Extrinsic Muscles of the Air-bladder in Some Siluroidae and the "Elastic Spring" Apparatus of Others Subordinate to the Voluntary Production of Sound? What is, According to Our Present Knowledge, the Function of the Weberian Ossicles?   59
  J. Anat. Physiol., **29,** 109–139, 205–229 (1894–1895)

4   BRIDGE, T. W., and A. C. HADDON: Note on the Production of Sounds by the Air-bladder of Certain Siluroid Fishes   115
  Proc. Roy. Soc. London, **55,** 439–441 (1894)

5   TOWER, R. W.: The Production of Sound in the Drumfishes, the Sea-Robin and the Toadfish   118
  Ann. N. Y. Acad. Sci., **18,** 149–180 (1908)

6   BURKENROAD, M. D.: Sound Production in the Haemulidae   149
  Copeia, No. 1, 17–18 (1930)

7   BURKENROAD, M. D.: Notes on the Sound-Producing Marine Fishes of Louisiana   151
  Copeia, No. 1, 20–28 (1931)

Contents

PART III: THE POST-WAR PERIOD

Editor's Comments on Papers 8 Through 11 — 162

8 DOBRIN, M. B.: Measurements of Underwater Noise Produced by Marine Life — 164
   Science, **105**, 19–23 (1947)

9 KNUDSEN, V. O., R. S. ALFORD, and J. W. EMLING: Underwater Ambient Noise — 168
   J. Mar. Res., **7**(3), 410–429 (1948)

10 FISH, M. P., A. S. KELSEY, JR., and W. M. MOWBRAY: Studies on the Production of Underwater Sound by North Atlantic Coastal Fishes — 188
   J. Mar. Res., **11**(2), 180–193 (1952)

11 MARSHALL, N. B.: The Biology of Sound-producing Fishes — 202
   Symp. Zool. Soc. London, No. 7, 45–60 (1962)

PART IV: MECHANISMS OF SWIM BLADDER SOUND PRODUCTION

Editor's Comments on Papers 12 Through 16 — 220

12 SKOGLUND, C. R.: Neuromuscular Mechanisms of Sound Production in *Opsanus tau* — 222
   Bio. Bull., **117**, 438 (1959)

13 PACKARD, A.: Electrophysiological Observations on a Sound-producing Fish — 223
   Nature, **187**(4731), 53–54 (1960)

14 TAVOLGA, W. N.: Mechanisms of Sound Production in the Ariid Catfishes *Galeichthys* and *Bagre* — 225
   Bull. Amer. Mus. Nat. Hist., **124**, 3, 5–30 (1962)

15 HARRIS, G. G.: Considerations on the Physics of Sound Production by Fishes — 280
   Marine Bio-Acoustics (W. N. Tavolga, ed.), Pergamon Press 1964, pp. 233–247

PART V: BIOLOGY AND COMMUNICATION

Editor's Comments on Papers 16 Through 19 — 296

16 TAVOLGA, W. N.: The Significance of Underwater Sounds Produced by Males of the Gobiid Fish, *Bathygobius soporator* — 298
   Physiol. Zool. **31**, 259–271 (1958)

17 GRAY, G.-A., and H. E. WINN: Reproductive Ecology and Sound Production of the Toadfish, *Opsanus tau* — 311
   Ecology, **42**, 274–282 (1961)

18 WINN, H. E., J. A. MARSHALL, and B. HAZLETT: Behavior, Diel Activities, and Stimuli That Elicit Sound Production and Reactions to Sounds in the Longspine Squirrelfish — 320
   Copeia, No. 2, 413–425 (1964)

19 WINN, H. E.: The Biological Significance of Fish Sounds *Marine Bio-Acoustics* (W. N. Tavolga, ed.), Pergamon Press 1964, pp. 213–230	333
Author Citation Index	**351**
Subject Index	**355**
About the Editor	**365**

CONTENTS BY AUTHOR

Alford, R. S., 168
Bridge, T. W., 115
Burkenroad, M. D., 149, 151
Dobrin, M. B., 164
Emling, J. W., 168
Fish, M. P., 188
Gray, G.-A., 311
Haddon, A. C., 115
Harris, G. G., 280
Hazlett, B., 320
Kelsey, A. S., Jr., 188

Knudsen, V. O., 168
Marshall, J. A., 320
Marshall, N. B., 202
Mowbray, W. M., 188
Packard, A., 223
Skoglund, C. R., 222
Sörensen, W., 59
Tavolga, W. N., 3, 47, 225, 298
Tower, R. W., 118
Winn, H. E., 311, 320, 333

SOUND PRODUCTION
IN FISHES

Part I

REVIEW AND RECENT ADVANCES

Editor's Comments
on Papers 1 and 2

1   **TAVOLGA**
    Excerpts from *Sound Production and Detection*

2   **TAVOLGA**
    Recent Advances in the Study of Sound Production in Fishes

Reprinted here in part is a review that appeared in 1971 in Volume 5 of *Fish Physiology* (edited by W. S. Hoar and D. J. Randall), through the courtesy of Academic Press, Inc. The second article is a review, written especially for this volume, of significant research in the field during the period 1971 through early 1976.

1

Copyright © 1971 by Academic Press, Inc.

Reprinted from pp. 135–162, 183–191, 192–204 of *Fish Physiology*, W. S. Hoar and D. J. Randall, eds., Academic Press, Inc., New York, 1971, pp. 135–205.

SOUND PRODUCTION AND DETECTION

William N. Tavolga

I. INTRODUCTION

The field of aquatic bioacoustics has grown rapidly in many directions, involving many allied areas of research. Major recent reviews of the subject include those by Moulton (1963), Protasov (1965), and Tavolga (1965). Three international symposia on aquatic bioacoustics have been held and their proceedings published (Cahn, 1967; Tavolga, 1964a, 1967a).

Sound is probably the most effective channel for long-range communication under water, and it has become clear over the past 20 years that many fishes utilize this channel. The mechanisms of sound production and the sounds themselves have formed an active area of research, aided by recent technical developments in underwater acoustics. Although sound production may be restricted to some as yet unknown fraction of all fish species, it is apparent that all fishes must be capable of receiving acoustic stimuli. Sound detection in an aquatic medium presents certain problems not normally encountered by terrestrial organisms, e.g., the separation of pressure from displacement detection seems to be characteristic of aquatic animals.

II. SOUND PRODUCTION

A. Historical Background

Prior to the 1940's published reports appeared sporadically on sounds produced by fishes and other aquatic animals. Aristotle and Pliny (cited by Moulton, 1963) gave several examples, but all such reports up through the nineteenth century were based upon what the unaided human ear could detect. Some primitive techniques used by fishermen possibly since prehistoric times have included pipes, oars, and other objects that would transmit underwater sounds to the ear. Moulton (1963) reviewed several instances of fishermen who rely on detecting the presence of desirable fish by listening for them.

An important nineteenth century contribution was the report of Dufossé (1874), in which many instances of sound production by fish were described, including several marine species. In addition to such essentially descriptive studies (Geoffroy St. Hilaire, 1829; Smith, 1927; von Ihering, 1930; Dijkgraaf, 1932; and others cited by Moulton, 1963, and Tavolga, 1965), there was an early interest in the mechanisms by which fish produced sounds. Agassiz (1850) suggested the swim bladder as a sonic organ, and Moreau (1876) made certain deductions as to the sonic function of the swim bladder in the sea robin, *Trigla,* based upon anatomical studies. An extensive anatomical and experimental investigation of swim bladder sonic mechanisms was reported by Tower (1908), in which he was probably the first to suggest that the vibration frequency of the bladder was equivalent to the fundamental frequency of the emitted sound. The production of sound by certain catfishes was reported by Sørensen (1894), and he provided a highly detailed description of the skeletal and muscular apparatus. Although these structures, such as the "elastic spring" in catfishes, had been described earlier by Müller (1842, 1843), their function in sound production was not known until Sørensen published his thesis in 1884, and his conclusions were confirmed by Bridge and Haddon (1894). In the following 40 years, accounts were published sporadically on further investigations of sonic mechanisms in fishes, notably those of Smith (1905) on sciaenids, Greene (1924) on

Porichthys, and Hardenberg (1934) on *Therapon.* In addition, structures such as stridulating teeth, fin spines, and other hard parts were found to be involved in sound production in many species (Sørensen, 1894; Burkenroad, 1930, 1931). Among small aquarium species, the croaking gourami, *Trichopsis vittatus,* is probably the best known sound producer (Stampehl, 1931; Reickel, 1936; Meder, 1953; J. A. Marshall, 1963).

The behavioral significance of fish sounds also occupied the interest of many investigators. Dufossé (1874) remarked on the possible communicative functions of these sounds. Most workers reported these sounds to be a sign of alarm or fright (Greene, 1924; Burkenroad, 1930, 1931), although the fact that some fish sounds, notably of sciaenids, are associated with the spawning season or with schooling had apparently been known to fishermen since ancient times.

As is often the case, major scientific and technological advances occur as a byproduct of the search for more efficient means of making war. The field of marine bioacoustics serves as a good example. Motivated by a concern for detection of submarines and other means of undersea warfare, new and efficient mechanisms for detection of underwater sound were developed during World War II. Hydrophones and their associated electronic and recording equipment also proved capable of detecting sounds produced by undersea animals. Shortly after the war, reports were made public that sounds produced by undersea animals often formed an ambient, interfering noise (Loye and Proudfoot, 1946; Knudsen *et al.,* 1948). Some progress was made in identification of sonic species and their seasonal occurrences (Dobrin, 1947, 1948), and in the use of sonics as a tool in the study of marine ecology (Johnson, 1948).

An important contribution to this field is represented by the reports of Fish *et al.* (1952) and Fish (1954). These publications described sounds produced by a wide variety of marine fishes, representing many families, and showed that sound production was much more common than had been previously supposed. Most of the examples given were recorded under artificial, aquarium conditions, and the sounds emitted were often from animals under duress. Such reports demonstrated that many fishes were potentially capable of producing sounds. Spectral analyses of the sounds showed a significant amount of species distinctiveness.

In 1953, Kellogg published a bibliography of sounds of marine organisms. Although not complete, this listing included a large majority of references available at that time. Over the past 15 years, the number of relevant articles and books issued totals at least 10 times the 53 listed by Kellogg. In addition, several reviews of the field have been published, notably by Backus (1958), Maliukina and Protasov (1960), N. B.

Marshall (1962), Moulton (1963), Protasov (1965), Schevill *et al.* (1962), Schneider (1961), and Tavolga (1960, 1965, 1967c). Two international symposia on marine bioacoustics covered many aspects of the field, including sound production in fishes, cetaceans, and other marine organisms (Tavolga, 1964a, 1967a).

A serious problem encountered in obtaining recordings of aquatic animals under natural conditions is that the listener is often working blindly and thus the identification of the sound producer often becomes impossible. A major approach to the solution of this problem has been to accompany sonic observations with visual observations by means of underwater television. A unique installation of this sort is now located off Bimini, Bahamas (Steinberg *et al.*, 1962; Steinberg and Koczy, 1964; Kronengold *et al.*, 1964). This system has been undergoing constant improvement and modification and has provided a large body of data on sonic species of fish (Cummings *et al.*, 1964, 1966; Steinberg *et al.*, 1965).

B. Underwater Acoustics

Any study of sound production or sound detection in fishes necessitates an understanding of the acoustic properties of water as a medium. Since water is much denser than air, the velocity of sound in water is almost 1500 meters/sec while in air it is about 330 meters/sec. In air, sound velocity is affected slightly by humidity, temperature, and barometric pressure. In water, temperature and pressure are independent variables in shallow areas, while at greater depths pressure affects the temperature. The curves that relate sound velocity to depth, therefore, become quite complex (Albers, 1965; Tschiegg and Hays, 1959; MacKenzie, 1960) (Fig. 1). Salinity increases sound velocity, and in the oceans sound velocity may attain 1540 meters/sec.

As a corollary to this almost fivefold difference in sound velocity, c, between air and water, the wavelengths, λ, of underwater sounds are almost five times the length of those in air for the same frequency, f, i.e., $\lambda = c/f$.

Since water is about a thousand times denser than air, more input energy is required to initiate the propagation of sound in water. However, once the sound is propagated, the acoustic energy will be transmitted faster in water. This transmission is further enhanced by the reflection of sound from the water surface (up to 99.9% is reflected back), from the sea bottom, and from interfaces that are formed by layers of water at different temperatures (Vigoreux, 1960; Albers, 1965).

If we measure the sound level under ideal conditions, that is, with

6. SOUND PRODUCTION AND DETECTION

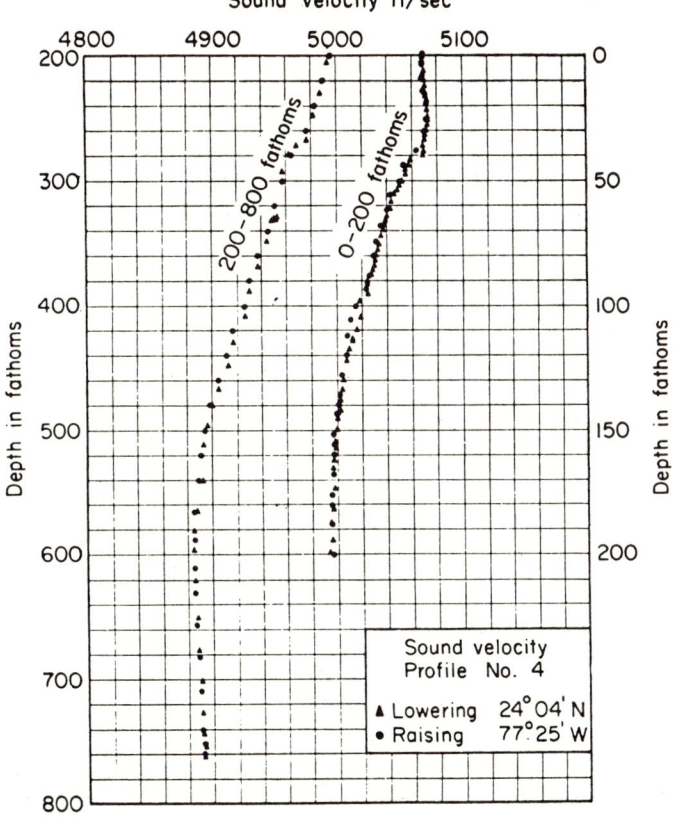

Fig. 1. Velocity of sound in the sea. The two series of determinations show that velocity varies from about 5070 ft/sec (1546 meters/sec) at the surface, to about 4800 ft/sec (1464 meters/sec) at depths below 500 fathoms (915 meters). After Tschiegg and Hays (1959).

the sea at dead calm and with no vessels or sound-producing animals about, there is still an ambient noise whose pressure is about 0.18 to 0.20 dyne/cm². This is usually expressed in terms of decibels (dB) with respect to a reference level of 1 dyne/cm² [= 1 μb (microbar)], and, in this case, the noise level would be about 15 dB below this reference level. In air, the reference level is usually taken to be 0.0002 μb, since this is the standard threshold of human hearing for a frequency of 1000 cps (Hz). This reference level has little objective meaning in underwater acoustics. Thus the 1-μb level is now almost universally used as the reference level in water and all the sound pressure values given in this report will be in decibels re 1 μb (abbreviated to dB μb) (Table I).

Table I
Comparative Chart of Approximate Acoustic Pressure Levels of Common Sounds in Air and in Water[a]

Sounds in air	Acoustic pressure (dB)	Sounds in water
Jet aircraft takeoff (at 75 meters)	60	Underwater dynamite explosion (at 100 meters)
Threshold for human aural discomfort (discomfort at 1000 Hz)	50	25 hp outboard motor (at 15 meters)
Loud auto horn (at 1 meter)	40	Toadfish boat-whistle sound (at 1 meter)
Small propeller aircraft (at 5 meters)	30	Rough sea (state 6)
New York subway train (at 10 meters)	20	Large chorus of marine catfish
Noisy business office	10	Noise of ships in busy harbor
Home high fidelity set	0	Large chorus of snapping shrimp (at 100 meters)
Average conversation (at 1 meter)	−10	Calm sea (state 0)
Private business office	−20	Squirrelfish hearing threshold (at 800 Hz)
Average residence	−30	
	−40	Threshold of hearing of ostariophysine fishes
Quiet country residence	−50	
Quiet whisper	−60	
Human hearing threshold (at 1000 Hz)	−70	
	−80	

[a] The reference point is set at 0 dB = 1 μb (= 1 dyne/cm^2). To convert to a reference point of 0.0002 μb add 74 dB.

Ambient noise in the sea normally includes the sounds produced by wave motion on the surface, friction of moving water currents against the bottom and against each other, the noise of shipping traffic, and, superimposed on all that, the noises of marine animals (Wenz, 1962, 1964) (Fig. 2). The average level of ambient sea noise is about 10 or 15 dB μb.

In air, sound is usually defined as a more or less periodic form of compression waves that can be detected by the human ear. The frequency range, again with reference to human hearing, is normally considered to be from 20 to 20,0000 Hz. At 20 Hz, the sound is "felt" rather than heard, and most people cannot detect sounds above about 16,000 Hz (= 16 kHz). This is a subjective and anthropomorphic definition of sound and does not apply to acoustic energy in water. Since water is highly resistant to compression, the propagation of sound in water usually involves particle displacement as well as compression. This displacement is partic-

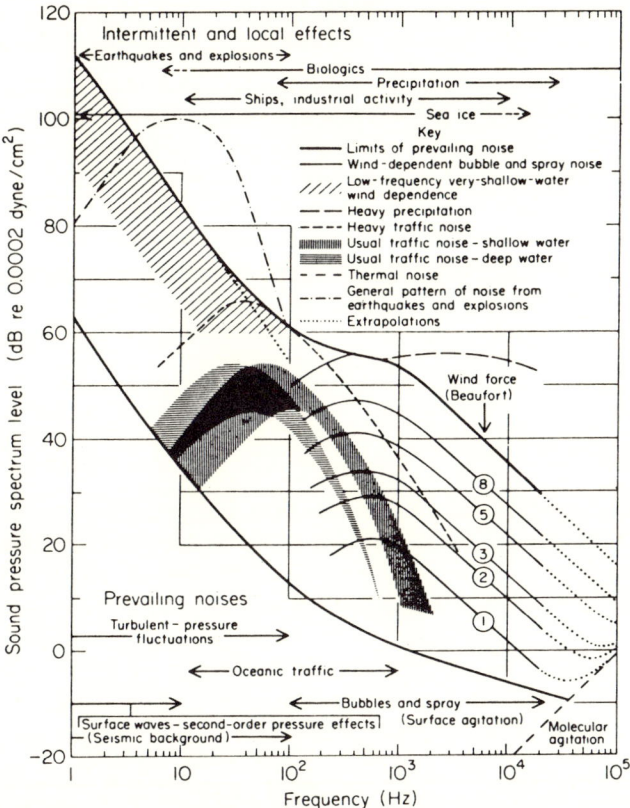

Fig. 2. Typical ambient noise spectra in the sea. Horizontal bars show approximate band of influence of various sources. Sound pressure level on the ordinate is given in decibels with reference to 0.0002 dyne/cm²; to convert to reference level used in this paper subtract 74 dB from all values. After Wenz (1964), with permission of Pergamon Press.

ularly evident at close range to the sound source, and the phenomenon has been termed the "near-field" effect, as opposed to the "far-field" compression waves. The relationship between these two forms of acoustic energy has been discussed extensively by van Bergeijk (1964, 1967a), Harris (1964), and Harris and van Bergeijk (1962). Under most conditions in the field and in aquarium tanks, both near- and far-field energy occur together and are difficult to separate with standard equipment, since a hydrophone is basically a pressure transducer and will respond to compression produced at its surface as a result of particle displacement, i.e., to near-field energy as well as to far-field energy. The energy propagated by the fins of a fish, the flow of water along the body of a moving fish, and even the currents of water flow in rivers and seas are essentially acoustic phenomena, and a line of demarcation between such energy and that of a distinctly audible hoot of a toadfish is difficult to draw. Some investigators have referred to these extremely low frequency or steady state displacements and pressures as "unsound" or "pseudosound" (Parvulescu, 1964, 1967; Ffowcs-Williams, 1967).

It is apparent, therefore, that in water the usual definitions of sound are not entirely applicable, and the distinction between rheotaxis and hearing is not clear. As evidenced by the problems and discussions at a recent conference on the lateral line of fishes (Cahn, 1967), the concern over underwater bioacoustics has extended far below what is ordinarily considered the audiofrequency range.

C. Sonic Mechanisms in Fishes

There appears to be little relationship between the morphology of sound-producing mechanisms and phylogenetic position in fishes. At the present time, however, only a few hundred out of more than 20,000 species have been clearly identified as sound producers. Many species, such as certain deep-sea forms (N. B. Marshall, 1967), appear to possess the means for sound production but have not yet been recorded. Three general types of sonic mechanisms are present in fishes: stridulatory, hydrodynamic, and swim bladder. Stridulatory sounds are produced by friction of teeth, fin spines, or bones. Hydrodynamic sounds result from swimming movements, especially during rapid changes of direction or velocity. The swim bladder acts as a sound projector when it is vibrated by contiguous or attached muscles.

1. Stridulatory Mechanisms

A large majority of teleost fishes possess opposing patches of denticles in the pharynx and are at least potentially capable of sound production during feeding. Some species stridulate pharyngeal denticles in connection with other activities such as alarm and territoriality. The best known sound producers of this type are members of the family Pomadasyidae—the grunts (Fig. 3). Burkenroad (1930) described the sounds and the mechanism in some detail for the white grunt, *Haemulon plumieri*. He also noted that if the swim bladder were deflated, the character of the sound became "dry" and lost its "gruntlike" quality. He concluded that the swim bladder acts as a "resonator." A similar arrangement has been described for other members of the family, and the margate fish, *Haemulon album*, has been observed to emit sounds, probably produced

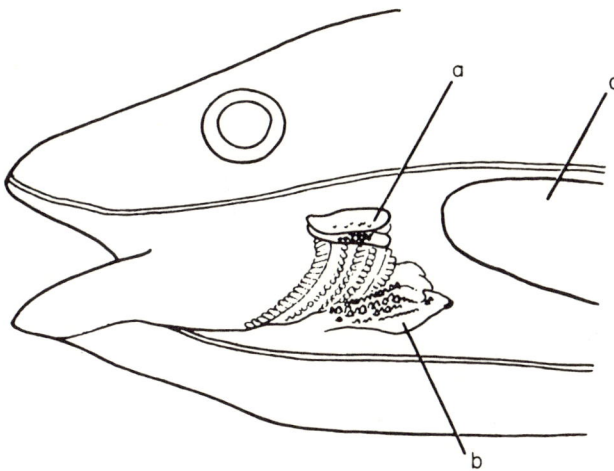

Fig. 3. Dissection of the pharyngeal region of a grunt, *Haemulon*, shows the location of pharyngeal denticles (a and b) and the swim bladder (c). After Tavolga (1965), with permission of the U. S. Naval Training Device Center.

by this mechanism, in connection with feeding and schooling behavior (Cummings *et al.*, 1966). The function of the swim bladder as a possible resonator was also demonstrated in triggerfishes (Salmon *et al.*, 1968).

The occurrence of pharyngeal tooth stridulation has been reported from a wide variety of species and families of teleostean fishes by Burkenroad (1931), Dobrin (1947), Dorai Raj (1960a), Fish (1954), Knudsen *et al.* (1948), Moulton (1958), Tavolga (1964b), and Taylor and Mansueti (1960). The courtship and territorial sounds of the croaking gourami, *Trichopsis vittatus*, are evidently produced by pharyngeal denticles (J. A. Marshall, 1963), and this small species offers many possibilities as an experimental animal in this field. It is also probable that the sounds of priacanthids are produced in this manner (Salmon and Winn, 1966).

Almost any predatory species of fish is likely to produce sounds when feeding, and even some herbivorous forms that browse on sessile plants and animals can emit sounds when crushing rocks and corals. Incisor types of teeth are capable of biting through the exoskeletons of crustacea and thus produce strong metallic sounds (Fig. 4). Given any food of moderate hardness, the action of teeth will produce sounds. Sometimes fish will gnash their teeth without the direct presence of food. Fish (1954) even listed some sharks and rays that produce sounds when feeding, despite the fact that elasmobranchs are not known to be sound producers in any specialized sense (Backus, 1963).

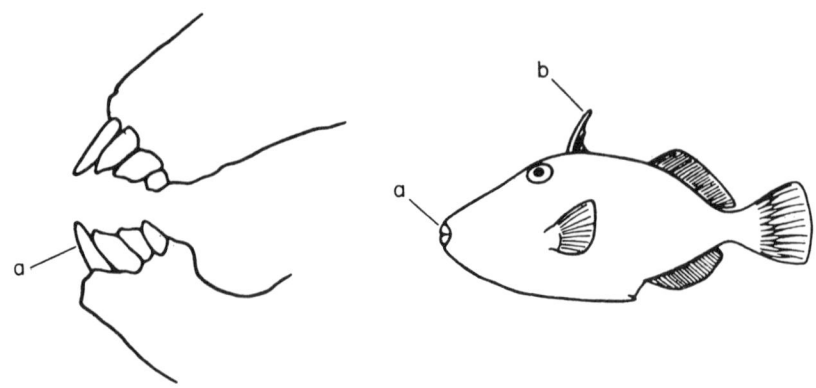

Fig. 4. The sharp incisorlike teeth (a) of the triggerfish, *Balistes,* can produce sounds during feeding. The erection of the dorsal fin spines (b) generates a stridulatory sound. After Tavolga (1965), with permission of the U. S. Naval Training Device Center.

Burkenroad (1931), Fish (1954), and Moulton (1958) listed a number of species that produce sounds by means of modified molariform teeth. Sound production of this type is particularly common among the members of the order Tetraodontiformes (puffers, filefish, etc.) and the family Scaridae (parrot fishes). Molariform teeth are also present in the family Sparidae (porgies), and Fish (1954) included the scup, *Stenotomus chrysops,* as a sound producer. Another common sparid is the pinfish, *Lagodon rhomboides,* whose sounds have been described by Burkenroad (1931) and Caldwell and Caldwell (1967). Knudsen *et al.* (1948) listed over a dozen additional species that were found to produce sounds when feeding.

Stridulation by the movement of special fin rays and spines has been reported in a few species. The sea catfish, *Galeichthys felis,* and the gafftopsail catfish, *Bagre marinus,* produce high-pitched squeaks when the enlarged pectoral fin spines are moved (Burkenroad, 1931; Tavolga, 1960), and several species of freshwater catfishes produce sounds in a similar fashion (Pfeiffer and Eisenberg, 1965; Sørensen, 1894; Mahajan, 1963). The erection of the specialized dorsal fin spines in triggerfish produces a stridulation (Schneider, 1961) (Fig. 4), and it is possible that low intensity sounds may be produced by sticklebacks, *Gasterosteus aculeatus* and *Apeltes quadracus,* when dorsal fin spines are moved (Fish, 1954).

Sounds produced by the friction of adjacent bones were reported from the clown fish, *Amphiprion,* by Schneider (1961, 1964a) and for the sea horse, *Hippocampus,* by Fish (1953, 1954). In both these cases, the probable mechanism is the stridulation of the posterior margin of the skull against some vertebral element. A similar mechanism was described for the pipefish, *Syngnathus louisianae*, by Burkenroad (1931). Klausewitz (1958) described sounds produced by the Indian loach, *Botia hymenophysa*, which he attributed to the stridulation of Weberian ossicles, although this function of the Weberian ossicles should be considered highly doubtful.

2. Swim Bladder Mechanisms—Indirect

As noted above, Burkenroad (1930) observed that pharyngeal denticle stridulation in the grunts (family Pomadasyidae) was affected by deflation of the swim bladder. In many such cases, the swim bladder may act as a resonator and thus change the quality of the emitted sound. Stridulation of bones of the pectoral girdle has been found in a triggerfish, *Balistes*, by Schneider (1961). Bone stridulation in a sculpin, *Myoxocephalus* (Fish, 1954), generates a sound at a point adjacent to the anterior end of the swim bladder. In a study on seven species of triggerfish, Salmon *et al.* (1968) found sounds were produced by the rubbing of pectoral fins against the body sides where skin thinly covered lateral evaginations of the swim bladder (Fig. 5).

The swim bladder has also been thought to function literally as a drum. Triggerfish, *Balistes*, were reported to produce a sound by beating their pectoral fins against the areas of the body wall that cover the swim bladder (Moulton, 1958), and some species of serranids appeared to beat their opercula in sound production (Fish, 1954; Tavolga, 1960). Schneider

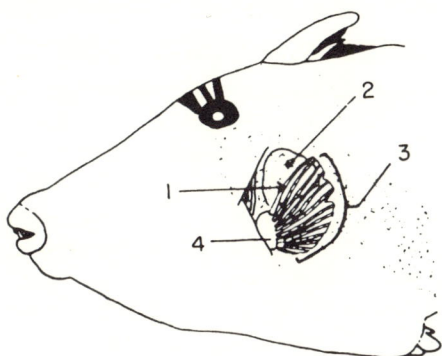

Fig. 5. Sound production in a triggerfish, *Rhinecanthus rectagulus*, takes place by rubbing the pectoral fins against the swim bladder: (1) pectoral fin spine, (2) drumming muscle, (3) pectoral fin rays, and (4) fleshy muscular lobe of pectoral fin. After Salmon *et al.* (1968).

(1961), however, stated that the drumming in triggerfishes was probably produced only out of water. Hazlett and Winn (1962) described the sonic mechanism of a serranid, *Epinephelus striatus,* as consisting of a pair of sonic muscles (see below).

3. Swim Bladder Mechanisms—Gas Expulsion

In some fishes, where there is a pneumatic duct between the swim bladder and esophagus, sounds can be detected when air bubbles are eructed. Fish (1954) and Dufossé (1874) reported this type of sound production in some eels and catfishes. A characid fish, *Glandulocauda inequalis,* is reported to produce sounds by gulping air at the water surface (K. Nelson, 1965).

4. Swim Bladder Mechanisms—Extrinsic Muscles

Among the best known of the sonic fishes are members of the drum and croaker family (Sciaenidae). The sonic mechanism consists of a pair of large muscles (sometimes present only in males) that are derived from the lateral body wall musculature (Fig. 6). These muscles are contiguous to the lateral walls of the swim bladder, and their vibrations occur in short bursts producing a series of drumlike beats or knocking sounds. There are a number of variations in the size and position of the sonic muscles among the species of the family, but the fundamental arrangement is the same in all the sonic forms (Dijkgraaf, 1947a; Fish, 1954; Schneider, 1961, 1967; Schneider and Hasler, 1960; Tower, 1908).

A type of sonic mechanism common to a wide variety of teleosts con-

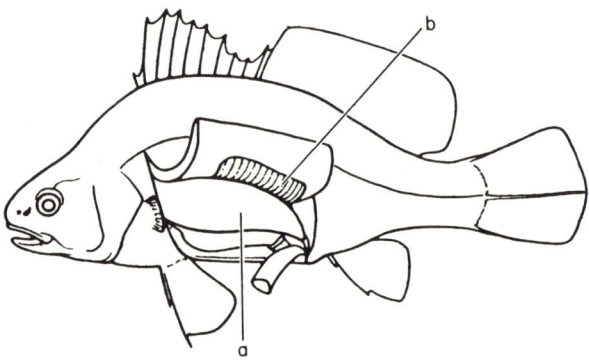

Fig. 6. The swim bladder (a) in the drumfish, *Aplodinotus,* is vibrated by a pair of sonic muscles (b). Redrawn from Schneider and Hasler (1960) after Tavolga (1965), with permission of the U. S. Naval Training Device Center.

sists of a pair of short muscles that originate on the occipital region of the skull and insert on the ribs and connective tissue that forms the anterior–dorsal surface of the swim bladder (Figs. 7 and 8). This mechanism has been described in some detail for the Nassau grouper, *Epinephelus striatus,* the squirrelfish, *Holocentrus rufus,* the tiger fish, *Therapon,* and a scorpaenid, *Sebasticus,* by Hazlett and Winn (1962), Winn and Marshall (1963), Schneider (1964b), and Dôtu (1951), respectively.

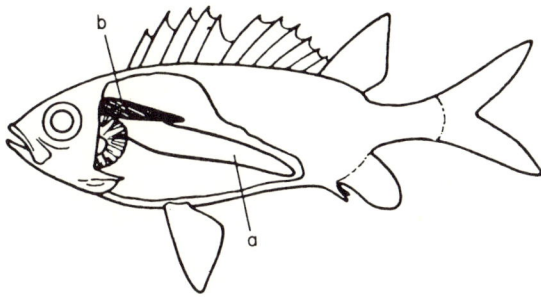

Fig. 7. The sonic muscles (b) in the squirrelfish, *Holocentrus,* extend from the rear of the skull and insert on the first pairs of ribs. The ribs are tightly laced to the swim bladder (a). After Tavolga (1965), with permission of the U. S. Naval Training Device Center.

A modification and specialization of the above mechanism is present in some of the catfishes (Siluroidea). First noted by Müller (1842, 1843), the catfishes commonly possess a fused shelf of bone over the anterior and dorsal surfaces of the swim bladder. The many variations of this arrangement were described by Bridge and Haddon (1889, 1893). In some species, this shelf of bone is extremely thin and springlike and has

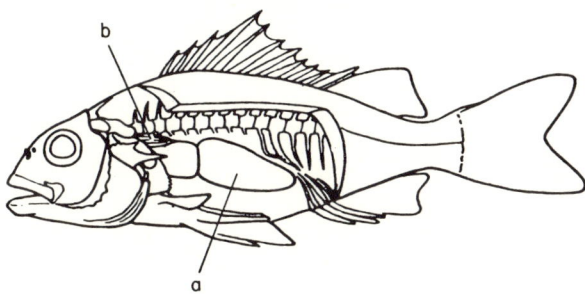

Fig. 8. The swim bladder (a) in the tigerfish, *Therapon,* is vibrated by a pair of sonic muscles (b). Redrawn from Schneider (1964b) after Tavolga (1965), with permission of the U. S. Naval Training Device Center.

been called the *"Springfederapparat"* (Müller, 1842) and "elastic spring" (Sørensen, 1894). This structure is derived from the transverse processes of the first few vertebrae and was first reported to function in sound production by Sørensen (1894) and later confirmed by Tavolga (1962). In marine catfishes, *Bagre marinus* and *Galeichthys felis,* the sonic mechanism has been described in some detail as consisting of a pair of sonic muscles that originate from the epiotic and neighboring bones of the skull and insert on the upper surfaces of the elastic spring. Vibration of these muscles causes the swim bladder to vibrate, and the elasticity of the bony shelf acts as the antagonist to the contraction of the sonic muscles (Tavolga, 1962).

Many species of fishes have been found to possess a pair of muscles connecting the skull and swim bladder, and, on the basis of morphology,

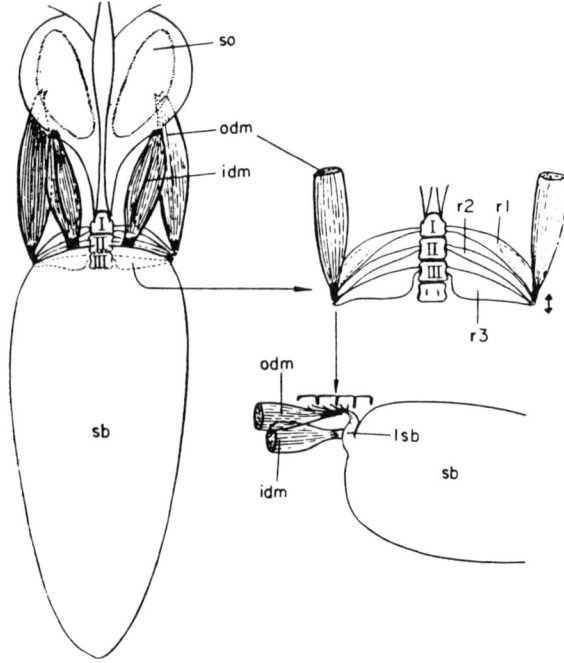

Fig. 9. Swim bladder and presumed sonic mechanism of a brotulid, *Monomitpus*. Left: The complete mechanism showing two pairs of muscles originating on the rear of the skull and inserting on the ribs; upper right: insertion points of the sonic muscles on the ends of the ribs; lower right; side view showing attachment of sonic muscles to ribs and swim bladder. Key: idm and odm, inner and outer drumming muscles; I, II, and III, vertebrae with corresponding ribs, r1, r2, r3; sb, swim bladder; lsb, ligament; so, saccular otolith. After N. B. Marshall (1967), with permission of Pergamon Press.

6. SOUND PRODUCTION AND DETECTION

it is probable that these are sonic muscles. Such structures have been found in many brotulids, macrourids, and other deep-sea forms (Walters, 1960; N. B. Marshall, 1967; Schneider, 1961) (Fig. 9). It is probable that future deep-sea exploration will identify these forms as sound producers.

5. Swim Bladder Mechanisms—Intrinsic Muscles

In some families of fishes the sonic muscles are completely attached to the walls of the swim bladder. In such cases, the bladder can be dissected out of the body cavity and can function as a sound-producing mechanism just by stimulation of the nerves leading to the sonic muscles (Fig. 10).

The sonic mechanism of the toadfish, *Opsanus*, is typical and probably the best known. The swim bladder and its role in sound production has been described by Tower (1908), Rauther (1945), Fish (1954), and Tavolga (1960, 1964b). The bladder is heart-shaped with its apex pointed

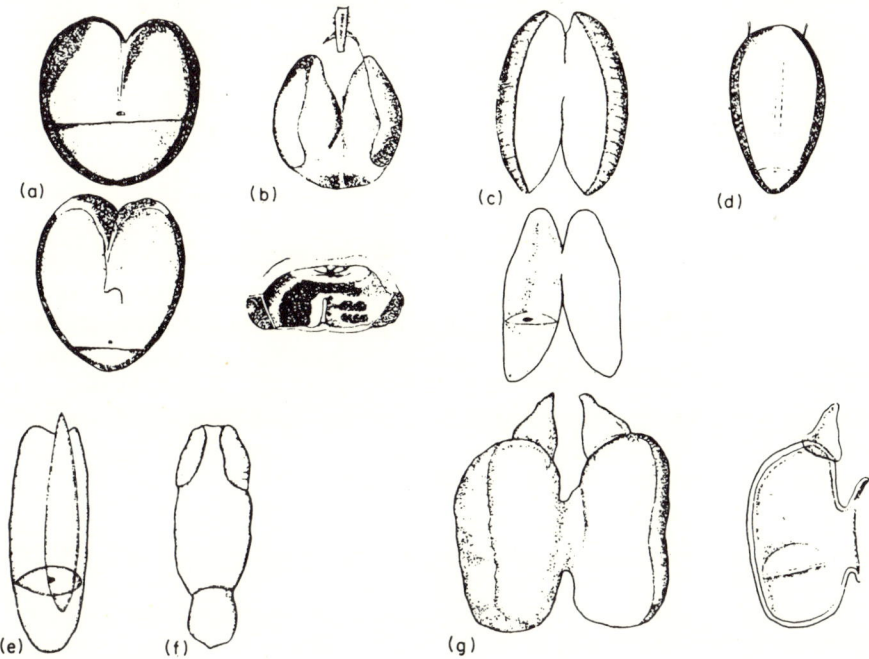

Fig. 10. Sound producing mechanisms in many species use a swim bladder with intrinsic sonic muscles: (a) *Opsanus* (Tower, 1908), (b) *Porichthys* (Greene, 1924), (c) *Prionotus* (Tower, 1908), (d) *Trigla* (Rauther, 1945), (e) *Sebasticus* (Dôtu, 1951), (f) *Zeus* (Dufossé, 1874), (g) *Dactylopterus* (Dufossé, 1874). After Schneider (1967), with permission of Pergamon Press.

posteriorly. Two broad bands of muscle are firmly attached along the lateral surfaces, with the fine, striated fibers running obliquely (Fig. 10a). The interior of the bladder contains an extensive vascular network and a pair of "red glands." A thin transverse membrane with a small sphincter separates the secretory and absorptive chambers. The remarkable gas exchanging properties of this organ has been extensively investigated by Fänge (1966) and Fänge and Wittenberg (1958).

The first experimental studies on the sonic function of the swim bladder of toadfish were those of Tower (1908). His observations, confirmed by Tavolga (1964b), showed that the fundamental frequency of the emitted sound is exactly equivalent to the vibration frequency of the sonic muscles.

Thus far, although differences in the character of the cell exist among populations and different species of toadfishes, no structural differences in the swim bladder mechanism have been found to account for these variations in pitch, duration, and quality (Tavolga, 1958c, 1964b). In *Porichthys*, a genus of fish closely related to the toadfishes, the sonic mechanism appears to be virtually identical (Greene, 1924) (Fig. 10b).

The sea robins (Triglidae) possess essentially the same mechanism of sound production as the toadfishes except for the shape of the swim bladder. In the sea robins (*Trigla* in Europe; *Prionotus* on our Atlantic coast), the swim bladder is composed of two dirigible-shaped chambers side by side, and attached to one another by a narrow passageway (Figs. 10c and 10d). The arrangement of the sonic muscles and their function appears to be similar to that of the toadfish (Moulton, 1960a; Moreau, 1876). A family related to the sea robins is the Dactylopteridae—the flying gurnards. Although little is known of their ability for sound production, the mechanism seems to be identical to that of the sea robins (Tower, 1908; Fish, 1954; Tavolga, 1964b) (Fig. 10g). Simultaneous electromyograms and sound recordings in the midshipman, *Porichthys notatus*, showed that sound pulses and muscle potentials corresponded in a one-to-one fashion (Cohen and Winn, 1967).

6. Mechanics of Swim Bladder Sound Production

If we consider the swim bladder and its associated skeletal and muscular systems as an underwater loudspeaker, the energy source for the emitted sound is the contraction of specialized muscles. In no case is there a muscular antagonist to the action of the sonic muscles. In the catfishes, the "elastic spring" maintains tension on the sonic muscles and returns the fibers to their normal condition after each contraction (Tavolga, 1962). In the squirrelfishes and groupers, the sonic muscles insert on the first pairs of ribs. These muscles contract against the tension of hypaxial muscles, vertebral ligaments, and connective tissue lacings to the roof of the swim bladder (Hazlett and Winn, 1962; Winn and Marshall, 1963). In the sea robins and toadfishes, where the sonic muscles are intrinsic to the swim bladder, the elasticity of the bladder walls and the internal pressure of the distended bladder will tend to return the muscle fibers to normal length after each contraction. In the sciaenids, the return of the sonic muscles to normal shape after contraction is a function of the swim bladder tension, in part, and probably a function of the elasticity of the lateral body wall musculature. In catfishes, Tavolga (1962) showed that the fundamental frequency of the sound is a direct translation of the frequency of contraction of the muscle. Packard (1960) recorded sounds and muscle action potentials simultaneously in the pigfish, *Congiopodus*, from New Zealand, and he found that the contraction frequency coincided with the sound frequency. A similar observation was made by Barber and Mowbray (1956) in the sculpin and Schneider (1964b, 1967) in *Therapon*. Winn and Marshall (1963) found that each sound from a squirrelfish, *Holocentrus rufus*, consisted of up to 5 pulses, about 10 msec apart. This corresponds to a fundamental frequency of about 100 Hz. Muscle action potentials coincided with the pulses.

Artificial stimulations of nerves leading to the sonic muscles can show that the muscles are capable of unusually rapid contraction and recovery cycles. In the gaff-topsail catfish, *Bagre marinus*, the sonic muscles took up to 12 sec to tetanize at 150 pulses/sec, and up to 3 sec at 200 pulses/sec (Tavolga, 1962). Using the same stimulating equipment as described earlier (Tavolga, 1962), the nerves leading to the sonic muscles were stimulated in the toadfish, *Opsanus tau*, the slender sea robin, *Prionotus scitulus*, the squirrelfish, *Holocentrus ascensionis*, and the red hind, *Epinephelus guttatus*. *Opsanus* and *Prionotus* muscles were found to be most resistant to tetanization, but all could be stimulated at pulse frequencies over 100/sec without tetanization (Tavolga, 1964b).

Among vertebrates, the occurrence of such fast-acting muscles is unusual. Extrinsic eyeball muscles are known to reach 350 contractions/sec, but the majority become tetanized when stimulated at 50 pulses/sec. The sonic muscle of *Opsanus* has a contraction–relaxation cycle of 10 msec (Skoglund, 1959), which would theoretically limit its response to about 100 contractions/sec, but Skoglund's observations were limited to single twitches. Sonic muscles in the squirrelfish show fusion at frequencies above 200/sec, while normal white muscle begins to show summation at 50 and fusion at 100 pulses/sec (Gainer *et al.*, 1965). Electron microscopy showed the presence of an unusually developed sarcoplasmic reticulum which Fawcett and Revel (1961) have related to the muscle's fast-acting properties. Swim bladder muscles in nonsonic fish may also possess this reticulum and its associated fast-acting properties (Kilarski, 1964). In the catfishes, squirrelfishes, groupers, and sciaenids (drumfish and croakers), the sonic muscles are characteristically red in color as a result of a high degree of vascularization, and this may enhance the resistance to tetanization and fatigue. Histological examination of the sonic muscles of these species showed uniformly thin fibers and numerous polyaxonal innervations (Gainer and Klancher, 1965).

The innervation of the sonic muscles shows features common to many species. In the catfishes, the sonic muscle is supplied by a branch of the occipital nerve (Tavolga, 1962). Except in certain Ostariophysi, there are normally two pairs of occipital nerves in teleosts. Their homology is uncertain, but they are presently thought to be homologous to the hypoglossal (XIIth) cranial nerve of the tetrapods. Based upon dissections, stimulation experiments, and serial cross sections, the nerve supply to the sonic muscles was traced in the squirrelfish, toadfish, sea robin, and red hind. Both pairs of occipital nerves were found to innervate these muscles in all of the above species (Tavolga, 1964b). Except for the sciaenids, in which sonic muscles are supplied by spinal nerves (Schneider and Hasler, 1960), it appears that the innervation of the sonic muscles in widely divergent teleosts is the same and that the muscles in all these forms must be homologous structures, despite the gross differences in appearance and location (Tavolga, 1964b).

In the catfishes, if the swim bladder is damaged, deflated, or filled with water, the sound output of the sonic muscle and the elastic spring is greatly lowered, but the sound quality, i.e., its harmonic content, remains unchanged (Tavolga, 1962). Winn and Marshall (1963) found that in the squirrelfish, they could no longer detect the sound if the swim bladder was completely filled with water. Partial deflation of the bladder reduced the sound amplitude but did not affect the fundamental frequency or other properties (also true for other species: Salmon et al., 1968). In toadfishes, if the swim bladder was partially deflated, the spontaneous grunting sounds continued to have the same spectral characteristics but showed a reduced amplitude by as much as 20 dB (Tavolga, 1964b). It does not appear likely, therefore, that the resonating role of the swim bladder is an important one, especially since the excess internal pressure in the swim bladders of most fishes is low (Alexander, 1959a,b).

The pulsation of the bladder should be essentially that of a large air bubble and would be affected by the low compressibility of the medium and a variety of other physical factors. The net result is a damping of the resonance and a decrease in the resonant frequency. Based on the formula given by Meyer (1957), the resonant frequency of a bubble of air of radius 3 cm at atmospheric pressure would be about 100 Hz. If we treat the swim bladders of fishes as if they were spherical air bubbles, the bladder dimensions in most individuals of squirrelfish, red hind, etc., would give resonant frequencies in the 50–100-Hz range. Even such rough calculations indicate that the fundamental frequencies of most fish sounds are in the same order of magnitude as the resonant frequencies of the swim bladders. The resonance peak of fish swim bladders was found to be flat, with a maximum rise of about 3–4 dB (Tavolga, 1964b). Conceivably, this would enable the fish to gain some efficiency from resonance, but this gain would be quite small.

Harris (1964) investigated the properties of mathematical and physical models of vibrating swim bladders. He concluded that the pulsating air bubble, such as a swim bladder, would have the elastic and resonant qualities for a highly efficient underwater sound producer in the low frequency range that fishes usually produce. The toadfish, for example, develops sound pressures of up to 35 dB μb at distances within 5 meters, and many species produce sounds powerful enough to be heard distinctly above the sea surface. An underwater loudspeaker was constructed along the physical principles of a toadfish swim bladder and was found to be quite efficient in the 100–2000-Hz range (Tavolga and Wodinsky, 1963).

The mechanics of swim bladder sound production can be summarized as follows: The fundamental frequency of the sound emitted is a direct translation of the contraction frequency of the sonic muscles. The contractions of the sonic muscles produce volume and pressure changes in the swim bladder, and, therefore, the entire surface of the bladder pulsates. The fact that the natural frequency of the swim bladder as an air bubble is in the same order of magnitude as the vibration frequency of the sonic muscles undoubtedly aids the efficiency of the system, and the mechanical energy of the moving muscle is transmitted by way of air pressure changes to the entire bladder surface. These pulsations are transmitted through the tissues of the fish to the outside with little loss, since the fish is virtually transparent to water-borne sound. In acoustic and electronic terms, therefore, the swim bladder is an impedance-matching device analogous to the large surface cone of a high fidelity loudspeaker.

7. Hydrodynamic and Swimming Sounds

The movement of any object through the water will create displacement. Such displacements and compression waves may be rhythmic subsonic vibrations when produced by the fins and body of a swimming fish. Such phenomena have been classified as hydrodynamic sounds by Shishkova (1958a,b) and Moulton (1960b). Moulton also considered the possibility that some of the sounds produced by swimming fishes are of internal origin, i.e., produced by muscular and skeletal movements within the animal.

The most intense of these sounds are emitted when the animal turns rapidly or changes its velocity. In his review of fish propulsion, Nursall (1962) pointed out that the main thrust in a swimming fish arises from transverse movements of the contralateral waves. Turning is accomplished by a unilateral wave as the head of the fish turns, and the caudal portion is used as a fulcrum. It is clear that lateral displacement of the medium will arise during straight-line swimming, and these will tend to be rhythmic since the body movements are rhythmic. The turning of the head will produce a strong displacement of the medium. These displacements will, of course, result in compression waves, which can be detected as sound by most hydrophones.

Skudrzyk and Haddle (1963) defined flow noise as turbulences and concomitant pressure fluctuations produced by the motion of a body through water. This is primarily a near-field phenomenon and, as such, would not be propagated over large distances. Although the above study was concerned with the noise generated by the motion of rigid objects, such as vessels, it is obvious that surface turbulence and eddies would be produced by swimming fish.

The locomotion of fish, therefore, produces pressure and displacement effects in three possible ways: the more or less rhythmic effects of undulatory movement; the turbulence generated by flow noise; and, possibly, internally generated locomotor sounds.

D. Characteristics of Fish Sounds

It has long been recognized that a verbal description of a sound is inadequate and often misleading. This problem has been especially troubling to observers attempting to describe sounds of animals such as birds and amphibians. With the advent of high fidelity tape recording equipment, the problem of preserving the data seemed to be solved, although precautions still had to be taken against artifacts resulting from overloaded amplifiers, incorrect equalization, and other electronic difficulties. Photographs of oscilloscope tracings of sounds were frequently used, and these analyses are still especially valuable for short, rapidly pulsed sounds. Fish (1954) presented her data on sounds of fishes in the form of frequency analysis graphs that were made with an octave band filter.

An important improvement over the octave band filter was the development of the sound spectrograph. This instrument is now well known in many areas of acoustic research, including virtually all phases of bioacoustics. In brief, the sound spectrograph takes a sample of the sound and produces a graph of frequency against time. Other displays such as frequency against relative intensity are also possible. Many examples of its use in bioacoustics can be found in Lanyon and Tavolga (1960) and Busnel (1963). Spectrographic displays of various fish sounds are shown in practically all current publications in this field.

As applied to underwater bioacoustics, the use of the sound spectrograph has been carefully evaluated by Watkins (1967). The capabilities and limitations of this instrument are often not fully understood by biologists. The display of apparent harmonics, for example, can result from repetition of pulses, and other complexities can be introduced if the sound consists of short, repeated bursts of pulses, with each pulse consisting of a brief tone or complex of tones. Often the so-called fundamental frequency is actually a pulse repetition rate. In addition, the verbal descriptions of bioacoustic phenomena have had to be standardized, and attempts at this standardization resulted in glossaries compiled by Broughton (1963) and Bondesen and Davis (1966), although only a minority of the terms they listed apply to the sounds of fishes.

A serious source of difficulty in making original recordings of animal sounds is the effect of the reverberations, reflections, absorptions, and other acoustic phenomena in the environment. Field recordings are always plagued by such problems, including the presence of background noise. Such problems are particularly troublesome in underwater recordings. The spectral and other characteristics of sea noise and man-made noise have been summarized by Wenz (1964), but the acoustic properties of the ocean bottom, surface, suspended particles, and air bubbles are still under intensive study (Albers, 1965; Richardson, 1957). The problems of recording sounds of captive fishes are even more complex, and the acoustic field generated in an aquarium tank, for example, virtually defies analysis (Parvulescu, 1964, 1967). Any spectral analysis of an underwater sound, therefore, must be cautiously interpreted, especially if the acoustic conditions are not controlled and not specified (Schneider, 1967; Tavolga, 1965).

These difficulties are especially evident in attempts to identify unknown sound sources. Sounds of marine animals are often recorded without any visual confirmation of the source of the sound, and it is tempting to make comparisons between such field recordings and recordings of known species in captivity. This problem was discussed in detail by Tavolga (1965).

Tables of data on pitch and other characteristics of fish sounds are given in papers by Fish *et al.* (1952) and Fish (1954). The tables are based primarily on recordings made under laboratory conditions, and, in many cases, the stimulus for sound production was some noxious stimulus such as electric shock. These data, therefore, must be treated cautiously and specific conclusions as to sonic characteristics of a given species need to be corroborated with field recordings under natural conditions. Much of the following discussion is based upon reviews by Fish (1954, 1964), N. B. Marshall (1962), Moulton (1958, 1963), Schneider (1967), Tavolga (1960, 1964b, 1965), and Winn (1964).

1. Stridulatory Sounds

Sounds produced by the grating of pharyngeal denticles, jaw teeth, fin rays, or bones are essentially nonharmonic in structure, i.e., they do not resolve, in a spectrogram, into a series of horizontal parallel bars at harmonic intervals. Such sounds contain many harmonically unrelated frequencies (Fig. 11).

Two general categories of stridulatory sounds can be distinguished on the basis of predominant pitch. High frequency stridulations are usually produced by the gnashing of jaw teeth and, sometimes, the patches of pharyngeal denticles during feeding. The component frequencies extend continuously from below 100 to over 8000 Hz. Predominant frequencies are in the 1000–4000-Hz range, and durations are extremely variable.

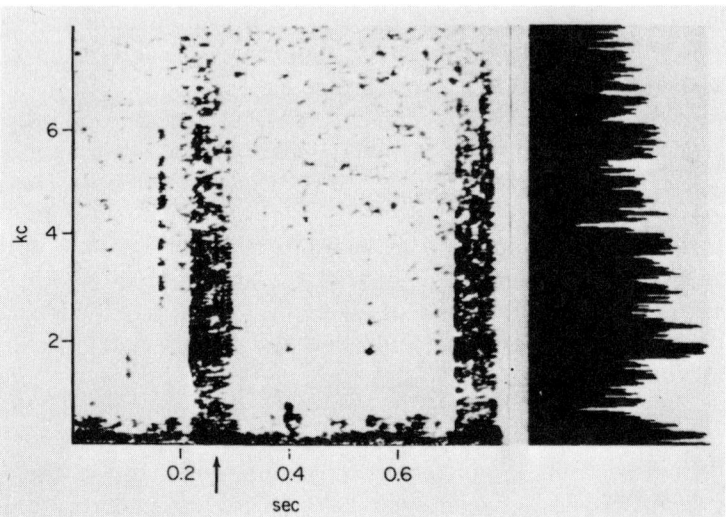

Fig. 11. Spectrogram of stridulatory sounds produced by a jack, *Caranx*. A frequency-intensity section shown on the right was taken at the point indicated by the arrow. This is characteristic of short, broad-band pulsed sounds. After Tavolga (1965), with permission of the U. S. Naval Training Device Center.

Such sounds have been described verbally as rasps, scratches, clicks, chirps, and scrapes.

The second category of stridulatory sounds are those in which the swim bladder plays some part in determining the quality of the sound. Verbally, the sounds have been called grunts, croaks, thumps, knocks, etc. The frequency range is generally from 1000 to 8000 Hz, and the predominant frequencies are below 1000 Hz. Comparison of the analyses made by Fish (1954), Moulton (1958), and other authors shows a considerable divergence of data as to frequency range and predominant pitch. It is probable that the different types of equipment, different methods of eliciting the sonic behavior, and, most important, the different conditions under which recordings were made can account for the discrepancies in the data.

2. Swim Bladder Sounds

Sounds produced by the vibration of muscles around or attached to the swim bladder are usually recognizable by their harmonic structure. If accompanied by stridulation, the harmonics can sometimes be partially masked in a spectrogram, but to the ear these sounds seem vibrant and possess a tonal quality. A number of parameters of such sounds can be defined and measured by the use of spectrographic and oscillographic analysis. The fundamental and predominant frequencies usually can be determined. If the sound is a sustained call, its duration usually varies within narrow limits. Often the sound consists of a number of rapidly repeated pulses, and in such instances it is necessary to know the duration of each pulse, the pulse repetition rate, and the usual number of pulses within the sound complex.

The number and relative strengths of harmonics is extremely variable. In studies on the sounds of marine catfish (Tavolga, 1962), it became evident that the harmonic components could be varied by changing the character of the surroundings, the type of hydrophone, the distance from the second source, and the recording and analyzing levels. The hypothesis was advanced that the actual emitted sound was virtually a pure tone.

In the process of listening to the boat whistle calls of toadfish, *Opsanus tau* (Fig. 12), Tavolga (1964b) recorded a long series of sounds emitted by single individuals that remained in the same position for many hours. Several environmental variables were found to affect the number and relative intensities of harmonics, e.g., the distance of the hydrophone from the sound source was of prime importance. Schneider (1967) found that the construction material of an aquarium tank affected the quality of the recorded sounds of *Therapon*.

The midshipman, *Porichthys notatus*, is also a member of the toadfish

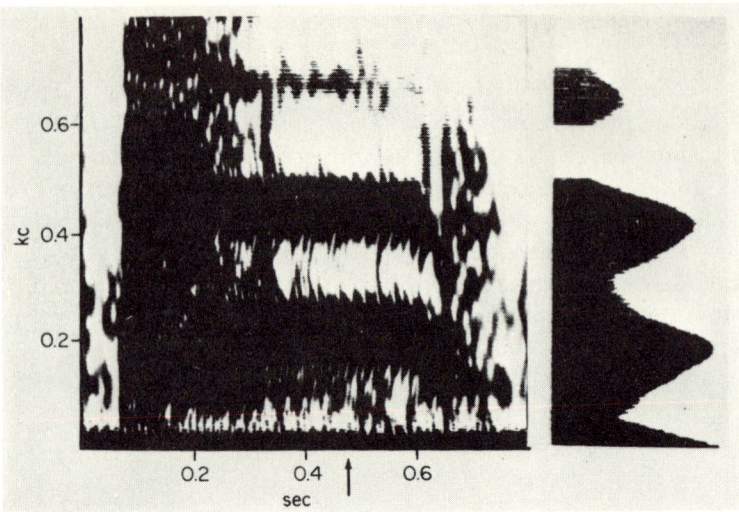

Fig. 12. Spectrogram of a boat-whistle sound of a toadfish, *Opsanus*. The fundamental frequency of this harmonic sound is about 250 Hz, and the frequency-intensity section on the right was taken at the point indicated by the arrow. After Tavolga (1965), with permission of the U. S. Naval Training Device Center.

family (Batrachoididae) and is known as a sound producer (Greene, 1924). Sounds recorded both in the field and in the laboratory were either short grunts or long buzzes (Cohen and Winn, 1967). Both types of sounds consist of bursts of sound pulses. To use the terminology recommended by Watkins (1967), the pulse frequencies were 175–200 Hz, the pulse repetition rates were about 50/sec, and the burst durations varied from about 77 msec (grunts) up to 3 sec (buzzes). The short bursts sometimes occurred in trains.

Unlike the toadfish, the majority of fish swim bladder sounds are short pulses, with a fundamental frequency of from 75 to 100 Hz. This is true for the members of the Sciaenidae, the largest family of sonic fishes, including croakers and drumfish. Fish (1954) described the sounds of several species, primarily under captive conditions, and Kellogg (1955) presented several recorded examples. The sounds of the squirrelfish, *Holocentrus* (Winn and Marshall, 1963) (Fig. 13), triggerfishes of several species (Schneider, 1961; Moulton, 1958; Vincent, 1963a), some of the groupers and sea basses (Fish, 1954; Tavolga, 1960; Hazlett and Winn, 1962), the sea catfish (Tavolga, 1962) (Fig. 14), and many other species fall into this acoustic category.

Some species produce sounds of lower fundamental frequency, as, for example, the codfish and haddock with a fundamental of 40–50 Hz (Brawn, 1961; A. D. Hawkins and Chapman, 1966). Others exhibit higher

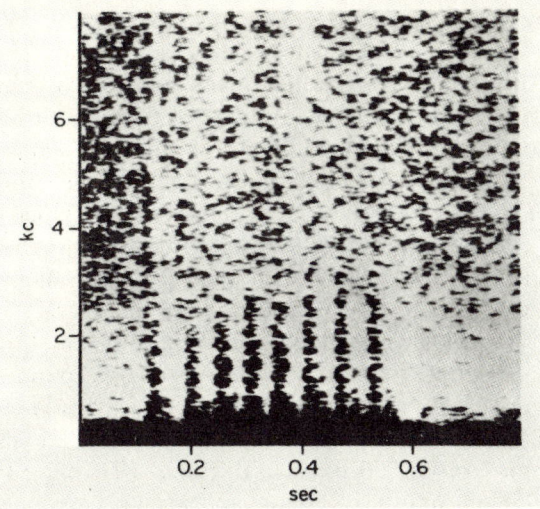

Fig. 13. Spectrogram of a burst of eight sound pulses produced by a squirrelfish, *Holocentrus*. The fundamental frequency within each pulse is about 75–100 Hz. After Tavolga (1965), with permission of the U. S. Naval Training Device Center.

frequencies. The yelps of the gaff-topsail catfish range over 250 Hz (Tavolga, 1962), and the calls of toadfish may reach above 300 Hz (Tavolga, 1958c).

Some figures on the duration of swim bladder sounds in various species

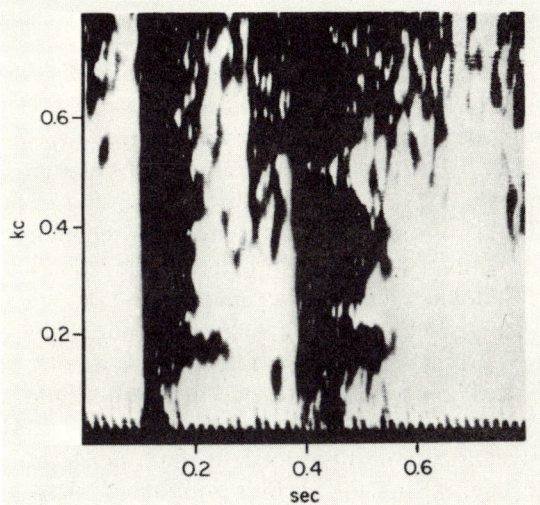

Fig. 14. Spectrogram of two gruntlike sounds produced by a marine catfish, *Galeichthys*. These are short pulses with a fundamental frequency of about 200 Hz. After Tavolga (1965), with permission of the U. S. Naval Training Device Center.

were given by Moulton (1958), Tavolga (1960), and Winn (1964) in spectrograms. In the majority of species, the sounds consist of single pulses with a duration of 20–100 msec each. Some forms characteristically produce a train of such pulses, and the repetition rates are probably species-specific. Sounds of the black grouper, *Mycteroperca bonaci,* usually occur in volleys of 4 or 5 pulses each, while other species normally emit one sound pulse at a time (Tavolga, 1960). Squirrelfish, *Holocentrus,* produce rattling volleys of up to 20 pulses in quick succession (Winn and Marshall, 1963). Many species of croakers and drumfish also produce such trains of pulses, and Winn (1964) classified such emissions as "fixed-interval signals." He proposed that this kind of temporal patterning can serve as a primitive means of communication.

Long, sustained tones are unusual. The boat-whistle sounds of toadfish vary from 350 to 450 msec in length, and occasional yelps of the gafftopsail catfish may reach 500 msec in duration (Tavolga, 1960).

The low-pitched pulses of many species are extremely difficult to distinguish from one another on the basis of acoustical characteristics alone. Harmonic structure is greatly affected by the conditions of recording and the equipment used. There is little known as to the consistency with which certain species produce characteristic pulse trains. The problem of identification of sound sources, therefore, is one that will require considerably more data than are now available.

3. Hydrodynamic and Swimming Sounds

The character of sounds produced by the motion of fishes through the water has only recently been recognized and described. Thus far, Moulton (1960b) has been the only investigator to report any spectral analyses on such sounds. He has found that the sounds are nonharmonic with the dominant frequencies extending far below 100 Hz.

The main sound output from individual or schooling fish occurred when there was a rapid change in direction or speed. The sound resembled a low roar or that of a wooden mallet striking the side of a boat under water. Such sounds could be detected from single predatory fish such as jacks or barracuda (Fig. 15). Sounds of veering schools of sardines, herrings, and anchovies were of lower amplitude and appeared to have more high frequency components (Tavolga, 1964b).

The pressure fluctuations produced by flow noise are affected by a wide variety of factors such as surface roughness, shape, and velocity. The predominant frequencies, upon spectral analysis, are in the range below 500 Hz, and the frequencies below 100 Hz are least affected by changes in the above variables (Skudrzyk and Haddle, 1963).

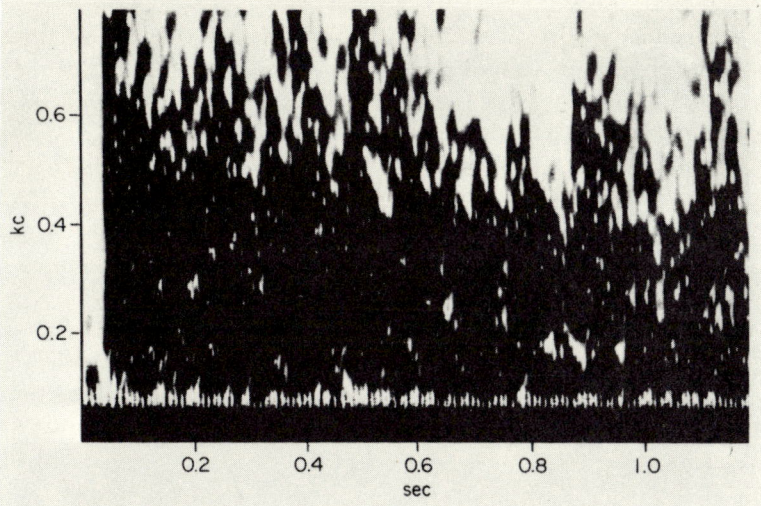

Fig. 15. Spectrogram of a hydrodynamic sound produced by a small school of jacks, *Caranx*. The sound begins almost explosively and contains many harmonically unrelated frequencies, mostly below 200 Hz. After Tavolga (1965), with permission of the U. S. Naval Training Device Center.

All of these hydrodynamic phenomena generate primarily near-field displacements and, with the usual pressure-sensitive hydrophones, can be detected only at short range and when emitted at a high intensity.

4. Sounds of Unknown Mechanisms

Tavolga (1956, 1958a,b, 1960) reported sounds produced by several species of small tidal zone fishes—gobies and blennies. These were low frequency thumps with a fundamental well below 100 Hz and with no harmonic distribution of frequencies. Acoustically these sounds resemble hydrodynamic pulses. Kinzer (1961) and Protasov *et al.* (1965) described similar sounds from European species of gobies. Tavolga (1960) proposed that a possible sonic mechanism here might be an explosive ejection of water through the gill slits.

Romanenko and Protasov (1963) found that the beluga sturgeon, *Huso huso*, produced at least four kinds of sounds: (1) whistles with a dominant frequency of about 3800 Hz, (2) a broad-band hissing, (3) sharp pulses with a dominant frequency of 100–125 Hz, and (4) short clicks with acoustic energy predominant around 4000 Hz and possibly higher.

The electric ray, *Torpedo marmorata*, was reported to produce low frequency grunts (below 100 Hz) in connection with its electrical discharges (Vincent, 1963b). Despite efforts to electrically isolate the hydrophone, the acoustic nature of these signals is still doubtful. If true, this report could become the only definite record of an elasmobranch producing sounds other than hydrodynamic ones.

[Material has been omitted at this point.]

IV. ACOUSTIC COMMUNICATION IN FISH

The channels available for animal interactions are: photic (visual), mechanical (including tactual and acoustical), thermal, chemical (gustatory and olfactory), and electrical (limited to certain fishes). Information transmitted along any of these channels can be broadly classified as: long range vs. short range and directional vs. nondirectional. Each of these channels has certain limitations in an aquatic medium as compared with a terrestrial environment.

The photic channel is severely limited in water (Dietrich, 1963), especially in seawater in regions of high planktonic concentration. The probability is that the range for effective vision in the marine environment is generally less than 1 meter, and in areas of high turbidity this effective range may be reduced to only a few centimeters.

The chemical channel is potentially an effective one in water, because of the large range of substances that are easily suspended or dissolved in water. This channel, however, is slow and nondirectional, and the source of the stimulus can be located only by means of some kinesislike movements in which the animal simply moves about until it finds areas of progressively higher stimulus concentrations.

Although useful in a terrestrial environment, the thermal channel is virtually unavailable to aquatic animals, especially to ectothermic forms such as fishes. Water absorbs heat rapidly, and a thermal gradient attenuates much too fast for any effective reception as a stimulus in an interaction.

Both ac and dc electrical fields are readily set up in water, especially in saltwater, and many species of fishes, in addition to the well-known electrical forms, are now known to be able to detect electrical potentials produced by other organisms. How widely this energy channel is used in interaction among fishes still remains to be studied.

Aside from the short-range, direct contact function of tactile receptors, the mechanical channel offers several advantages for interactions and communication among fishes. Acoustic energy under water is effective as pressure and as displacement. Both the inner ear and the lateral line are essentially displacement sensitive, but the inner ear receives near-field displacements from the nearby swim bladder. The swim bladder acts as a transducer for pressure waves and transforms them into local near-field effects (Harris and van Bergeijk, 1962; van Bergeijk, 1964). Pure pressure waves are efficiently propagated in water and this form of acoustic energy is probably the most rapid and effective channel for long-range interactions. At short range, directional orientation to a sound source by fishes can take place in the near field, probably by means of the lateral line. The majority of sounds produced by swim bladder mechanisms in fishes range from 50 to 100 Hz, and this would provide an effective near-field range of from about 50 to 3 meters, respectively. The acoustic channel, therefore, is one of long effective range and high information content for interactions among aquatic animals, especially fishes.

It is clear that fishes interact with each other in a variety of ways, using various sensory modalities, and the interactions occur in many different behavioral contexts. Some of these interactions have been called "communication," but the exact definition of animal communication is not always clear. Some authors, for example, have favored a broad definition that could conceivably include all kinds of interactions, while others restricted the use of the term "communication" to intraspecies interactions (Frings and Frings, 1964). Attempts have been made to find factors in common between human communication and that of other animals (Marler, 1961; Sebeok, 1965). Many of the approaches to animal communication involve cross-phyletic comparisons with little regard for the differences in phyletic position and level of organization of the organisms compared.

The concept of levels of organization is basic to biological science, and this concept is particularly applicable to behavior. Schneirla (1953) postulated a hierarchical arrangement of levels of behavioral integration with each level defined by qualitatively different organization and development of integrative systems. Tavolga (1969, 1970) applied the concept of levels of organization to the definition of communication. Communication is not a single phenomenon; rather, there are many kinds of communication. There are many kinds of interactions that are characteristic of different levels of organization, and only some of these levels involve communication in the strict sense. Interactions among fishes occur at various levels of complexity and organization. The understanding of these interactions must depend upon an appreciation of the level involved in each case and the developmental history of the behavior.

Levels of interaction among organisms were defined by Tavolga (1969, 1970) as follows:

Vegetative: Interactions by virtue of physical presence alone, as well as through growth or tropism.

Tonic: Interactions resulting from the more or less continuous processes fundamental to species-typical development and function, e.g., homeostasis.

Phasic: Interactions that result from discontinuous, more or less

6. SOUND PRODUCTION AND DETECTION 185

regular stages or events in the development of an organism. This would include some primitive types of sex discriminatory and food discriminatory behavior.

Signal: Interactions in which receiver and/or emitter utilizes some specialized structures. Specialized forms of behavior would also be involved.

Symbolic and Language: Interactions based on high levels of nervous integration, as in primates, including man.

The term "communication," therefore, could include any one or more levels in the above heirarchy of interactions. Restriction of the scope of the term now becomes practical and meaningful. Signal, symbolic, and language levels would comprise communication, while the vegetative, tonic, and phasic levels would include primitive forms of interactions (Tavolga, 1969). For communication to take place any one of the following conditions should be met: (1) The emitter must possess a specialized stimulus-producing mechanism (chemical, morphological, or behavioral); (2) the stimulus must occupy a narrow portion of the available spectrum of the channel (frequency range, duration, patterning, chemical specificity); and (3) the receiver must possess specialized receptors and respond in a specific manner.

Most fishes can react with each other on the tonic level. Excretory wastes, normal mucous secretions, carbon dioxide output, and other metabolic byproducts are readily diffused through water and can affect other animals in some physiological fashion. Body shapes, basic color patterns, normal locomotion, and other manifestations of homeostasis and maintenance behavior can result in both intra- and interspecies interactions.

Phasic level interactions are also common in fishes. The reaction to *Schreckstoff* is an example, even though it can sometimes be species-specific. Many predator–prey interactions are on this level, as well as much of aggressive and territorial behavior. Schooling is primarily a phasic level interaction, and some aspects of reproductive behavior are also phasic.

Although there is little evidence on signal level interactions in fishes, indications are that the exchange of specific signals may be important in reproductive behavior, especially during the prespawning, and parental care stages. The courtship stage is probably most crucial, since species and sex discrimination takes place at this time, and a continuing series of attracting interactions is necessary to hold the pair in proximity until the proper physiological state for spawning is reached. Such interactions can also stimulate the members of the pair to develop into spawning condition. Specialized reproductive activities, therefore, are the main interaction types among fishes that fall into the category of communication as defined here.

Many of the sounds described and cataloged by Fish (1954, 1964) and her co-workers were elicited from captive animals under various unspecified conditions of duress, including administration of electric shock. It is not known what behavioral significance these sounds could have. Moreover, the electric shock method may in some cases produce a neuromuscular response rather than a true behavior. Any animal with a swim bladder is potentially capable of sound production if the body wall muscles are abruptly stimulated to contract. Interpretations of biological significance, therefore, can only be made on the basis of field recordings or of relatively unrestrained animals under experimental conditions.

Some stridulatory sounds, such as those of grunts (Pomadasyidae) and other forms that grind their pharyngeal teeth, have been recorded only in air or from specimens grasped under water in an aquarium (Burkenroad, 1930). The fin ray stridulations of marine catfish are produced when the animal is captured and pulled out of water (Tavolga, 1960). Such sounds are very unlikely to have any biological significance. Many of these stridulatory sounds have most of their acoustic energy at frequencies above the hearing range of most fishes and thus should be considered unlikely to have any behavioral significance.

The gnashing sounds of feeding should be within the hearing range of fishes, but these sounds have not yet been shown to have any communicative value. It is quite possible that individuals hearing other fish eating may soon learn to come for food to the source of the sound. Hydrodynamic sounds may also play a role in the life of both a predator and a prey species. Sharks have long been known to approach the vicinity of a wounded fish, and D. R. Nelson and Gruber (1963) proposed that much of this attraction is the result of low frequency acoustic stimuli. Young lemon sharks, *Negaprion brevirostris*, will not only approach but will even attack a hydrophone that is emitting pulses of broadband noise, and similar hydrodynamic sounds were found to attract sharks (Banner, 1968). A detailed description of field studies on the effects of sounds on free-ranging sharks shows dramatic results (Myrberg *et al.*, 1969), but it is not yet clear what the specifications and parameters of an adequately attractive stimulus are. Not only sharks but also other predatory fishes have been shown to be attracted to the source of low frequency sound pulses, especially those hydrodynamic disturbances associated with active predation (Richard, 1968). Such interactions are probably on the phasic level, since the stimuli produced by the prey animals result from simple locomotor behavior, and the responses of the predators are nonspecific approaches.

A number of sounds are produced in some stressful situation. Groupers, squirrelfish, jacks, and a variety of other species have been described as doing so (Fish, 1954; Moulton, 1958; Tavolga, 1960). Approach of foreign objects, a poke of a stick, etc., are examples of such stimuli. Whether the sound is indicative of alarm, fright, anger, annoyance, or other emotional state is not known. When the approach of another fish stimulates sound production, then there is a likelihood of the sound functioning as a warning call from a territory holder toward an intruder, as in groupers (Moulton, 1958; Tavolga, 1960), squirrelfish (Winn et al., 1964), toadfish (Tavolga, 1960; Gray and Winn, 1961), pinfish (Caldwell and Caldwell, 1967), triggerfish (Salmon et al., 1968), and several other forms listed by Winn (1964). Most of these interactions are probably tonic or phasic in level, but the data on toadfish (Winn, 1967) indicate a more specific signal level communication.

Since ancient times, the relation of certain kinds of fish sounds to spawning behavior has been known. This is especially true of members of the croaker and drumfish family (Sciaenidae). In many parts of the world, including the Florida coast, fishermen commonly listen for the deep rattling sound of drumfish during their spring spawning migration. Schneider and Hasler (1960) described the spawning sounds of the freshwater drum, *Aplodinotus*. The European sciaenid *Corvina* has been reported by Dijkgraaf (1947a) and by Protasov and Aronov (1960) as producing sounds in connection with spawning behavior.

The codfish, *Gadus callarias*, and the haddock, *Melanogrammus aeglefinus*, have been reported to produce low frequency sounds during the spawning season (Brawn, 1961; Hawkins and Chapman, 1966).

The boat-whistle sounds of the toadfish, *Opsanus*, were thought to be related to spawning or to territorial behavior (Fish, 1954; Fish and Mowbray, 1959; Tavolga, 1958c, 1960). Gray and Winn (1961) and Winn (1964, 1967) have presented data showing that the sounds appear to be emitted by males only and serve to attract females to nesting sites. Winn (1967) also demonstrated the interactions between sounding territorial toadfish.

Positive evidence of the function of fish sounds in reproductive behavior have come from studies on freshwater minnows (Delco, 1960; Winn and Stout, 1960; Stout, 1963), on a characid (K. Nelson, 1965), on tidal zone species of gobies and blennies (Tavolga, 1956, 1958a,b), and on certain cichlid fishes (Myrberg et al., 1965). Experiments involving playback of sounds to the fish were used to establish the fact that sounds produced by courting males could attract females. It is probable that many of the instances of sounds used in reproductive behavior represent signal level communication, but needed data on specificity of the signal are often lacking.

Although no sex differences appear to exist in the morphology of the sonic mechanisms of the Batrachoididae, only male toadfish, *Opsanus tau*, are known to emit boat-whistle sounds (Gray and Winn, 1961; Winn, 1964), and only males of the midshipman, *Porichthys notatus*, produce any sounds at all (Cohen and Winn, 1967).

Some species appear to produce sounds when schooling or aggregating. Aggregations of sea robins produce a characteristic staccato call (Moulton, 1956), and nighttime schools of marine catfish form large choruses (Tavolga, 1960).

In some long-term observations of local populations, Breder (1968) found that the sea catfish, *Galeichthys felis*, aggregate and produce choruses of the "percolator" sounds from April to October, with a lull in July and August. These choruses usually start after 5 P.M. and cease before 11 P.M. Optimum water temperatures for chorus formation were from 74° to 89°F, and increased chorusing was noted during new moon periods.

Breder (1968) also observed that the boat-whistle sounds of the toadfish, *Opsanus beta*, were normally heard in March, April, August, September, and October, usually most frequently when the *Galeichthys* choruses were most vigorous. The repetition rate of the toadfish calls was found to be temperature dependent, averaging 0.93/min at 74°F and 1.92/min at 83°F. Toadfish calls of this Florida west coast species disappeared at water temperatures below 73°F and above 91°F.

As in the catfish (above), many species seem to show a daily rhythm in their sonic behavior. Winn *et al.* (1964) demonstrated this cyclic activity in the squirrelfish. This species, as well as many others, tend to show dawn and dusk peaks of sonic activity, possibly correlated with feeding or territorial movements. Cummings *et al.* (1964) found a cyclic occurrence, with dawn and dusk peaks, for many sounds of biological origin as monitored by a shore-based hydrophone system.

Since so many of the sounds produced by different species are similar, Winn (1964) proposed that the coding of information may be through temporal patterning. With the exception of a few species, like the toadfish, that produce harmonic tones, most fish sounds are short pulses with a fundamental near 100 Hz. There are, however, distinct differences among species in the grouping and repetition rates of these pulses. It is conceivable, according to Winn, that species discrimination and the communication of information as to emotional state (i.e., alarm, territorial defense, and prespawning behavior) is by way of temporal coding. This type of coding would be characteristic of the signal level of communication.

The presence of echolocation abilities among fishes have yet to be demonstrated. Griffin (1955) described a single instance in which a deep-sea species (no identified) was recorded and echoes of its cries could be detected bouncing back from the ocean floor. On this basis, Griffin proposed that fishes could conveivably utilize echoes of their own sounds for orientation. These observations, however, are in need of confirmation.

V. PROBLEMS AND PROSPECTS FOR THE FUTURE

The study of underwater bioacoustics is an interdisciplinary field and has captured the interests of physicists, acousticians, engineers, psychologists, physiologists, and scientists in many other specialities. As noted earlier, it is also a rapidly expanding field and as a by-product of the expansion, numerous new areas of research are unfolding. Some of these are basic and some applied, and it was pointed out by Galler (1967) that this field has served as a good example of the interaction between basic and applied research.

Ever since the development of ASDIC and Sonar, the effectiveness of echo-ranging equipment in detecting marine life has shown a steady increase. Although not directly related to sound production and detection by fishes, the utilization of these techniques for detection of fishes has moved to increasing degrees of precision (Hester, 1967; Cushing, 1967; Weston, 1967) and deserves mention here. Sonar has already proved its value in assisting the commercial fisherman, and it promises to be an excellent tool for the study of fish locomotion and the behavior of fish schools.

The utilization of sound to guide fish has only recently been applied commercially. The bibliography compiled by Moulton and Backus (1955) showed that these techniques for guiding or attracting fishes have been utilized by fishermen in many parts of the world. Some of the techniques are primitive and of ancient origin (Busnel, 1959; Wolff, 1966). Recent attempts have been made, using modern electronic equipment, to attract fishes by playback of their feeding sounds, and some success in this area has been reported (Hashimoto and Maniwa, 1967). Preliminary studies on guiding migrating salmon by means of acoustic stimuli have also shown promise for the future (Vanderwalker, 1967). The recent observations by Richard (1968) demonstrated the feasibility of attracting a variety of predatory fishes by means of low frequency sound. For the successful development of these techniques, it is clear that more basic information is still required on the hearing range of many of the commercially important species, as well as more information on the behavioral significance of sounds produced by these fishes.

Although the technique of listening for sounds produced by fishes is known as an art among fishermen in several areas of the world (Moulton, 1963), the use of this method in major commercial fisheries has yet to be tested. Some preliminary studies have been attempted, using a combination of sonobuoy and telemetry equipment (Hashimoto et al., 1960). Considering the technology now available, it should be feasible to detect and identify fish sounds electronically, although much needed information on identifying characteristics of commercially desirable species is still lacking.

It is no longer sufficient to present another instance of a fish making sounds, and such information needs to be correlated with the behavioral context of the sound. Long-term recordings and observations, like those supplied by the acoustic–video installation of the Institute of Marine Science at Bimini, Bahamas (Kronengold et al., 1964), will be increasingly necessary. Bioacoustic observations by small submersibles hold much promise for the future of the field (Backus et al., 1968), and some scuba divers are already becoming equipped with listening devices. In this connection, the identification of many hitherto unspecified field contacts will become possible. Behavioral studies, however, will have to be supplemented by laboratory observations and experiments where environmental conditions can be controlled.

A problem that has troubled both field and laboratory investigators has been the exact specification and description of the acoustic stimulus. Watkins (1967) pointed out the pitfalls of analyzing equipment, but, further, it is extremely difficult to separate the two forms of energy that usually exist together in an underwater sound field: pressure and displacement. Most hydrophones are basically pressure transducers and will detect pressure changes produced by a near field as well as those of a far field. The needs of biologists for a small, but sensitive, displacement transducer have not yet been satisfied (Tavolga, 1967c), and this is another area in which an interdisciplinary approach to the problem is necessary.

The problem of frequency discrimination by fish is an intriguing one. If future evidence demonstrates the applicability of a place theory to fish hearing, it may necessitate a reexamination of the concept of the critical band.

6. SOUND PRODUCTION AND DETECTION   191

The mechanisms of sound production in fishes need further elucidation, especially in relation to the remarkable fast-acting properties of the sonic muscles. If these properties, as well as those of common innervation and embryonic origin, are found to be as widely distributed among the diverse families of fishes as is presently apparent, then questions begin to arise as to the evolutionary origin of sound production.

Data are now being accumulated at a rapid rate on the behavioral significance of fish sounds. Acoustic interactions and communication among fishes are evidently more common than was thought only about 20 years ago. Both laboratory and field investigations are now establishing the behavioral contexts for such interactions in many species. One area of this study that has yet to be examined is the ontogeny of these interactions. A strong trend in modern animal behavioral study is toward an understanding of the development of behavior. It is no longer appropriate to simply label a behavior as innate or learned, but it is necessary to investigate the physiological, experiential, genetic, and other antecedents of the behavior as it develops in the individual (Schneirla, 1965). One of the few ontogenetic behavioral studies in fishes has been on the development of schooling. Shaw (1960, 1961) showed the possible relationship of the optomotor response in larval fishes to aggregation and schooling. Investigations of the development of sonic behavior in fishes are clearly indicated and desirable.

The future of the field of fish bioacoustics is exciting, and areas for research are available to workers from many disciplines. Many basic problems still exist, but at the same time the application of the basic information is feasible and has a great potential toward improving our understanding of aquatic resources in general.

[Material has been omitted at this point.]

REFERENCES

Agassiz, J. L. (1850). Manner of producing sounds in catfish and drumfish. *Proc. Am. Acad. Arts Sci.* **2**, 238.

Albers, V. M. (1965). "Underwater Acoustics Handbook—II." Penn. State Univ. Press, University Park, Pennsylvania.

Alexander, R. M. (1959a). The physical properties of the swim bladder in intact Cypriniformes. *J. Exptl. Biol.* **36**, 315–332.

Alexander, R. M. (1959b). The physical properties of the swim bladders of fish other than Cypriniformes. *J. Exptl. Biol.* **36**, 347–355.

Backus, R. H. (1958). Sound production by marine animals. *J. Underwater Acoustics* **8**, 191–199.

Backus, R. H., Craddock, J. E., Haedrich, R. L., Shores, D. L., Teal, J. M., Wing, A. S., Mead, G. W., and Clarke, W. D. (1968). *Ceratoscopelus maderensis*: Peculiar sound-scattering layer identified with this myctophid fish. *Science* **160**, 991–993.

Barber, S. B., and Mowbray, W. H. (1956). Mechanism of sound production in the sculpin. *Science* **124**, 219–220.

Bondesen, P., and Davis, L. I. (1966). Sound analysis within biological acoustics. Terms and definitions. *Natura Jutlandica* **12**, 235–239.

Brawn, V. M. (1961). Sound production by the cod (*Gadus callarias* L.). *Behaviour* **18**, 239–255.

Breder, C. M., Jr. (1968). Seasonal and Survival occurrences of fish sounds in a small Florida bay. *Bull. Am. Museum Nat. Hist.* **138**, 325–378.

Bridge, T. W., and Haddon, A. C. (1894). Notes on the production of sounds by the air bladder of certain siluroid fishes. *Proc. Roy. Soc.* **55**, 439–441.

Broughton, W. B. (1963). Method in bio-acoustic terminology. Glossarial index. *In* "Acoustic Behaviour of Animals" (R.-G. Busnel, ed.), pp. 3–24 and 824–910. Elsevier, Amsterdam.

Burkenroad, M. D. (1930). Sound production in the Haemulidae. *Copeia* pp. 17–18.

Burkenroad, M. D. (1931). Notes on the sound-producing marine fishes of Louisiana. *Copeia* pp. 20–28.

Busnel, R.-G. (1959). Etude d'un appeau acoustique pour la pêche, utilisé au Sénégal et au Niger. *Bull. Inst. Franc. Afr. Noire* **A21**, 346–360.

Busnel, R.-G., ed. (1963). "Acoustic Behaviour of Animals." Elsevier, Amsterdam.

Caldwell, D. K., and Caldwell, M. C. (1967). Underwater sounds associated with aggressive behavior in defense of territory by the pinfish, *Lagodon rhomboides*. *Bull. S. Calif. Acad. Sci.* **66**, 69–75.

Cohen, M. J., and Winn, H. E. (1967). Electrophysiological observations on hearing and sound production in the fish, *Porichthys notatus*. *J. Exptl. Zool.* **165**, 355–370.

Cummings, W. C., Brahy, B. D., and Herrniknd, W. F. (1964). The occurrence of underwater sounds of biological origin off the west coast of Bimini, Bahamas. *In* "Marine Bio-Acoustics" (W. N. Tavolga, ed.), pp. 27–43. Pergamon Press, Oxford.

Cummings, W. C., Brahy, B. D., and Spires, J. Y. (1966). Sound production schooling, and feeding habits of the margate, *Haemulon album* Cuvier, off North Bimini, Bahamas. *Bull. Marine Sci.* **16**, 626–640.

Cushing, D. H. (1967). The acoustic estimation of fish abundance. *In* "Marine Bio-Acoustics" (W. N. Tavolga, ed.), Vol. 2, pp. 75–91. Pergamon Press, Oxford.

Dietrich, G. (1963). "General Oceanography." Wiley (Interscience), New York.

Dijkgraaf, S. (1932). Über Lautäusserungen der Elritze. *Z. Vergleich. Physiol.* **17**, 802–805.

Dijkgraaf, S. (1947a). Ein Töne erzeugender Fisch in Neapler Aquarium. *Experientia* **3**, 493–494.

Dobrin, M. B. (1947). Measurements of underwater noise produced by marine life. *Science* **105**, 19–23.

Dobrin, M. B. (1948). Recording sounds of undersea life. *Trans. N. Y. Acad. Sci.* [2] **11**, 91–96.

Dorai Raj, B. S. (1960a). On the production of underwater sound by *Therapon jarbua*. *Current Sci.* (India) **29**, 277–278.

Dôtu, Y. (1951). On the sound producing mechanisms of a scorpaenid fish, *Sebasticus marmoratus*. *Kyushu Imp. Univ. Dept. Agr. Bull. Sci.* **13**, 286–288.

Dufossé, M. (1874). Recherches sur les sons expresifs que font entendre les poissons d'Europe. *Ann. Sci. Nat.* [5] **19**, 1–53; **20**, 1–134.

Fänge, R. (1966). Physiology of the swimbladder. *Physiol. Rev.* **46**, 299–322.

Fänge, R., and Wittenberg, J. B. (1958). The swim bladder of the toadfish (*Opsanus tau* L.). *Biol. Bull.* **115**, 172–179.

Fawcett, D. W., and Revel, J. P. (1961). The sacroplasmic reticulum of a fast-acting fish muscle. *J. Biophys. Biochem. Cytol.* **10**, Suppl., 89–109.

Ffowcs-Williams, J. E. (1967). Flow noise. *In* "Underwater Acoustics" (V. M. Albers, ed.), Vol. 2, pp. 89–102. Plenum Press, New York.

Fish, M. P. (1953). The production of underwater sound by the northern seahorse, *Hippocampus hudsonius*. *Copeia* pp. 98–99.

Fish, M. P. (1954). The character and significance of sound production among fishes of the western North Atlantic. *Bull. Bingham Oceanog. Coll.* **14**, 1–109.

Fish, M. P. (1964). Biological sources of sustained ambient noise. *In* "Marine Bio-Acoustics" (W. N. Tavolga, ed.), pp. 175–194. Pergamon Press, Oxford.

Fish, M. P., and Mowbray, W. H. (1959). The production of underwater sounds by *Opsanus* sp., a new toadfish from Bimini, Bahamas. *Zoologica* **44**, 71–76.

Fish, M. P., Kelsey, A. S., Jr., and Mowbray, W. M. (1952). Studies on the production of underwater sound by North Atlantic coastal fishes. *J. Marine Res.* **11**, 180–193.

Frings, H., and Frings, M. (1964). "Animal Communication." Ginn (Blaisdell), Boston, Massachusetts.

Gainer, H., and Klancher, J. E. (1965). Neuromuscular junctions in a fast-contracting fish muscle. *Comp. Biochem. Physiol.* **15**, 159–165.

Gainer, H., Kusano, K., and Mathewson, R. F. (1965). Electrophysiological and mechanical properties of squirrelfish sound-producing muscle. *Comp. Biochem. Physiol.* **14**, 661–671.

Geoffroy St. Hilaire, I. (1829). Histoire naturelle des poissons du Nil. *In* "Déscription de l'Égypt," 2nd ed., Vol. 24, pp. 141–338.

Gray, G.-A., and Winn, H. E. (1961). Reproductive ecology and sound production of the toadfish, *Opsanus tau*. *Ecology* **42**, 274–282.

Greene, C. W. (1924). Physiological reactions and structure of the vocal apparatus of the California singing fish. *Am. J. Physiol.* **70**, 496–499.

Griffin, D. R. (1950). "Underwater Sounds and the Orientation of Marine Animals—A Preliminary Survey," Tech. Rept. No. 3, Proj. NR 162–429. ONR and Cornell University.

Griffin, D. R. (1955). Hearing and acoustical orientation in marine animals. *Deep-Sea Res.* **3**, Suppl., 406–417.

Harris, G. G. (1964). Considerations on the physics of sound production by fishes. In "Marine Bio-Acoustics" (W. N. Tavolga, ed.), pp. 233–247. Pergamon Press, Oxford.

Hashimoto, T., and Maniwa, Y. (1967). Research on the luring of fish shoals by utilizing underwater acoustical equipment. In "Marine Bio-Acoustics" (W. N. Tavolga, ed.), Vol. 2, pp. 93–104. Pergamon Press, Oxford.

Hashimoto, T., Nishimura, M., and Maniwa, Y. (1960). Sonobuoy, detection of fish by sonobuoy. Bull. Japan. Soc. Sci. Fisheries 26, 245–249.

Hawkins, A. D., and Chapman, C. J. (1966). Underwater sounds of the haddock, Melanogrammus aeglefinus. J. Marine Biol. Assoc. U. K. 46, 241–247.

Hazlett, B. A., and Winn, H. E. (1962). Sound producing mechanism of the Nassau grouper Epinephelus striatus. Copeia pp. 447–449.

Hester, F. J. (1967). Identification of biological sonar targets from body-motion Doppler shifts. In "Marine Bio-Acoustics" (W. N. Tavolga, ed.), Vol. 2, pp. 59–74. Pergamon Press, Oxford.

Horton, J. W. (1959). "Fundamentals of Sonar." U. S. Naval Inst., Annapolis, Maryland.

Johnson, M. W. (1948). Sound as a tool in marine ecology, from data on biological noises and the deep scattering layer. J. Marine Res. 7, 443–458.

Jones, F. R. H., and Pearce, G. (1958). Acoustic reflexion experiments with perch (Perca fluviatilis Linn.) to determine the proportion of the echo returned by the swim bladder. J. Exptl. Biol. 35, 437–450.

Kellogg, W. N. (1953). Bibliography of the noises made by marine organisms. Am. Museum Novitates 1611, 1–5.

Kellogg, W. N. (1955). "Sounds of Sea Animals," Vol. 2, Florida. Folkways Record Album No. FPX 125, Folkways Record and Service Corp., New York.

Kilarski, W. (1964). The organization of the sarcoplasmic reticulum in skeletal muscles of the swim-bladder in the burbot (Lota lota L.). Acta Biol. Cracov., Ser. Zool. 7, 161–168.

Kinzer, J. (1961). Über die Lautäusserungen der Schwarzgrundel Gobius jozo. Aquar. Terrar-kunde 7, 7–10.

Klausewitz, W. (1958). Lauterzeugung als Abwehrwaffe bei der hinterindischen Tiger-schmerle (Botia hymenophysa). Natur Volk 88, 343–349.

Knudsen, V. O., Alford, R. S., and Emling, J. W. (1948). Underwater ambient noise. J. Marine Res. 7, 410–429.

Kronengold, M., Dann, R., Green, W. C., and Loewenstein, J. M. (1964). An acoustic-video system for marine biological research. Description of the system. In "Marine Bio-Acoustics" (W. N. Tavolga, ed.), pp. 11–26. Pergamon Press, Oxford.

Lanyon, W. E., and Tavolga, W. N., eds. (1960). "Animal Sounds and Communication," Publ. No. 7. Am. Inst. Biol. Sci., Washington, D. C.

Loye, D. F., and Proudfoot, D. A. (1946). Underwater noise due to marine life. J. Acoust. Soc. Am. 18, 446–449.

Mackenzie, K. V. (1960). Formulas for the computation of sound speed in sea water. J. Acoust. Soc. Am. 32, 100.

Mahajan, C. L. (1963). Sound producing apparatus in an Indian catfish Sisor rhabdophorus Hamilton. J. Linnean Soc. London, Zool. 44, 721–724.

Maliukina, G. A., and Protasov, V. R. (1960). Hearing, "voice" and reactions of fish to sounds. Usp. Sovrem. Biol. 50, 229–242.

Marler, P. (1961). The logical analysis of animal communication. J. Theoret. Biol. 1, 295–317.

Marshall, J. A. (1963). Sound production and reproductive behavior of the croaking gourami, Trichopsis vittatus. Am. Zoologist 3, 231.

Marshall, N. B. (1951). Bathypelagic fish as sound scatterers in the ocean. *J. Marine Res.* **10**, 1–17.
Marshall, N. B. (1962). The biology of sound-producing fishes. *Zool. Soc. London, Symp.* 45–60.
Marshall, N. B. (1967). Sound-producing mechanisms and the biology of deep-sea fishes. *In* "Marine Bio-Acoustics" (W. N. Tavolga, ed.), Vol. 2, pp. 123–133. Pergamon Press, Oxford.
Meder, E. (1953). Trichopsis vittatus, der knurrende Gurami. *Aquar. Terrar. Z.* **6**, 1–2.
Meyer, E. (1957). Air bubbles in water. *In* "Technical Aspects of Sound" (E. G. Richardson, ed.), Vol. 2, pp. 222–229. Elsevier, Amsterdam.
Midttun, L., and Hoff, I. (1962). Measurements of the reflection of sound by fish. *Rept. Norweg. Fishery Invest.* **13**, 1–18.
Moreau, A. (1876). Recherches expérimentales sur les functions de la vessie natatoire. *Ann. Sci. Nat.* [6] **4**, 1–85.
Moulton, J. M. (1956). Influencing the calling of sea robins (*Prionotus* spp.) with sound. *Biol. Bull.* **111**, 393–398.
Moulton, J. M. (1958). The acoustical behavior of some fishes in the Bimini area. *Biol. Bull.* **114**, 357–374.
Moulton, J. M. (1960a). The acoustical anatomy of teleost fishes. *Anat. Record* **138**, 371–372.
Moulton, J. M. (1960b). Swimming sounds and the schooling of fishes. *Biol. Bull.* **119**, 210–223.
Moulton, J. M. (1963). Acoustic behaviour of fishes. *In* "Acoustic Behaviour of Animals" (R.-G. Busnel, ed.), pp. 655–687. Elsevier, Amsterdam.
Moulton, J. M., and Backus, R. H. (1955). Annotated references concerning the effects of man-made sounds on the movements of fishes. *Fisheries Circ., Dept. Sea Shore Fisheries, Maine* No. 17, 1–8.
Müller, J. (1842). Beobachtungen über die Schwimmblase der Fische, mit Bezug auf einige neue Fischgattungen. *Arch. Anat. Physiol.* pp. 307–329.
Müller, J. (1843). Untersuchungen über die Eingeweide der Fische. *Abhandl. Kgl. Akad. Wiss. Berlin* pp. 109–170.
Myrberg, A. A., Jr., Kramer, E., and Heinecke, P. (1965). Sound production by cichlid fishes. *Science* **149**, 555–558.
Nelson, K. (1965). The evolution of a pattern of sound production associated with courtship in the characid fish, *Glandulocauda inequalis*. *Evolution* **18**, 526–540.
Nursall, J. R. (1962). Swimming and the origin of paired appendages. *Am. Zoologist* **2**, 127–141.
Packard, A. (1960). Electrophysiological observations on a sound-producing fish. *Nature* **187**, 53–54.
Parrish, B. B., Hemmings, C. C., Chapman, C. J., and Hawkins, A. D. (1968). (no title). *In* "Fisheries of Scotland," Report for 1967, p. 99. Edinburgh.
Pfeiffer, W., and Eisenberg, J. F. (1965). Die Lauterzeugung der Dornwelse (Doradidae) und der Fiederbartwelse (Mochokidae). *Z. Morphol. Oekol. Tiere* **54**, 669–679.
Protasov, V. R. (1965). "Bioakustika Ryb." Akad. Nauk, Moscow.
Protasov, V. R., and Aronov, M. I. (1960). On the biological significance of sounds of certain Black Sea fish. *Biofizika* **5**, 750–752.
Protasov, V. R., Tsvietkov, V. I., and Rashcheperin, V. K. (1965). Acoustic signals emitted by *Neogobius melanostomus* Pallas. *Zh. Obshch. Biol.* **26**, 151–160.
Rauther, M. (1945). Über die Schwimmblase und die zu ihr in Beziehung tretenden

somatischen Muskeln bei den Triglidae und anderen Scleroparei. *Zool. Jahrb. Abt. Anat.* **69,** 159–250.

Reickel, A. (1936). *Trichopsis vittatus* (Cuvier & Valenciennes) (Croaking or purring gourami). *Aquarium* **4,** 257–259.

Richardson, E. G. (1957). Propagation of sound in the atmosphere and the sea. *In* "Technical Aspects of Sound" (E. G. Richardson, ed.), Vol. 2, pp. 1–30. Elsevier, Amsterdam.

Romanenko, E. V., and Protasov, V. R. (1963). On the sounds of the beluga. *Priroda* **1963,** 118–120.

Salmon, M., and Winn, H. E. (1966). Sound production by priacanthid fishes. *Copeia* pp. 869–972.

Salmon, M., Winn, H. E., and Sorgente, N. (1968). Sound production and associated behavior in triggerfishes. *Pacific Sci.* **22,** 11–20.

Schevill, W. E., Backus, R. H., and Hersey, J. B. (1962). Sound production by marine animals. *In* "The Sea" (M. N. Hill, ed.), Vol. 1, pp. 540–566. Wiley (Interscience), New York.

Schneider, H. (1961). Neuere Ergebnisse der Lautforschung bei Fischen. *Naturwissenschaften* **48,** 513–518.

Schneider, H. (1964a). Bioakustische Untersuchungen an Anemonenfishen der Gattung Amphiprion (Pisces). *Z. Morphol. Oekol. Tiere* **53,** 453–474.

Schneider, H. (1964b). Physiologische und morphologische Untersuchungen zur Bioakustik der Tigerfische (Pisces, Theraponidae). *Z. Vergleich. Physiol.* **47,** 493–558.

Schneider, H. (1967). Morphology and physiology of sound-producing mechanisms in teleost fishes. *In* "Marine Bio-Acoustics" (W. N. Tavolga, ed.), Vol. 2, pp. 135–158. Pergamon Press, Oxford.

Schneider, H., and Hasler, A. D. (1960). Laute und Lauterzeugung beim Süsswassertrommler *Aplodinotus grunniens* Rafinesque. *Z. Vergleich. Physiol.* **43,** 499–517.

Schneirla, T. C. (1953). The concept of levels in the study of social phenomena. *In* "Groups in Harmony and Tension" (M. Sherif and C. Sherif, eds.), pp. 52–75. Harper, New York.

Schneirla, T. C. (1965). Aspects of stimulation and organization in approach/withdrawal processes underlying vertebrate behavioral development. *In* "Advances in the Study of Behavior" (D. S. Lehrman, R. A. Hinde, and E. Shaw, eds.), Vol. 1, pp. 1–74. Academic Press, New York.

Sebeok, T. A. (1965). Animal communication. *Science* **147,** 1006–1014.

Shishkova, E. V. (1958a). On the reactions of fishes to sounds and the spectrum of trawler noise. *Tr. Vses. Inst. Morsk. Ribn. Hozaist. Okeanograf.* **34,** 33–39.

Shishkova, E. V. (1958b). Notes and investigations on sounds produced by fishes. *Tr. Vses. Inst. Morsk. Ribn. Hozaist. Okeanograf.* **36,** 280–294.

Skoglund, C. R. (1959). Neuromuscular mechanisms of sound production in *Opsanus tau. Biol. Bull.* **117,** 438.

Skudrzyk, E. J., and Haddle, G. P. (1963). Flow noise, theory and experiment. *In* "Underwater Acoustics" (V. M. Albers, ed.), pp. 255–278. Plenum Press, New York.

Smith, H. M. (1905). The drumming of drum fishes (Sciaenidae). *Science* **22,** 376–378.

Smith, H. M. (1927). The so-called musical sole of Siam. *J. Siam Soc.* **7,** Nat. Hist. Suppl., 49–54.

Sørensen, W. (1884). "Om Lydorganer hos Fiske: En physiologisk og comparativeanatomisk Undersögelse." Copenhagen.

Sørensen, W. (1894). Are the extrinsic muscles of the air bladder in some Siluroidea and the "elastic spring" apparatus of others subordinate to the voluntary production of sound? *J. Anat. Physiol.* [9] **29**, 109–139, 205–229, 399–423, and 518–552.

Stampehl, H. (1931). Ctenops vittatus (der Knurrende Gurami). *Bl. Aquar.-u. Terrarienk.* **42**, 394–395.

Steinberg, J. C., and Koczy, F. F. (1964). An acoustic-video system for marine biological research. Objectives and requirements. In "Marine Bio-Acoustics" (W. N. Tavolga, ed.), pp. 1–9. Pergamon Press, Oxford.

Steinberg, J. C., Kronengold, M., and Cummings, W. C. (1962). Hydrophone installation for the study of soniferous marine animals. *J. Acoust. Soc. Am.* **34**, 1090–1095.

Steinberg, J. C., Cummings, W. C., Brahy, B. D., and MacBain (Spires), J. Y. (1965). Further bio-acoustic studies off the west coast of North Bimini, Bahamas. *Bull. Marine Sci.* **15**, 942–963.

Stout, J. F. (1963). The significance of sound production during the reproductive behavior of *Notropis analostanus* (Family Cyprinidae). *Animal Behaviour* **11**, 83–92.

Tavolga, W. N. (1956). Visual, chemical and sound stimuli as cues in the sex discriminatory behavior of the gobiid fish, *Bathygobius soporator*. *Zoologica* **41**, 49–64.

Tavolga, W. N. (1958a). The significance of underwater sounds produced by males of the gobiid fish, *Bathygobius soporator*. *Physiol. Zool.* **31**, 259–271.

Tavolga, W. N. (1958b). Underwater sounds produced by males of the blenniid fish, *Chasmodes bosquianus*. *Ecology* **39**, 759–960.

Tavolga, W. N. (1958c). Underwater sounds produced by two species of toadfish, *Opsanus tau* and *Opsanus beta*. *Bull. Marine Sci. Gulf Caribbean* **8**, 278–284.

Tavolga, W. N. (1960). Sound production and underwater communication in fishes. In "Animal Sounds and Communication" (W. E. Lanyon and W. N. Tavolga, eds.), Publ. No. 7, pp. 93–136. Am. Inst. Biol. Sci., Washington, D. C.

Tavolga, W. N. (1962). Mechanisms of sound production in the ariid catfishes *Galeichthys* and *Bagre*. *Bull. Am. Museum Nat. Hist.* **124**, 1–30.

Tavolga, W. N., ed. (1964a). "Marine Bio-Acoustics." Pergamon Press, Oxford.

Tavolga, W. N. (1964b). Sonic characteristics and mechanisms in marine fishes. In "Marine Bio-Acoustics" (W. N. Tavolga, ed.), pp. 195–211. Pergamon Press, Oxford.

Tavolga, W. N. (1965). "Review of Marine Bio-Acoustics. State of the Art: 1964," Tech. Rept. 1212-1. U. S. Naval Training Device Center, Port Washington, New York.

Tavolga, W. N., ed. (1967a). "Marine Bio-Acoustics," Vol. 2. Pergamon Press, Oxford.

Tavolga, W. N. (1967c). Underwater sound in marine biology. In "Underwater Acoustics" (V. M. Albers, ed.), Vol. 2, pp. 35–41. Plenum Press, New York.

Tavolga, W. N. (1969). Fishes. In "Animal Communication" (T. A. Sebeok, ed.), pp. 271–288. Indiana Univ. Press, Bloomington, Indiana.

Tavolga, W. N. (1970). Levels of interaction in animal communication. In "Evolution and Development of Animal Behavior" (L. R. Aronson, E. Tobach, D. S. Lehrman, and J. S. Rosenblatt, eds.), pp. 281–302. Freeman, San Francisco.

Taylor, M., and Mansueti, R. J. (1960). Sounds produced by a very young crevalle jack, *Caranx hippos*, from the Maryland seaside. *Chesapeake Sci.* **1**, 115–116.

Tower, R. W. (1908). The production of sound in the drumfishes, the sea robin and the toadfish. *Ann. N. Y. Acad. Sci.* **18**, 149–180.

Tschiegg, C. E., and Hays, E. E. (1959). Transistorized velocimeter for measuring the speed of sound in the sea. *J. Acoust. Soc. Am.* **31**, 1038–1039.

Vigoureux, P. (1960). Underwater sound. *Proc. Roy. Soc.* **B152**, 49–51.
Vincent, F. (1963a). Note préliminaire sur des émissions acoustiques chez *Balistes capriscus* L. *Bull. Centre Etudes Rech. Sci. Biarritz* **4**, 307–316.
Vincent, F. (1963b). Observations préliminaires sur un type d'émission acoustique chez *Torpedo marmorata* Risso. *Bull. Centre Etudes Rech. Sci. Biarritz* **4**, 317–327.
von Ihering, R. (1930). Sur la voix des poissons d'eau douce. *Compt. Rend. Soc. Biol.* **103**, 1327–1328.
Walters, V. (1960). The swimbladder of *Velifer hypselopterus*. *Copeia* pp. 144–145.
Watkins, W. A. (1967). The harmonic interval: Fact or artifact in spectral analysis of pulse trains. *In* "Marine Bio-Acoustics" (W. N. Tavolga, ed.), Vol. 2, pp. 15–43. Pergamon Press, Oxford.
Wenz, G. M. (1962). Acoustic ambient noise in the ocean—spectra and sources. *J. Acoust. Soc. Am.* **34**, 1936–1956.
Wenz, G. M. (1964). Curious noises and the sonic environment in the ocean. *In* "Marine Bio-Acoustics" (W. N. Tavolga, ed.), pp. 101–119. Pergamon Press, Oxford.
Weston, D. E. (1967). Sound propagation in the presence of bladder-fish. *In* "Underwater Acoustics" (V. M. Albers, ed.), Vol. 2, pp. 55–88. Plenum Press, New York.
Winn, H. E. (1964). The biological signficance of fish sounds. *In* "Marine Bio-Acoustics" (W. N. Tavolga, ed.), pp. 213–231. Pergamon Press, Oxford.
Winn, H. E. (1967). Vocal facilitation and the biological significance of toadfish sounds. *In* "Marine Bio-Acoustics" (W. N. Tavolga, ed.), Vol. 2, pp. 283–304. Pergamon Press, Oxford.
Winn, H. E., and Marshall, J. A. (1963). Sound-producing organ of the squirrelfish, *Holocentrus rufus*. *Physiol. Zool.* **36**, 34–44.
Winn, H. E., and Stout, J. F. (1960). Sound production by the satinfin shiner, *Notropis analostanus*, and related fishes. *Science* **132**, 222–223.
Winn, H. E., Marshall, J. A., and Hazlett, B. A. (1964). Behavior, diel activities, and stimuli that elicit sound production and reaction to sounds in the lonspine squirrelfish. *Copeia* pp. 413–425.

2
RECENT ADVANCES IN THE STUDY OF SOUND PRODUCTION IN FISHES

William N. Tavolga

This article was prepared expressly for this Benchmark volume.

It is evident by now that many fishes use sounds in intra- and inter-species interactions. Simply to add to the list of known sonic species is no longer of interest in and of itself. Descriptions of the sounds are no longer significant unless they are tied in with some behavioral observations. The questions in this field have shifted from the descriptive to the functional areas, i.e., communication theory and the analysis of the particular role sounds play in the behavioral repertoire of the species.

BEHAVIORAL CORRELATES OF SOUND PRODUCTION

Since the 1971 review (Tavolga, 1971a), there have been a number of significant contributions, several of these on species native to fresh water, a habitat that hitherto has been somewhat neglected in this field.

In two reports on cichlids, territorial and aggressive displays were found to accompany sound emissions in *Cichlasoma centrarchus* (Schwarz, 1974) and *Tilapia mossambica* (Lanzing, 1974). Sounds of *Cichlasoma* males appear to inhibit aggressive responses in conspecifics, and perhaps these sounds function as part of threat displays that establish dominance hierarchies. In six species of sunfish (*Lepomis,* Centrarchidae), sounds associated with courtship behavior were described from field observations (Gerald, 1971). Although the sounds were all basically similar, some species-typical characteristics were described.

Reporting on a series of studies on the bicolor damselfish (*Eupomacentrus partitus*), Myrberg (1972a, 1972b) used a combination of both field and laboratory observations and compiled a large amount of data based upon a sequential and correlational analysis of sound production and behavior. This species produced five different sounds, each correlated with a specific behavior pattern or color change associated with territorial and reproductive activities. One of the unique features of this study was the use of a video-acoustic installation with shore-based monitoring and recording equipment. A similar correlation of sound and behavior was reported from field observations on the dusky damselfish, *Eupomacentrus dorsopunicans* (Burke and Bright, 1972).

Through the use of an underwater habitat, Bright and Sartori (1972) recorded sound production and behavior in three species of squirrelfishes in the Bahamas (*Holocentrus coruscus, H. rufus,* and *Myripristis jacobus*). In addition to describing species differences among the sounds, they were able, through long term observation, to show diel cycles in sound production and the correlation of certain sounds with aggressive displays by territory-holding individuals.

In Pacific species of squirrelfishes (*Myripristis violaceus* and *M. pralinius*), Horch and Salmon (1973) demonstrated species-typical characteristics in their sounds.

Large choruses of sciaenid fishes have been well-known to produce a high level of ambient noise. One recent addition to the literature was a report by Fish and Cummings (1972) on the breeding chorus of orangemouth corvina (*Cynoscion xanthulus*). The recordings were made of an artificially introduced population in the Salton Sea in California, and the nighttime ambient noise level was increased by a factor of 50 dB resulting entirely from the sonic activity of these fish.

RESPONSES TO SOUND PLAYBACKS IN FISHES

The capacities of fishes to detect underwater sounds has been reviewed elsewhere (Tavolga, 1971a; Popper and Fay, 1973; and in another volume in this series).

The use of playbacks of sounds to fishes in both laboratory and field conditions provides one of the best means for obtaining data on behavioral correlates of the sounds. This is especially so since we now know the range of frequencies, intensities, and discriminatory capacities that many fishes possess. Specific responses, behavioral and sonic, were observed during conspecific sound playbacks in sunfish (Gerald, 1971), squirrelfish (Horch and Salmon, 1973), damselfish (Myrberg, 1972b; Myrberg and Spires, 1972), and toadfish (Fish, 1972).

Responses to sound playbacks in the damselfish confirmed the behavioral correlations that were made through observations of interactions among individuals (Myrberg, 1972b). Sounds of related species of damselfishes are similar in basic configuration, but individuals of the bicolor damselfish (*Eupomacentrus partitus*) responded more readily to sounds recorded from conspecifics than from other species (*E. leucostictus* and *E. planifrons*) (Myrberg and Spires, 1972). In this report, however, there were a significant number of responses to the sounds of other species, and, it might be added, "responses" were observed during control periods. The controls used here were simply periods of silence. Although the authors concluded that *E. partitus* could discriminate the sounds of its own species, the data show that the discrimination is strongly error-prone. In playback experiments done with sunfish (Gerald, 1971), the animals were clearly able to distinguish between recordings of conspecifics from recordings of music (Grofe's *Grand Canyon* Suite, for example).

SONIC COMMUNICATION IN FISHES

The toadfish (*Opsanus*) is probably the most extensively studied species from the point of view of sonic communication. Not only is the species common and widely distributed, but its "boat-whistle" sounds, produced by territory-holding males, are powerful and distinctive. Most fish sounds are remarkably similar in being low-pitched, short pulse bursts, but the toadfish call is one of the few with a clear tonal quality and a fundamental frequency well above 100 Hz. Even though it is rare to hear a toadfish boat-whistle in captivity, the habits of the animals make field studies feasible. They tend to be sedentary, and once located in an appropriate shelter (beer cans are highly preferred), the sounds of a given animal can usually be monitored for extended periods.

A significant recent contribution to toadfish communication was provided by Fish (1972), in which he did an extensive series of playback experiments. While it has been previously described that toadfish respond to playbacks of their own boat-whistle sounds, Fish was able to vary the animal's calling rate by altering the playback calling rate and, in effect, produce an antiphonic alternation between the animal and the playback. This confirmed the impression that antiphony exists among several animals that are within hearing distance of one another. In addition, Fish found that artificial 200-Hz tones were effective in eliciting the same responses as were the naturally recorded playbacks.

Although the rate of calling by a toadfish can be influenced by playback, there is no change in the animal's particular call characteristics, nor does the change in calling rate persist after playbacks cease. In addition, since the same changes in calling rate can be elicited by artificial approximations of the call, it is apparent that the animals do not discriminate between these. It may be that this lack of discrimination results from the fact that, under normal conditions, there are virtually no other tonal sounds in that frequency that exist in the toadfish's environment. Fine discrimination, therefore, would not be of any particular value. It is possible, furthermore, that this species is incable of much frequency discrimination, as has been demonstrated for several non-ostariophysine marine species (see the review by Tavolga, 1976a, in another volume in this series).

Winn (1972) reviewed most of the previous studies on the toadfish sounds, and went on to discuss communication in fishes in general. The toadfish response system is evidently quite broad, as long as the stimulus sound has a dominant frequency of about 200 –300 Hz, a duration of 150 –300 msec, and a repetition rate that gives enough time between sounds for the animal to respond. In reviewing his classification of fish sounds, Winn re-emphasized the importance of temporal patterning in fish communication systems. Fixed and variable interval sound bursts are most common to fishes, while harmonic-frequency calls, as in the toadfish, are rarely encountered.

The position of fishes with regard to their level of organization in communication is as yet unclear (Tavolga, 1974). Although many species appear to produce nonspecific sounds that provide a broad spectrum of information, i.e., general territorial aggressiveness or reproductive readiness, there are a few in which the sounds are related to highly specific activities and which, even in the strictest sense, can be defined as "signals." In order not to dilute the value of the term "communication," I suggested (Tavolga, 1974) that this concept be restricted to levels of organization in which signal exchange is clearly identifiable. In the strict sense, signals are characterized by being produced by specialized structures, with their energy output within a narrow spectrum of a single channel. The information content of a signal is coded, in the sense that the particular characteristics of the signal are closely linked to the emitter's condition. At this level, the specificity of signals is such that communication is generally limited to intraspecific interactions. The receiving system must be tuned to separate the signals from noise, the responses of the receiver are usually quite specific, and in intraspecific communication there must be a shared code in the exchange of signals.

The system of boat-whistles in the toadfish appears to fall well within the above definition of signals and communication (Winn, 1972), and the variety and specificity of sounds made by damselfish may permit the allocation of their system of sounds and responses to the level of communication (Myrberg, 1972a).

SONIC MECHANISMS IN FISHES

Since the publication of the 1971 review (Tavolga, 1971a, reprinted here in part), there have been only two reports on the mechanisms of sound production in fishes.

Sound production among cichlid fishes has only recently become known. Lanzing (1974) provided a detailed description of the sounds of *Tilapia mossambica*. The sounds are stridulatory, with emphasis well above 1 kHz, and are produced by the scraping of opposing patches of pharyngeal denticles.

Drumming sounds, with a fundamental of about 110 Hz, were reported from five species of piranhas (*Serrasalmus*) (Markl, 1971). The sonic mechanism was described as the swim bladder activated by synchronous contractions of a pair of attached drumming muscles. As in other sonic species, the sonic muscles have fast contractile properties and a high resistance to tetany.

In both the above reports, little information is available on the behavioral correlates of sound production.

CENTRAL NERVOUS CORRELATES OF SONIC BEHAVIOR

In recent years major advances have been made in the field of animal behavior through the development of techniques for stimulating and recording at specific sites in the central nervous system, particularly in the brain. Advances in this direction have been made in the field of fish sound production.

Although it has been readily possible to stimulate sound production by electric shock, it is not always clear that the sound so evoked is the same as a sound produced by mechanisms of central nervous origin. In fact, it is sometimes possible to evoke sounds from fish that normally do not produce any— simply by stimulating appropriate musculature in the region of the swim bladder. Demski and Gerald (1972) followed techniques used in studying birds and mammals, and were able to evoke sounds from toadfish that were almost identical with the harmonic boat-whistle sounds of territory-holding animals. The stimulus sites were in the medulla. In a later study, Demski, Gerald, and Popper (1973) and Demski and Gerald (1974) located a sonic motor center in the midbrain in the toadfish, and speculated on the possible homology of this brain area with comparable areas found in birds and amphibians.

One of the many interesting ramifications of these studies was the discovery that only male toadfish could be stimulated. Although the sonic mechanisms appear to be identical in both sexes, only males have ever been heard to produce the boat-whistle. The possible central nervous differences in sonic function would bear examination.

ACOUSTIC ORIENTATION IN FISHES

The entire problem of sound localization in fishes is reviewed elsewhere (Tavolga, 1976a, in another volume in this series). In connection with sound production, however, there are some reports of significance in which sounds have been found to be attractive to fishes.

The sounds of squirrelfish consist of staccato-like repetitions of low-frequency pulses. These sounds possess most of their energy in the near field (displacement), and other animals have been found to localize the sound sources. Popper, Salmon, and Parvulescu (1973) described some open bay observations in two species of Hawaiian squirrelfish (*Myripristis*), in which animals were seen to orient within a range of 3 meters to the source of playback of conspecific alarm calls. The concluded that the orientation was to the vector component of acoustic energy, i.e., the near-field displacement energy. This was further supported by behavioral and electrophysiological experiments in two other species of *Myripristis* (Horch and Salmon, 1973).

Most investigations on acoustic orientation in fishes have been primarily concerned with the receptor systems and the mechanisms through which fishes can localize the position of a sound source. This aspect is more appropriately reviewed elsewhere (Tavolga, 1976a). The function of fish sounds as an aid to their orientation, however, has only recently been considered, even as a possibility. The marine catfish *Arius felis** is a well-known sound producer, whose short, 100-Hz pulses are emitted almost constantly. In addition to the function of these sounds in aiding aggregation, the reflection and reverberation of the sounds contain some information about the presence and location of nearby obstacles (Tavolga, 1971b). Once it could be demonstrated that the information was available, through the interaction of sound output and its return, then the next step was to determine if, indeed, the animals make use of this information. Through experiments with transparent barriers and surgically blinded or muted fish, the ability of these catfish to use their sounds in short range (up to 10 cm) obstacle detection has been demonstrated (Tavolga, 1976b). Obstacle detection in this fish represents a simple and primitive echolocating system, at least relative to the highly evolved and specialized systems in dolphins and bats. The catfish acoustic orientation system is probably more akin to that of man, rather than to that of the more specialized forms, and therefore offers some opportunities for the investigation and analysis of echolocation on its most primitive level in the vertebrates.

*Formerly *Galeichthys*

Recent Advances in the Study of Sound Production in Fishes

SUMMARY

Since 1971, the advances in the study of sound production in fishes has been in three main areas.

1) A general and synthetic approach to the study of acoustic communication in fishes, through an appreciation of the level of organization involved and a comparison with other vertebrates.

2) The application of brain recording and stimulation techniques to the search for the neural correlates of acoustic behavior.

3) The discovery that at least one species of fish can use its own sounds to aid in acoustic orientation—a primitive form of echolocation, as it were.

REFERENCES

Bright, T. J., and Sartori, J. D. (1972). Sound production by reef fishes. *Hydro Lab J.*, **1**, 11–20.

Burke, T. E., and Bright, T. J. (1972). Sound production and color changes in the dusky damselfish. *Hydro Lab J.*, **1**, 21–29.

Demski, L. S., and Gerald, J. W. (1972). Sound production evoked by electrical stimulation of the brain in toadfish *(Opsanus beta). Anim. Behaviour,* **20**, 507–513.

Demski, L. S., and Gerald, J. W. (1974). Sound production and other behavioral effects of midbrain stimulation in free-swimming toadfish, *Opsanus beta. Brain, Behav., and Evolution,* **9**, 41–59.

Demski, L. S., Gerald, J. W., and Popper, A. N. (1973). Central and peripheral mechanisms of teleost sound production. *Amer. Zoologist,* **13**, 1141–1167.

Fish, J. F. (1972). The effect of sound playback on the toadfish. In *Behavior of Marine Animals,* vol. 2, H. E. Winn and B. L. Olla, eds., Plenum Press: New York, pp. 386–434.

Fish, J. F., and Cummings, W. C. (1972). A 50-dB increase in sustained ambient noise from fish *(Cynoscion xanthulus). J. Acoust. Soc. Amer.,* **52**, 1266–1270.

Gerald, J. W. (1971). Sound production during courtship in six species of sunfish (Centrarchidae). *Evolution,* **25**, 75–87.

Horch, K., and Salmon, M. (1973). Adaptations to the acoustic environment by the squirrelfishes *Myripristis violaceus* and *M. pralinius. Mar. Behav. Physiol.,* **2**, 121–139.

Lanzing, W. J. R. (1974). Sound production in the cichlid *Tilapia mossambica* Peters. *J. Fish. Biol.,* **6**, 341–347.

Markl, H. (1971). Schallerzeugung bei Piranhas (Serrasalminae, Characidae). *Ztschr. vergl. Physiol.,* **74**, 39–56.

Myrberg, A. A., Jr. (1972a). Ethology of the bicolor damselfish, *Eupomacentrus partitus* (Pisces: Pomacentridae): A comparative analysis of laboratory and field behaviour. *Anim. Behaviour Monogr.,* **5**, 199–283.

Myrberg, A. A., Jr. (1972b). Using sound to influence the behavior of free-ranging marine animals. In *Behavior of Marine Animals,* vol. 2, H. E. Winn and B. L. Olla, eds., Plenum Press: New York, pp. 435–468.

Myrberg, A. A., Jr., and Spires, J. Y. (1972). Sound discrimination by the bicolor damselfish, *Eupomacentrus partitus. J. Exp. Biol.,* **57**, 727–735.

Popper, A. N., and Fay, R. R. (1973). Sound detection and processing by teleost fishes: a critical review. *J. Acoust. Soc. Amer.,* **53**, 1515–1529.

Popper, A. N., Salmon, M., and Parvulescu, A. (1973). Sound localization by the Hawaiian squirrelfishes, *Myripristis berndti* and *M. argyromus. Anim. Behaviour,* **21**, 86–97.

Schwarz, A. (1974). Sound production and associated behaviour in a cichlid fish, *Cichlasoma centrarchus. Ztschr. Tierpsychol.,* **35,** 147–156.

Tavolga, W. N. (1971a). Sound production and detection. In *Fish Physiology,* vol. 5, W. S. Hoar and D. J. Randall, eds., Academic Press: New York, pp. 135–205.

Tavolga, W. N. (1971b). Acoustic orientation in the catfish, *Galeichthys felis. Ann. N.Y. Acad. Sci.,* **188,** 80–97.

Tavolga, W. N. (1974). Acoustic orientation in the sea catfish, *Galeichthys felis. Ann. N.Y. Acad. Sci.,* **188,** 80–97.

Tavolga, W. N. (1974). Application of the concept of levels of organization to the study of animal communication. In *Non-verbal Communication,* L. Krames, P. Pliner, and T. Alloway, eds., Plenum Press: New York, pp. 51–76.

Tavolga, W. N. (1976a). Recent advances in the study of fish audition. In *Sound Detection in Fishes,* W. N. Tavolga, ed., *Benchmark Papers in Animal Behavior,* Dowden, Hutchinson & Ross: Stroudsburg, Pa., pp. 37–52.

Tavolga, W. N. (1976b). Acoustic obstacle detection in the sea catfish (*Arius felis*). In *Sound Reception in Fish,* A. Schuijf, ed., Elsevier: Amsterdam. (In press).

Winn, H. E. (1972). Acoustic discrimination by the toadfish with comments on signal systems. In *Behavior of Marine Animals,* vol. 2, H. E. Winn and B. L. Olla, eds., Plenum Press: New York, pp. 361–385.

Part II

HISTORY AND EARLY DESCRIPTIONS

Editor's Comments
on Papers 3 Through 7

3 **SØRENSEN**
Are the Extrinsic Muscles of the Air-bladder in Some Siluroidae and the "Elastic Spring" Apparatus of Others Subordinate to the Voluntary Production of Sound? What Is, According to Our Present Knowledge, the Function of the Weberian Ossicles?

4 **BRIDGE and HADDON**
Note on the Production of Sounds by the Air-bladder of Certain Siluroid Fishes

5 **TOWER**
The Production of Sound in the Drumfishes, the Sea-Robin and the Toadfish

6 **BURKENROAD**
Sound Production in the Haemulidae

7 **BURKENROAD**
Notes on the Sound-Producing Marine Fishes of Louisiana

A perusal of the history of man's knowledge of underwater sonic animals shows it going back probably to prehistoric times, with little or no distinctions made between true fishes and other aquatic vertebrates such as whales and dolphins (see the section on historical background in Paper 1 in this volume). Most reports were based upon the sounds produced by a fish under extreme duress, i.e., hanging in air from a fishing line. Some observant fishermen, however, took to listening for their prey by using primitive stethoscope-like devices or by simply putting their heads beneath water. By the nineteenth century, however, some reports of significance on fish sounds began to reach the scientific literature.

During the march of Napoleon Bonaparte's armies through Egypt, biologists and other scientists were brought along to explore and report on the exotic fauna and flora. As part of Napoleon's "scientific" expedition in 1789, the eminent zoologist and contemporary of Georges Cuvier, Étienne Geoffroy St. Hilaire was to make collections of animals and plants to bring back to France. He remained there until 1801, at the time of the French evacuation of Alexandria. Based upon his vast collections, wrested by force from the British, who followed on Napoleon's heels, he published a twenty-four volume work: *Description de l'Egypt* (1821–1830). An excellent copy of this set is in the library of the American Museum of Natural History. Some chapters were written by the son, Isidore Geoffroy St. Hilaire, and in the sections on fish, specific mention was made of a "tetradon" (probably a puffer) that made noises by scraping its teeth and slapping its pectoral fins against the body. Geoffroy St. Hilaire not only reported on some meagre anecdotal natural history, but quoted at length (in typical Gallic fashion) recipes for the preparation of these fishes.

At one point, the recognized father of American ichthyology, J. Louis R. Agassiz, showed an interest in fish sounds. An often cited publication on the subject was a brief comment Agassiz made at the 365th meeting of the American Academy of Arts and Sciences on 6 August 1850 as follows:

"Professor Agassiz, in reply to a question of the President, stated that the shrill noise heard on suddenly drawing a catfish out of water is occasioned by the escape of air from the air-bladder through the pharynx; and, in reply to a remark of Dr. Gould, he stated that a somewhat similar explanation is applicable to the noise made by the drumfish when taken from the water, a fact recently ascertained by Dr. Holbrook." (*Proc. Amer. Acad. Arts & Sci.*, **2**, 238).

The state of the art was no better when William Sørensen published his thesis in 1884.* Although the original work was in Danish, a language not commonly used in scientific literature, he republished the major part of his studies in English in 1894 –1895. Most of that article is reprinted here, except for the last section, which deals almost entirely with the structure and function of the Weberian ossicles and not with fish sound production. Sørensen's work is a model of the careful, detailed description that was expected of a doctoral thesis of the time, and it also contains results of some experiments. In addition, there are many discursive passages in which the work of others is cited and criticized. Although pedantic and turgid in style, the writing is a typical example of thesis writing in the nineteenth century European university, which required a thesis to be a true, complete, exhaustive monograph. The modern American Ph.D. dissertation does not fare well in a comparison. The reader's attention should be drawn to Sørensen's ideas on the mechanisms of sound production, particularly in the case of the

---

*Sørensen, William. "Om Lydorganer hos Fiske. En Physiologisk og comparativanatomisk Undersøgelse." V. Thåning & Appels: Kjøbenhavn. 1884.

stridulatory sounds of certain catfishes. His conclusions on the role of vibrating muscles in swim bladder sounds are as accurate now as in 1884.

Evidently, Sørensen's original contribution was not really taken seriously by the "establishment," as represented by two eminent British zoologists, T. W. Bridge and A. C. Haddon. They misread or, more likely, missed reading, Sørensen's Danish, and criticised him for drawing functional conclusions without examining living material. In a short article (reprinted here), they had to retract their criticism, and they published this apology early in 1894, prior to the appearance of Sørensen's English translation. Sørensen's rather testy reply can be found in a footnote on page 214 of the paper reprinted here.

In a comparable study of functional anatomy, R. W. Tower provided probably the first evidence that the dominant output frequency of the toadfish swim bladder was a direct translation of the contraction frequency of the sonic muscles. Tower's 1908 report still stands as the accepted description of sonic mechanisms in fishes that use specialized muscles to vibrate the swim bladder.

In two short papers M. D. Burkenroad described sound production in several marine species, notably the grunts (Haemulidae). These fish, whose popular name describes their sound output, were known as such because of the sounds they emit when caught on hook-and-line. The fact that the fish were known to make sounds only under such traumatic conditions demonstrates the pitiful lack of our knowledge about the behavioral significance of fish sounds even in the 1930s. One observation of significance made by Burkenroad was to the effect that the air in the swim bladder is important in giving the sounds their resonant character.

3

Reprinted from *J. Anat. Physiol.*, **29**, 109–139, 205–229 (1894–1895)

ARE THE EXTRINSIC MUSCLES OF THE AIR-BLADDER IN SOME SILUROIDÆ AND THE "ELASTIC SPRING" APPARATUS OF OTHERS SUBORDINATE TO THE VOLUNTARY PRODUCTION OF SOUNDS? WHAT IS, ACCORDING TO OUR PRESENT KNOWLEDGE, THE FUNCTION OF THE WEBERIAN OSSICLES? A Contribution to the Biology of Fishes. By WILLIAM SÖRENSEN, Copenhagen.

Τῶν πόνων πωλοῦσιν ἡμῖν πάντα τ'ἀγάθ οἱ θεοί.

(Ἐπίχαρμος.)

HAVING seen from a memoir by Professors Bridge and Haddon, in the *Proceedings of the Royal Society*[1] (I*a*1), which was evidently the preliminary of a more extensive work, that these naturalists were engaged in studies of the anatomy of the Siluroidæ, and especially with that of the air-bladder and the modified vertebræ of these fishes, I took the liberty to forward, on February 9, 1891, to each of the said gentlemen a copy of two of my papers. I had treated in the former[2] of these papers the sound-producing organs in fishes; in the latter[3] the morphology of certain parts of the skeleton of fishes. In the former the Siluroidæ had precisely been the fishes from which I started my investigations. In the latter the conditions in the Siluroidæ, as the title of the paper shows, had been the principal object of my studies.

---

[1] I*a*: Bridge, T. W., and Haddon, A. C., "Contributions to the Anatomy of Fishes. I. The Air-Bladder and Weberian Ossicles in the Siluridæ" (*Proc. Roy. Soc. Lond.*, xlvi., 1890, No. 283, pp. 309–328). I*a*2: Bridge, J. W., and Haddon, A. C., "Contributions to the Anatomy of Fishes. II. The Air-Bladder and Weberian Ossicles in the Siluroid Fishes" (*Ibid.*, vol. lii., 1892, pp. 139–157.) The latter of these two papers I have read since the first sheet of this memoir of mine was in print.

[2] II*b* : Sörensen, William, *Om Lydorganer hos Fiske: En physiologisk og comparativ-anatomisk Undersögelse.* [On Sound-producing Organs in Fishes: A physiological and comparative anatomical examination.] Kjöbenhavn, 1884.

[3] III : Sörensen, William, *Om Forbeninger i Svömmeblæren, Pleura og Aortas Væg og Sammensmeltning deraf med Hvirvelsöjlen særlig hos Siluroiderne, samt de saakaldte Weberske Knoglers Morphologi.* [On Ossifications of the Air-Bladder, the Pleura, and the Wall of the Aorta, and their Fusion with the Vertebral Column, especially in the Siluroidæ, together with the Morphology of the so-called Weberian Ossicles.] *Avec un résumé en Français.* (*Danske Vidensk. Selsk. Skr. 6 R. Nat.-math. Cl. Bd.* vi. 2, Kjöbenhavn, 1890, pp. 67–152.)

Last summer Professors Bridge and Haddon published a voluminous work,[1] containing their anatomical studies of the Siluroidæ, a work which, on account of the evident diligence and ability displayed, as well as from the great number of fishes treated, will be for a long period a most important work in the anatomy of the organs in question.

Dr. Lütken, Professor of Zoology at the University of Copenhagen, has been so kind as to lend me his copy of this work, and I have thus been fortunate to make myself acquainted with its contents a long while before I should otherwise have been able to do so. It has been a great satisfaction to me to see that these two naturalists agree with me as to the morphological interpretation of the skeletal parts, as well in their morphological summary (pp. 224–261), as now and again in the special description of the fishes, and that they share the opinions I have suggested in my last work, of which they have evidently perused the "resumé en français." Nowhere, as far as I can see, have they urged any objections to the views I have set forth. However, I take the liberty to remark that the reports they have made of this paper of mine are not quite reliable,—a circumstance which is evidently due to their difficulty in interpreting the Danish language, nearly related though it be with the English. Thus they relate (p. 260) that I consider the "claustrum" as the neural spine of the first vertebra; the "stapes" (scaphium, Br. and Hadd.) as the neural arch of the first vertebra; the "incus" (intercalarium, Br. and Hadd.) as the neural arch of the second vertebra; the "malleus" (tripus, Br. and Hadd.) as the rib of the third vertebra. But such is not exactly my opinion of these skeletal parts. As to the "claustrum," on the contrary, I have been careful not to call it a neural spine, and I have shown that at the first 3 or 4 vertebræ in the Ostariophyseæ[2] (in other Physostomi only at

[1] Ib: Bridge, T. W., and Haddon, A. C., "Contributions to the Anatomy of Fishes. II. The Air-Bladder and Weberian Ossicles in the Siluroid Fishes" (*Phil. Trans. Roy. Soc. Lond.*, vol. clxxxiv., 1893, B., pp. 65-333).

[2] The name given to all families furnished with the Weberian ossicles by Sagemehl (p. 22). [IV: Sagemehl, M., "Beiträge zur vergleichenden Anatomie der Fische. III. Das Cranium der Characiniden nebst allgemeinen Bemerkungen über die mit einem Weberschen Apparat versehenen Physostomen-Familien" (in *Morphol. Jahrbuch.* x., Leipzig, 1885, pp. 1–119). With great force Sagemehl (*ib.*, p. 9) denies the existence of the Weberian ossicles in the Gymnarchidæ, which Bridge and Haddon count among the Ostariophyseæ. I have not examined the *Gymnarchus niloticus*.]

PRODUCTION OF SOUNDS BY AIR-BLADDER OF SILUROIDÆ. 111

the first vertebra) there exists a separate ossicle, which sometimes forms part, sometimes not, of the spinal canal, and which, as far as I can judge, is homologous with the ossa imparia in the Acipenser.

After having mentioned the Weberian ossicles in each of the families belonging to the Ostariophyseæ, I have, to facilitate the survey, compiled the following table (III. p. 105):—

	Characini.	Cyprinoidei.	Cobitini.	Gymnotini.	Siluroidæ.
Claustrum.	The os commissurale of the 1st vertebra.			Is wanting.	The os commissurale of the 1st vertebra is often wanting.
Stapes.	The neural arch of the 1st vertebra.				
Incus.	The neural arch of the 2nd vertebra + ossified ligament.			Only ossified ligament.	
Malleus.	The rib of the 3rd vertebra + the basal part[1] of the rib + ossified air-bladder + ossified ligament.	The rib of the 3rd vertebra + ossified air-bladder + ossified ligament.			In Clarias and Plecostomus the rib of the 3rd vertebra + ossified air-bladder + ossified ligament. In the other genera [known to me] the rib of the 3rd vertebra + the basal part[1] of the rib + ossified air-bladder + ossified ligament.
Os suspensorium mihi [sc. vesicæ natatoriæ].	The basal part[1] of the rib of the 4th vertebra + ossified air-bladder.				Shares this function with other bones.

To whosoever has read, and were it only this table, it will be quite evident that my views as to the morphology of these

[1] Or the processus transversus.

ossicles are recorded by Professors Bridge and Haddon in a very superficial manner.

In all the (nine) genera of the Siluroidæ which I have examined, the "incus" in full-grown specimens consists of ossified ligament; but Bridge and Haddon have made the very interesting discovery that this ossicle forms a smaller or larger part of the neural canal in some genera (*e.g.*, Macrones, Liocassis, Pseudobagrus, Bagroides—Cryptopterus, Callichrous), in which respect these genera approach to the other families of the Ostariophyseæ.

When, however, Bridge and Haddon say (I*b*, p. 261) :—

"A further question arises as to whether, in addition to a modified neural arch, the intercalarium [incus, Web.] did not originally include an element comparable to a transverse process. We are inclined to think that it did, and that the horizontal process of the ossicle, when present, represents the modified transverse process of the second vertebra. In its origin from the neural arch or ascending process, the horizontal process conforms precisely to the contiguous transverse processes of the fourth and fifth vertebræ, which spring in exactly the same way from the neural arches, and not from the centra of their respective vertebræ."

I may answer this question in the negative, because in fact I have already done so—a circumstance which, however, the authors had not observed, as they had not made themselves acquainted with the *Danish* part of my paper in question, in which is set forth the documentation of my morphological views. 1°. In the other families of the Ostariophyseæ, at the 2nd vertebra, there exists, besides the "incus" when present, a real[1] transverse process; nay, in the Gymnotini this process even carries a rib. 2°. A transverse process (or rib) in fishes always springs from the centrum of the vertebra, and never from the neural arch; when it seems to do so, it is only so in appearance. And to abide by the Siluroidæ, one glance at the figure (*Tb*. I. fig. 10) which I have given of this part of the vertebral column in the foetus of *Galeichthys feliceps* also proves how the transverse processes of the (real) 4th and 5th vertebræ spring from the centra of their respective vertebræ, while the proximal end of the "incus" is fixed in the wall of the neural channel. And, starting from this foetus, I have proved (III. pp. 101, 102) that

[1] I must add this designation, as on the 1st vertebra in the Cyprinoidei there exists a transverse process, which is false.

when the transverse processes of the normal vertebræ in the Siluroidæ *seem* to spring both from the centrum and the arch of their vertebra, this is due to a later ossification of the ligaments which—as in other fishes—unite the transverse processes (or ribs) with the neural arch. And I may add that when the modified transverse processes of the (real) 4th and 5th vertebræ seem to spring from the neural arches, this is equally due to a secondary transformation during the growth (in the very young animal), as briefly intimated in my paper (pp. 103–105). The " horizontal process " of the " incus " is ossified ligament.

Professors Bridge and Haddon say (I*b*, p. 249), that in Clarias " the inferior limb " of the " post-temporal " " becomes quite rudimentary, and loses its usual articulation with the basi-occipital." On the contrary, I have declared that this " inferior limb " does not exist at all in this genus. This might seem but a very slight difference. To be sure, the difference of expression is very slight, but the difference as to the interpretation of the fact in question is anything but slight. For, as I have shown,—a fact also mentioned by Bridge and Haddon,—this " inferior limb " is a very essential point of the skeletal structure of the Siluroidæ. But what is this " inferior limb "? In most of the Teleostean Fishes the " pectoral girdle "—as is well-known—consists of three bones, called by Cuvier the " suprascapula," the " scapula," and the " humerus." As is equally well known, the uppermost of these bones is, as a rule, joined with two bones of the skull, viz., with the os squamosum (or temporale), and with the epioticum (or paroccipitale). The bone in the middle, in the Siluroidæ coalesced with the uppermost one to form the " suprascapula " (Cuvier), or the " post-temporal " (Br. and Hadd.), is—a fact[1] less well known—in most Fishes, also in Amia and Polypterus, united, as a rule,

---

[1] For this reason I have had the ligament drawn in all the figures I have published, where it was possible to do so. The only author who has mentioned it is, as far as I know, Mettenheimer, C., *Disquisitiones Anatomico-comparativæ de membro piscium pectorali*, Berolini, 1847. This ligament is also ossified in Dactylopterus (where it acts the same part as in the Siluroidæ), Aulostoma, and Ostracion (but not in Batrachus tau). This also appears to be the case in Lepidosiren, Protopterus, and Ceratodus. But as I have dissected none of these genera, I dare not state with certitude that the bone ("the first rib," Günther) which unites the first vertebra or the basis-cranii with the "shoulder-girdle" is homologous with the said ligament.

with the occipitale basilare, or uncommonly, as in the Cyprinoidei and the Gadoidæ, with the centrum of the 1st vertebra. But while the junction between the "suprascapula" and the epioticum is brought about by means of a slender process of the suprascapula — *i.e.*, an ossified ligament — in most cases the junction between the lower end of the scapula and the occipitale basilare, or the centrum of the 1st vertebra, is brought about by means of a ligament. For it is but rarely the case that the latter is ossified: in the Cyprinoidei a shorter or longer part of it is ossified; and as the ossification takes its beginning from the proximal end, it has the appearance of being a "transverse process" of the 1st vertebra; but in nearly all the Siluroidæ (viz., with the exception of Clarias) this ligament is a more or less considerable bone, the strength of which, as well as the manner in which it is connected with the occipitale, is in relation to the size of the first ray of the pectoral fin, and whether the "suprascapula" is connected with the transverse process of the (real) 4th vertebra or no. Ten years ago I had already pointed out this fact in my first paper (II*b*, pp. 3 and 21). And in both my papers I have proved that in Clarias there exists no connection whatever between the "suprascapula" and the occipitale basilare, because the said ligament has not been able to form itself, the accessory gill-cavity, in which the air-breathing dendritic organs are inclosed, being placed where the ligament should be, and has therefore, if I may say so, supplanted it. In return, the pectoral girdle has in this genus been strengthened by the helmet (the dermoidal bones) having attained to a size unparalleled in all other Siluroidæ except when the large (2nd) ray of the dorsal fin is a very effective weapon—in Clarias it is no weapon at all, but a mere weak ray, and the preceding "rudimentary" ray is here wanting.

If the authors had known the true nature of this "inferior limb" of the post-temporal, and the importance of the degree of the development of this bone in these animals, they would never have been "tempted to think that the post-temporal plates of Macrones and the allied genera might represent a form of 'elastic spring' mechanism" (p. 245). For, neither from a morphological nor from a physiological point of view, have these two things anything to do with each other. But, as I shall prove in the following pages, the authors had, in their Memoir in the *Transactions* of the Royal Society, completely overlooked my first paper.

I.

Are the extrinsic muscles of the air-bladder in some Siluroidæ and the "elastic spring" apparatus of others, subordinate to the voluntary production of sounds?

> Πάντα δὲ ταῦτα τὴν δοκοῦσαν φωνὴν ἀφιᾶσι . . . τὰ δὲ τοῖς ἐντὸς τοῖς περὶ τὴν κοιλίαν. Πνεῦμα γὰρ ἔχει τούτων ἕκαστον, ὃ προστρίβοντα καὶ κινοῦντα ποιεῖ τοὺς ψόφους.
> —*Aristotle.*

While Professors Bridge and Haddon agree with me in the opinions I have propounded as to the morphology of the skeletal elements treated in my second paper (III.), I cannot, I am sorry to say, pride myself on this concurrence as to the physiology of the air-bladder in the Siluroidæ treated in my first paper (II*b*). In order to show this, I shall take the liberty to quote some remarks by the said authors in their detailed account of the "Physiology of the Air-bladder and Weberian Ossicles in the Siluridæ" (I*b*, pp. 261–303). Page 269:—

"In addition to the various other methods by which voluntary sounds are produced in different fishes, the air-bladder not unfrequently shares in the function of phonation. Such sounds are either produced by the vibration of the internal annular diaphragm (Moreau), or by the vibration of certain extrinsic muscles (Dufossé[1]), the air-bladder in the latter case intensifying the sound produced by acting as a resonator. Dufossé (*loc. cit.*) is also of opinion that some Ostariophyseæ (*e.g.*, some Cyprinidæ and one or two Siluridæ) produce breathing noises ('les bruits de souffle') by the expulsion of gas from the air-bladder through the ductus pneumaticus, and it has been suggested [2] that the grunting sounds emitted by Clarias have

---

[1] The first paper of this author here referred to. (V*a*: Dufossé, *Recherches sur les bruits et les sons expressifs que font entendre les Poissons d'Europe* . . . *Annales d. Sci. Nat. 5 Sér.*, T. xix. Paris, 1874. Art. No. 5.)

[2] By whom?—Day, to whom Bridge and Haddon refer in a paper entitled "Instincts and Emotions in Fishes" (*Jour. Linn. Soc.*, xv., 1881, pp. 31–58) only says as follows:—"Sir Emerson Tennent observed that a Siluroid fish (Clarias) found in the lake at Colombo is said by the fishermen to make a grunt

a similar origin. *The possibility that the Weberian ossicles have any thing whatever to do with phonation, either in the Siluridæ or in other Ostariophyseæ, is very remote,*[1] and need be but briefly considered."

Page 270: "We are strongly inclined to the opinion that although sounds may indirectly have their origin in the air-bladder, *they have no relation to it other than as accidental accompaniments in the exercise of its normal hydrostatic function,*"—with the following footnote: "For these reasons, and *in the absence of definite experimental evidence*, we cannot at present accept Sörensen's *ingenious* theory that the extrinsic muscles of the air-bladder in the Pimelodinæ, and the 'elastic-spring' apparatus of other Siluridæ, are solely subordinate to the voluntary production of sounds." "In one example cited above (Clarias) it is almost certain that the grunting sound which the fish is said to make could not be caused *by the voluntary expulsion of gas from the air-bladder*, inasmuch as this organ is not only rudimentary, but almost completely encapsuled by bone. Eliminating such doubtful examples of the association of the air-bladder with phonation in a few Siluridæ and Cyprinidæ, it may be urged with regard to the rest that the comparative[2] rarity of *well authenticated instances of the production of voluntary sounds*, the absence of extrinsic muscles in all but a few genera (Pimelodinæ), and *the want of internal vibratory diaphragms, or other obviously vocal structures, are quite sufficient to prove that the air-bladder takes little or no part in this function, at all events, by any of the ordinary methods known in other Fishes.*"

P. 296: "In the great majority of the Ostariophyseæ the escape of air from the air-bladder through the ductus pneumaticus apparently takes place only as the result of the expansion of the contained gases under the influence of diminished hydrostatic pressure, although it is possible that the rate of overflow may in some way be regulated. In some few Siluridæ, however, there does seem to exist a special

---

under water when disturbed." And in saying so he is almost literally quoting Sir Emerson Tennent, whose words run as follows (*Ceylon*. Fifth edition, vol. ii., Lond. 1860, p. 470):—"The fishermen assert that a fish, about five inches in length, found in the lake at Colombo, and called by them 'Magoora,' makes a grunt when disturbed under water." Bridge and Haddon, indeed, quote a paper by Day bearing the very same title, and to be found in the *Transactions of the Zool. Soc.*, vol. xv., 1880; but nowhere in the *Transactions* of this Society, neither in vol. xv. nor in the volume published in 1880, there exists such a paper. I therefore presume that they refer to the paper which I have mentioned, the more so as *the references in the physiological division of their work*, with the exception of the references to their own papers, to those of Ramsay Wright, and to Günther's "Introduction," are wanting in precision.

[1] The italics in this and the following quotations are mine.

[2] The Siluroidæ are, among all families of Fishes, the one which, before the time of Dufossé, counted the greatest number of species known as sound-producing—both Platystoma Orbignyanum, Pseudaroides clarias, and 2-3 species of the genus Doras, for instance, were known as such by Cuvier and Valenciennes. The statements of naturalists as d'Orbigny and Charles Darwin, Professors Bridge and Haddon, will not, I suppose, design as being not "well authenticated."

mechanism *by which, under certain conditions, the air-bladder may be subjected to considerable compression, and the air which it contains either forcibly expelled, or greatly reduced in volume by condensation* [Moreau!].[1] This mechanism presents two important modifications, viz., the "elastic spring apparatus" and the powerful extrinsic muscles of the Pimelodinæ."

Pp. 297–298:—"The mobility and elasticity of the transverse process which forms each spring will certainly give to the lateral portions of the anterior wall that capacity for sharing in the distension of the anterior chamber which is prevented in all other Siluridæ by the absolute rigidity of the processes in question, but it is at the same time, equally clear that the 'elastic spring' apparatus cannot possibly give the fish any power of directly compressing the air-bladder, except under certain conditions, viz., when the anterior chamber becomes distended through the diminution of pressure which occurs in movements of ascent, coincidently with the forward movement of the two springs as the result of the voluntary or reflex contraction of their protractor muscles. Under such circumstances the [elastic spring,] mechanism potentially acquires the power to modify the capacity of the air-bladder, for the subsequent relaxation of the muscles will at once enable the springs, through the force of their own recoil, to exert their full strength in compressing both the air-bladder and its gaseous contents." And to the words "gaseous contents" is added in the form of a footnote:—"Should this view of the mode of action of the 'elastic spring' apparatus prove correct, *it will be difficult to see how the mechanism can have anything to do with the production of voluntary sounds*, as suggested by Sörensen, *inasmuch as the Fish would only be able to exercise its vocal powers under conditions involving pressure reduction during ascent from a deeper to a more superficial level. Under such conditions only does it seem likely that the contained gases would be expelled with sufficient force to produce any definite or characteristic sounds.*"

P. 300:—"The Extrinsic Muscles of the Air-bladder in the Pimelodinæ.—A function substantially similar to that of the 'elastic spring' apparatus may, in all probability, be assigned to the powerful

---

[1] The authors have perused Moreau's paper (VI. Moreau, A.: "Recherches expérimentales sur les fonctions de la vessie natatoire": *Annales d. Sci. Nat.*, 6 Sér., T. iv. Paris, 1876. Art. No. 8), but they have not *studied* this most excellent memoir. Otherwise, they would have noticed his remarks (p. 52) which may regard also the theory of Joh. Müller about the function of the "elastic spring" apparatus in some Siluroidæ: "L'attention des savants n'avait pas été encore fortement appelée sur les phénomènes de la contraction musculaire sur la fatigue que le travail musculaire engendre et par conséquent sur l'invraisemblance d'efforts aussi prolongés et aussi énergiques que ceux que suppose la théorie traditionelle." For the very same objection may be urged against the above-cited opinion, nay, against all suggested by Bridge and Haddon on the function of the "compressor" muscles and the "elastic spring" mechanism. In the passages of these authors which I am going to quote on the following pages, I take the liberty to add the word ["Moreau!"] to similar remarks.

compressor muscles of the Pimelodinæ.[1] *These muscles cannot possibly have any share in dilating the air-bladder* and rarefying the contained gases in order to facilitate ascent, but it would certainly seem that they enable these particular Siluridæ to exercise a still more effective control over its distension, *inasmuch as the muscles are apparently able to compress the air-bladder at all times*, although more effectively, no doubt, when the latter is more or less distended [Moreau !]. By the contraction of these muscles during rapid movements of ascent the tendency to over-distension on the part of the air-bladder will be promptly counteracted, while a forcible expulsion of gas through the pneumatic duct would enable the Fish to speedily adjust its volume and specific gravity to a new plane of least effort at the more superficial level. . . . . In both series of Fishes it is extremely interesting to recall the existence of a special arrangement by means of which the compression of the air-bladder, either by the action of the 'elastic springs' or by the contraction of special *compressor* muscles, is prevented from imparting a too violent shock to the Weberian mechanism, and more especially to the fluids and sensory epithelia of the internal ear. . . . . . *The extreme difficulty of attempting to arrive at a satisfactory solution of* the various problems arising out of the physiology of the air-bladder, *through anatomical data alone*, is again forcibly illustrated, for *it is impossible entirely to exclude the possible relation of the extrinsic muscles of the Pimelodinæ to the function of sound production, and it may also be the case, although perhaps less likely, that the same reservation will also apply to the 'elastic spring' mechanism. That a violent expulsion of air from the air-bladder should produce definite sounds is extremely probable*, but how far such sounds can be considered as related to the primary function of these muscles, or as merely accidental concomitants to it, must for the present remain an open question. Sörensen has adopted the former suggestion, and regards both the *compressor* muscles and the 'elastic spring' mechanism as being subordinate to sound production. Nevertheless, *in the absence of confirmatory experimental evidence*, we still think it worth while to direct attention to an alternative interpretation of the function of these structures, which is at least as consistent with their morphology as any other view at present suggested. We have elsewhere (p. 270 and p. 298) suggested certain difficulties, which, in our opinion, are *serious* objections to Sörensen's views on this point."

To everybody who has perused these quotations from the work of Professors Bridge and Haddon, it will now be evident that these authors have set forth the following suggestions :—

1. *That the function of the "protractor" muscles of the "elastic spring" mechanism in some Siluroidæ consists in pulling forward*

[1] I beg to direct the attention of the reader to the fact, that, as will appear from the above quotation, the authors are of opinion that the effect produced by the protractor muscles of the "elastic spring" apparatus is nearly quite the reverse of the effect produced by the extrinsic muscles of the Pimelodina.

the springs, in order to enable the air-bladder to distend itself in front, and that the effect which they produce is therefore nearly quite the reverse of the effect produced by the extrinsic muscles, the so-called " compressor " muscles, of some other Siluroidæ.

2. *That all " internal vibratory diaphragms or other obviously vocal structures " are completely wanting in the air-bladder of these fishes.*

3. *That without confirmatory experimental evidence I have propounded a " theory," nay, an " ingenious " one, of the " elastic spring " mechanism and the extrinsic muscles being subordinate to the production of sounds.*

4. *That these sounds are produced by " a violent expulsion of air from the air-bladder."*

5. *That the authors have entirely crushed this theory.*

I am quite willing to acknowledge that they have completely succeeded in doing so. This victory has but one deficiency: that what they have succeeded in conquering, are—the windmills of Montiel! For the opinion which they impute to me, I never had nor suggested.

But why, then, do they impute this theory to me? Firstly, because they had entirely overlooked my book, *Om Lydorganer hos Fiske*, and had not formed an idea as to the contents of this work by perusing its " explicatio figurarum "[1]; and secondly, because, even if they themselves had not been fully aware of the fact, they were under the influence of the first paper, written in modern times, " On the Origin of Sounds produced by Fishes "[2]— a paper by no means worthy of the eminent genius of its author.

In this paper Joh. Müller says:—

" Im Munde jedes Fisches können, wenn er sich in der Luft befindet, Lufttöne entstehen, gleichviel ob er eine Schwimmblase besitzt oder nicht, ob die Schwimmblase geschlossen ist oder einen Luftgang in den Mund besitzt. Dagegen kann bei einem Fische, der unter Wasser tönt, an Lufttöne nur dann gedacht werden, wenn er einen Luftgang der Schwimmblase besitzt und wenn dieser hinreichend weit ist, *um Luft plötzlich auszutreiben.*"

[1] The names of the Siluroidæ, the air-bladders of which are represented among the figures, are, however, recited in their memoir.
[2] VII. Müller, Joh.: Ueber die Fische, welche Töne von sich geben, und die Entstehung dieser Töne (*Archiv f. Anat. u. Physiol.* Berlin, 1857, pp. 249-279).

For as Horace says :—

> Quo semel est imbuta recens, servabit odorem
> Testa diu.

If the authors had perused the Latin "explicatio figurarum" in my paper, they would have observed that wherever I have represented muscles, which make the air-bladder act as a sound-producing organ (if only one pair of such muscles do exist), I have designated each of these muscles as "musculus, cujus contractione *sonat* vesica natatoria," whether the fishes mentioned be furnished[1] with a ductus pneumaticus or no.[2] But how could it have been possible to suggest it as my opinion that (extrinsic or intrinsic) muscles of an air-bladder, *without* a pneumatic duct, might ever be able to produce sounds by "a violent expulsion of the air from the air-bladder"? Or, how could I ever—if I had but the slightest notion of the meaning of the Latin words—say, "*sonat* vesica natatoria," if the contraction of the muscles were to effectuate the expulsion of air from the air-bladder?

But if the authors had only read a short remark of mine in "les Comptes rendus,"[3] or in the "Annals,"[4] where it had been translated from "les Comptes rendus," they would have seen that what I have suggested on the production of sounds by means of the air-bladder in the Siluroidæ[5] in question, is quite different from what they impute to me; for in the passage quoted from the "Annals," my words have been thus translated:[6]—"In the Siluroidæ[5] the anterior portion of the swimming-bladder is drawn alternatively forward and backward by the contraction and relaxation[7] of the muscles. During these movements the air, in passing across the incomplete transverse septa, sets the

---

[1] Pseudaroides clarias (fig. 45); Synodontis schal (fig. 48).

[2] Diodon hystrix (fig. 49); Batrachus tau (fig. 60); Micropogon undulatus (fig. 61).

[3] IIa: Sörensen, William, "Sur l'appareil du son chez divers Poissons de l'Amérique du Sud" (*Compt. rend. de l'Acad. d. Sci.*, T. lxxxviii. Paris, 1879, pp. 1042-43).

[4] *Annals a. Mag. of nat. hist.* 5 Ser., vol. iv. London, 1879, pp. 99–100.

[5] Doras maculatus, Platystoma Orbignyanum?, Pseudaroides clarias.—The remark also refers to some genera of the Characini.

[6] I readily agree that, on account of its brevity, this summary is not quite clear. As to the manner in which the production of sounds itself is operated, I do not think, however, that it can easily be misunderstood.

[7] I have not succeeded in finding the correct words to express my idea, as may be seen later on from the quotation of my book.

latter in vibration, and the sound is produced. The height, or rather the depth, is in direct proportion to the rapidity of the vibration of the springs."

And if the authors had known a most valuable paper by the late M. Dufossé[1] on the air-bladder as a sound-producing organ, in a fairly considerable number of Fishes (16 species, 7 genera), all without a pneumatic duct, they would have seen that my "theory" was quite in accordance with the result at which this skilful French author had arrived through his examinations—not in his study, but—of living animals. Then, certainly, the authors would not have said that the production of sound, according to my " theory," was not effectuated " at all events by any of the ordinary methods known in other Fishes." And then they would not have spoken about "the want of internal vibratory diaphragms . . . . ." *inasmuch as they themselves have described and represented such diaphragms (the " transverse septa") in a considerable number of Siluroidæ—even to the number of six pairs!*

However, even if the opinion which I had suggested was in accordance with the result obtained by Dufossé in his examinations of living fishes, Professors Bridge and Haddon might be entitled to call my opinion a "theory" if, what they evidently supposed me to have done, I had stayed in Copenhagen and "construirt" my "theory" "aus dem Inneren meines Bewusstseins." For that is exactly the way by which they have arrived at the opinions which they have suggested on the function of the air-bladder and the Weberian ossicles in the Siluroidæ.

So I must ask the reader to observe the full title of my book, *On Sound-Producing Organs in Fishes, a Physiological and Comparative Anatomical Research*, my intention being to intimate, in as few words as possible, that my results had been obtained in both ways—vivisection with regard to some forms, and examination of dead animals with regard to others.

Now, my examinations of Siluroidæ and Characini happen to be the basis of my paper, as I *commenced* these studies at the confluence of the Riacho del Oro with the Rio Paraguay.

[1] V*b*: Dufossé, "Recherches sur les bruits et les sons expressifs que font entendre les Poissons d'Europe . . ." (*Annales d. Sci. Nat.* 5 Sér., T. xx. Paris, 1874. Art. No. 3).

It must now be investigated whether my opinion about the function of the air-bladder and the extrinsic muscles or the "elastic spring" mechanism be a "theory" or no.

About *Doras maculatus* (Cuv. et Val.), I say in my paper, after having described (pp. 85–87) the structure of the air-bladder (*vide* figs. 1 and 2), as well as the "elastic spring" mechanism and its muscles, and after having pointed out the fact that the "malleus" (in this genus) is also a spring, p. 88, as follows:—

"*Observations on the Production of Sounds.*—When the belly of the recently caught fish is opened, and the intestines with their append-

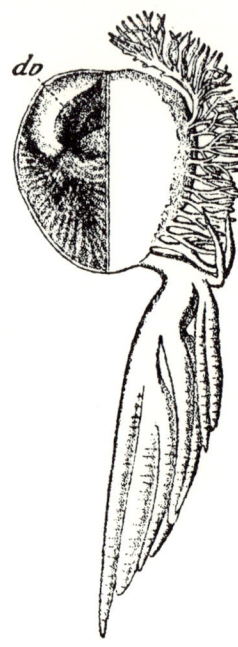

Fig. 1.—*Doras maculatus*, Cuv. et Val. The air-bladder of a specimen, measuring 50 cm. of length, seen from below. Diminished thrice. To the right the ventral wall of the air-bladder has been removed. The cæcal diverticula are not represented on this side. *dv*, one of the circular bony plates (fused with the hinder part of the muscular spring, the transverse process of the (real) 4th vertebra) in the fore end of the wall of the air-bladder. In a living specimen it is not so prominent in proportion to the remaining part of the air-bladder.

ages are taken out quickly, so that the air-bladder is laid open, it may be very easily observed that the air-bladder is in a state of convulsive vibration, at the same time as sounds are produced.[1] This sound

[1] A part from the sounds produced by the movements of the fins.

is a very deep, growling tone, which is so intense that *it is still to be heard very distinctly at a distance of* 100 *feet* [1] when the fish is *out of the water*. In contradistinction from the sound produced by the movements of the pectoral fins, the sound produced by the air-bladder is not discordant, and therefore it is not disagreeable to the ear. As far as I have been able to catch—I am sharp of hearing, but I have some difficulty in distinguishing notes—the air-bladder only commands one tone, but this tone may be more or less strong as it pleases the fish. If you move your fingers backwards and forwards on the air-bladder, you will soon perceive that the vibrating movement, arising at the same time as the sound, is strongest in front, especially near the muscular springs, and likewise that the muscles inserted upon the similar plates of these springs are contracted at the same time as the sound arises. If the muscles are cut asunder, the sound [2] is no more produced. If a small hole is made in the air-bladder (when the muscles are uninjured) the sound does not grow much fainter, but if the hole is enlarged the sound loses considerably in strength. If the

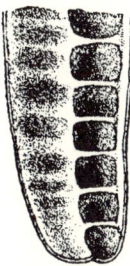

Fig. 2.—*Doras maculatus*, Cuv. et Val. The same specimen. The end of one of the large (posterior) cæcal diverticula of the air-bladder. It is opened, so as to permit a survey of its cellular structure. Natural size.

air-bladder is removed, the sound grows much fainter, but is still audible; then it is exclusively produced by the vibration of the springs. By direct observation I have not been able to prove that the beams of the principal chamber of the air-bladder, or the incomplete transverse septa of its cæcal diverticula, contribute to the production of the sound; but if the reader will compare this with what I am going to communicate in the following pages by *Pseudaroides*, I am of opinion that it will prove *without doubt* that the cæcal diverticula of the air-bladder, on account of the incomplete internal transverse septa (fig. 2), are most intimately concerned in the function of intensifying the sound by means of the air passing forwards and backwards above the septa. By a minute examination it may be observed that the foremost of the bony scutes on the side of the body shares in the vibration when the sound is produced. I presume that the

---

[1] A Danish foot is a little longer than an English.
[2] Apart from the sounds produced by the movements of the fins.

function of the ligament [1] serving to connect this plate with the circular plate of the muscular spring, does not only consist in transmitting the sounding vibrations of the air-bladder to the water, but also in preventing a too violent reaction of the spring when the muscle is relaxed."

After having described the air-bladder and its extrinsic muscles (p. 92) in *Platystoma Orbignyanum* ? Val.[2] and *Pseudaroides clarias*, Bl., and having proved the "malleus" (the "tripus," Br. and Hadd.) to be a spring, I continue (p. 93) as follows :—

"*Observations on the Production of Sounds.*—When the air-bladder of the living animal is laid open, it is very easy to perceive that the muscles in question are contracted at the same time as a strong, deep, growling sound is produced, while the wall of the air-bladder is set into a strong, vibrating motion. The majority of the specimens I have examined of Pseudaroides were not above a total length of 25 to 35 cm. Hence the walls of the air-bladder were not so thick as to prevent me from distinguishing, without opening the air-bladder, the internal transverse septa in the shape of darker transverse stripes; therefore, I was able to observe very distinctly that, during the emission of the sound, they were swinging (or being moved) very quickly to and fro (fig. 3).

This fact is sufficient to prove that they play a very important part in serving to intensify the sound, by means of the air vibrating above their edges from one chamber to another. If a small hole is made in the air-bladder of a Platystoma, the intensity of the sound is not diminished in any remarkable degree. But if a fissure, however small, is made in the air-bladder, the sound grows distinctly fainter, and at last quite ceases, even if the muscles are still in action.

As far as I have been able to observe, as is the case in Doras, only one contraction of the muscles takes place every time a sound is produced. This sound always lasts for a certain time, grows fainter toward the end, but suddenly ceases. On the nature of the sound the same may be said as I have stated about Doras. The sound produced by a Platystoma [3] is audible at a distance of more than 20 feet if the animal is on shore."

---

[1] This ligament connects the distal circular plate of the "elastic spring" with the foremost, hardly visible, of the thorny dermoidal plates, which in this genus are placed on the side of the body. A corresponding ligament I have not found in any other genus of the Siluroidæ furnished with the "elastic spring" mechanism.

[2] Professor Lütken, who was then so kind as to determine the Siluroidæ which had been the object of my examinations, was not able to determine with exactitude this species, as I had not brought home any uninjured specimens. But at any rate it is *not* Pl. fasciatum, as related by Bridge and Haddon (*Ib*, p. 118). For, according to Joh. Müller, it is easy to distinguish the air-bladder in this species from that which I have examined.

[3] The specimens examined generally had a length of 1 metre, or somewhat less.

I ought to have added that the sound is no more produced when the muscles are cut. Whether, by this operation, I have cut the pair of small muscles (M. "tensores tripodium," Br. and Hadd.) which Bridge and Haddon have always found along with the large extrinsic muscles I do not know, as I have not seen these muscles; I suppose it to be so.

As it was of some importance in connection with the subsequent comparative-anatomical part of my book, I added:—
"From the description of the air-bladder of these three genera it is evident that, particularly in Platystoma and Doras, it is so 'cellular' as is but rarely found in fishes; nay, *Doras maculatus*

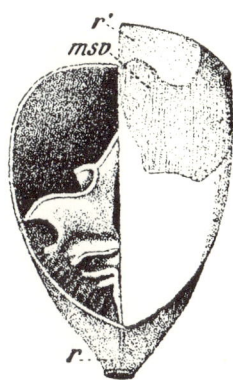

Fig. 3.—*Pseudaroides clarias*, Bl. The air-bladder of a specimen measuring 39 cm. of length, seen from below. To the right the ventral wall of the air-bladder has been removed, so as to permit the survey of the internal transverse septa, which divide its lateral halves into compartments. (By this operation a small part of the septa has been removed.) Diminished twice. *msv*, one of the muscles, which make the air-bladder act as a sound-producing organ; *r*, the kidneys behind the air-bladder, cut off from behind; *r'*, the pronephros.

seems to be of all fishes (perhaps with the exception of Gymnarchus) the one whose air-bladder contains the greatest number of 'cells.'"

As may be observed, this is in perfect accordance with the words of Bridge and Haddon (Ib, p. 236):—

"It may be remarked that cæcal appendages are very characteristic of those Siluroids in which an elastic spring 'apparatus' is present."[1] And in this connection it is well worth noticing what the authors add:—"A branching condition of the air-bladder, with the branches

[1] Not, however, in Synodontis, Malapterurus, and Euanemus.

ending in cæcal extremities, is very common in certain Physoclist genera. . . ."—See Günther,[1] p. 144–145.

For the genera mentioned in this place by Dr. Günther belong, with the exception of Polynemus, to the Sciænoidei, which are known to produce sounds by means of the air-bladder.

I myself was also struck with a similar very obvious idea, as may be seen from the following lines, a repetition in nearly the same words of what I wrote ten years ago (II*b*, pp. 107–109):—" Being engaged in the above-mentioned investigations at the mouth of the Riacho del Oro, where it enters Rio Paraguay, I was quite unacquainted with what had been published on this subject. I only knew that *Cottus scorpius*,[2] *Trigla (gurnardus)*, and *Dactylopterus volitans* are sound-producing. But having arrived at the result that the muscles of the air-bladders, which I had examined, serve to throw the air-bladder into sound-producing vibrations, and that a divided [3] air-bladder offered an improved organ for the production of sounds, I consulted V. Carus (and Gerstaecker's) *Handbuch der Zoologie*, which book was my only literary resource in South America, to see if any sound-producing fishes were mentioned there. This, it is true, was not the case; but as to Dactylopterus, the author

---

[1] XIII. Günther, A. C. L. G., "Introduction to the Study of Fishes," Edinburgh, 1880. The authors here also refer to Günther's *Catalogue of Fishes in the British Museum*, vol. v. But as this volume only contains Physostome fishes, I do not understand this reference.

[2] I have shown (II*b*, p. 78) that this fish produces sounds by stridulating between the præoperculare and the hyomandibulare. Dufossé (V*b*, pp. 91–103) and Professor L. Landois (in a book, *Thierstimmen*, Freiburg in Breisgau, 1874, published by his brother, Dr. H. Landois) judge the sound emitted by this fish to be due to the vibration of certain muscles. But while Dufossé designs these muscles as ". . . des muscles qui . . . font partie des régions inférieures des appareibs : hyoïdien, branchial et pharyngidien," Landois is of opinion that these muscles are the muscles of the shoulder-girdle. I have had no opportunity to resume this investigation after the perusal of the book of Landois.

[3] In the air-bladder of the Characini, examined by me, there is no further division but the well known one into an anterior and a posterior chamber; in *Platystoma orbignyanum* (?) and *Pseudaroides clarias* (see fig. 3 above) the air-bladder is divided into one large anterior chamber and five to six pair of chambers, growing gradually smaller behind and communicating with each other and with the first chamber by means of, comparatively speaking, large apertures along the lateral wall of the air-bladder; in *Doras maculatus* (see figs. 1 and 2 above) it is but partially divided by a longitudinal incomplete septum, and provided with numerous finger-like cæcal diverticula, among which the largest are incompletely divided in cells.

says (p. 539), "Schwimmblase getheilt, mit Muskeln;" and about Trigla (p. 538), "Schwimmblase wie bei Prionotus:" "Schwimmblase meist mit seitlichen Muskeln und in zwei Theile gespalten." Without knowing that this fact had already been stated by Dufossé,[1] I drew the obvious conclusion that Trigla and Dactylopterus also produce their sounds by means of the air-bladder, and that Prionotus was a sound-producing fish. In no other genera except these three muscles (extrinsic or intrinsic) were mentioned. But from what was stated about the form of the organ in the different genera, I arrived, by means of a vague conclusion, it is true, at the mere preliminary supposition with regard to twenty-six other genera or families, that the air-bladder might act as a sound-producing organ.

I am not going to fatigue the reader with a detailed account, but I shall confine myself to remarking that this idea proved correct,[2] so that, by means of my own investigations as well as by those of my predecessors, and more especially those of Dufossé, I was able to establish the following general thesis (II*b*, p. 182):—"*Where the air-bladder is a sound-producing organ, this organ is rendered the more effective, the thicker its walls are; the stronger or more elastic they are; the more it is divided into chambers* (and it is without any importance whether this division is brought about by external diverticula or by internal septa); *the more vigorous the musculature is,* and, in consequence, *the more closely it is connected with the skeleton,* provided the musculature be not intrinsic muscles. When these conditions are found together, we may determine an air-bladder as an organ producing sounds. As yet, however, with the restriction, *that the air-bladder must be either closed or furnished with a long, narrow, thin-walled pneumatic duct.*

For when the pneumatic duct is short and wide, and opens into the œsophagus with a fissure-like aperture, the *possibility* is

---

[1] The paper of Dufossé, dealing with this matter, was published in 1874, when I was a soldier, and I left Denmark in 1876, having never before been engaged in independent anatomical examinations of fishes.

[2] Among the twenty-six genera (or families) species producing sounds were known, or are now known: the Sciænoidei, Batrachus, Polypterus, Ceratodus, Lepidosiren, and Protopterus; muscles (extrinsic or intrinsic) were known in Therapon, Holacanthus, the Sciænoidei, Batrachus, Amblyopsis, Heterotis, Gymnarchus, Amia, Polypterus, and Lepidosiren.

not excluded, that this pneumatic duct may be capable of admitting the passage of the atmospheric air to and from the air-bladder, and in this case, therefore, it is possible that the latter may operate as an organ of respiration.

Besides the fishes, of which the anatomy is known to me through autopsy (Polypterus and Acipenser), information may be found in literature that the pneumatic duct is short and wide in Protopterus, Ceratodus, Spatularia, Arapaima (Sudis); Gymnarchus, Heterotis, Lepidosteus, Lepidosiren (in which four genera, according to the examinations of Hyrtl, the air-bladder is provided with muscles), as well as in Amia (in which, according to Franque, the air-bladder is provided with muscles). Now, it is true that several authors, on account of the anatomical data, have declared the air-bladder to be a lung in the following genera:—Arapaima (Sudis), Polypterus, Lepidosteus, Amia, Ceratodus, Protopterus, and Lepidosiren. But though it has, indeed, been observed that these fishes—with the exception of Polypterus[1]—are inhaling atmospheric air, still it is worth remarking that there exists no proof whatever[2] of the atmospheric air, respired by these fishes, being carried down into the air-bladder, as is absolutely necessary if the air-bladder is to be considered an organ of respiration.

In my book (II*b*, pp. 183–204) I have proved, as far as I can judge, that the reasons for attributing to the air-bladder the function of a lung, are anything but satisfactory. As I am here addressing myself to English-speaking naturalists, I shall confine myself to mention the Ceratodus and Protopterus, as these fishes have been repeatedly treated in English literature. The idea of attributing to the air-bladder the function of a lung arises from the *Lepidosiren paradoxa*, whose gills are so few and so small that most probably they cannot be sufficient to provide for the respiration of the animal. When this remarkable fish

[1] A sister of mine, Mrs Ida Leschly, who spent some years in Egypt, obtained information about the biology of this fish from the Egyptian fishermen. They unanimously stated this fish to live at the bottom of the river, and none of them ever saw an "Bichir" rising to the surface of the water to respire (whereas they were fully aware of the fact that such was the habit of the genus Clarias). One fisherman declared that he had heard it producing sounds in the water, but never out of it. Out of the water it is said to live only a few hours, though it is very tenacious of life.

[2] I must here remark that I have not followed these questions since 1884.

was treated in literature, no air-breathing fishes were known except the Cobitis; hence it was quite natural that, in the discussion on the Lepidosiren, no attention was paid to this isolated fact, the more so as this discussion turned chiefly upon its systematic position,—for, as is well known, it was considered by many naturalists as belonging to the Batrachia. No wonder, therefore, that the authors immediately agreed on the suggestion, that on the "lung," as it was called, was ingrafted the function of a respiratory organ. However, the probability that the air-bladder is really acting as a lung in this animal is somewhat confirmed by Dr Bohls [1]—

"Die Bedeutung der Lunge als Respirationsorgan kennzeichnet sich bei den lebendfrisch geöffneten Tieren durch die hellrote Farbe, die sie dem arteriell gewordenen Blut verdankt." When the author adds: "Die von mir gehaltenen *knurrten* beim Anfassen, ein Laut, der erzeugt wurde durch Auspressen der Luft aus den engen Kiemenöffnungen."

This explanation of the manner in which the sounds are produced appears to me rather doubtful.

The question was then proposed if the air-bladder was not to be considered a respiratory organ in other fishes, where it is cellular. Joh. Müller, meeting with an air-bladder of this kind in some Siluroidæ and in Erythrinus, thought fit [2] to reduce the question to a more definite form, and he answered it to the effect that the air-bladder in the Lepidosiren was to be considered a respiratory organ because of its receiving ("dark") blood from and returning ("light") blood to the heart, but that this was not the case either in the Polypterus, the organ in this fish receiving blood from the fourth gill-vein, and returning blood to the liver-veins, or in the Erythrinus, where it received blood from the arteries of the body, and returned blood to the veins of the body. This distinction, which is obviously founded on relations in the superior Vertebrata, was generally adopted, and is still considered decisive—thus Moreau calls it "les justes remarques." And yet it is incorrect, nevertheless. For the

[1] Bohls, "Mittheilungen über Fang und Lebensweise von Lepidosiren aus Paraguay," *Aus. d. Nachricht d. k. Ges. d. Wiss.*, Göttingen, 1894, No. 2.

[2] VIIIc: Müller, Joh., "Untersuchungen über die Eingeweide der Fische: Schluss der vergleichenden Anatomie der Myxinoiden" (*Abh. d. k. Akad. d. Wiss. zur Berlin*, A. d. J. 1843, Berlin, 1845, pp. 109-170).

physical principle of respiration is, that the blood enters into so close a relation with the air (atmospheric air, or air absorbed in water) that by a diosmosis a change of matter may be effected between the air in the blood and that without. And it is of no consequence whether the respiring capillaries are ramifications of an artery or of a vein. Consequently, an animal is able to respire with any part of its body provided that this part can enter into contact with the air: the outer skin, the cavity of the mouth, the gill-cavity, the intestines, the gills, or the lung. That the skin in Fishes shares in the respiration has already been shown by Spallanzani,[1] Provençal and Humboldt.[2] To what degree it shares in this function has not yet been proved, as far as I know; probably it does so in a rather great measure. But, on the other hand, respiration cannot take place when the air which, having been in contact with the respiring part of the body, has been rendered incapable of continuing the respiration, is not replaced by fresh air—that is, provided the respiring superficies is placed inside the body (gills, lung, intestines), by mechanical respiration. Whether an internal organ is an organ of respiration may therefore be decided by examining if it is the seat of mechanical respiration, which is, to a certain extent, constant, and whose continuation is necessary for the animal's existence, at least under certain circumstances. If we confine ourselves to the air-bladder, it has already been proved by Delaroche at the beginning of this century, and still more conclusively by Moreau, that there exists a constant exchange between the air in the air-bladder and the air in the blood,[3]

---

[1] Spallanzani, L., *Mémoires sur la respiration*, traduits par J. Senabier, Genève, 1803, pp. 113–114.

[2] IX. Provençal et Humboldt, "Recherches sur la réspiration des poissons," p. 392 (*Mémoires de physique et de chimie, de la société d'Arcueil*, T. II., Paris, 1809, pp. 359–404).

[3] The quantity of carbonic acid is variable. According to Moreau, in Perca, when normal, scarcely 1 per cent. is found, and in Barbus ½ per cent.; but the author, who unfortunately gives no details of his examinations, says in general: "Les proportions d'acide carbonique sont généralement au-dessous de 10 per cent. et même au-dessous de 5 pour 100." ✝ Schultze, Fr. ("Ueber den Gasgehalt der Schwimmblase einiger Süsswasserfische Deutschlands." *Pflüger's Archiv. f. d. gesammte Physiologie*, V. Bonn, 1872, p. 48–52), found in Barbus 1·4–4 per cent., and in Tinca 3·9–5·4 per cent. Schultze pretends to be, in 1872, the first author who has found the carbonic acid in the air-bladder. Without being thoroughly versed in this question, I know several authors previous to Schult

and yet it is no organ of respiration. It cannot be called so unless its air is renewed by mechanical respiration. If that be the case, it is an organ of respiration, no matter whence its blood comes and where it goes. Not to mention that (according to Hyrtl) the accessory gill-cavity in Saccobranchus receives some vessels from (and returns to) the adjoining parts of the body, and that (according to Jobert) the intestinal tube in Callichthys receives blood from the aorta, the venæ cavæ, and the vena porta renalis, there is another fact, pretty well known, which has no reference to the Fishes, which proves the untenability of the distinction set up by Joh. Müller. It is a well-known fact that the inwardly smooth hinder part of the lung in Snakes and in certain Sauria receives blood from the aorta. In accordance with the distinction of Joh. Müller, this fact is constantly interpreted to the effect that the hind part of the

who have made this observation. Moreau found carbonic acid in several Fishes ("Sur l'air de la vessie natatoire des poissons." In *Compt. rend. d. l'Acad. d. Sci.*, T. LVII., Paris, 1863, pp. 816–20). Humboldt found 2 per cent. in Exocœtus (*Reise in die Aequinoctial-Gegenden des neuen Continents*, I. Stuttgart, 1859, p. 179) and 4 per cent. in Poecilia Bogotensis (Humboldt et Aimé Bonpland: *Recueil d'observations de zoologie et d'anatomie comparée* . . ." T. II., Paris, 1833, p. 155). Provençal et Humboldt, in 1809, found 5·2 per cent. in Cyprinus carpio (IX., p. 401), and in 1789 Fourcroy found carbonic acid in Cyprinus carpio ("Observations sur le gaz azote contenu dans la vessie natatoire de la carpe" . . . in *Annales de Chimie*, I., Paris, 1789, pp. 47–51). If the author had consulted Cuvier, *Leçons d'anat. comp.*, he would have found (Edit. 2, T. VIII., p. 724) two of the authors mentioned, to whom I have referred here. When he says with regard to the two authors Biot and Erman, with whose papers he has made himself acquainted: "Dass die älteren Beobachter fälschlich einen vollständigen Mangel von CO_2 behauptete, eine Thatsache, die ihnen allerdings unerklärlich schien, die sie aber als solche hinnahmen," then he must have perused their papers rather superficially. For Biot ("Mémoire sur la nature de l'air contenu dans la vessie natatoire des Poissons," in *Mém. de phys. et de chim. d. l. soc. d'Arcueil*, T. I., Paris, 1807, pp. 254–281) says, p. 259: "Je n'avois pas . . . les moyens nécessaires pour mesurer exactement la quantité d'acide carbonique, . . . mais je me suis du moins assuré que cette quantité est fort petite." Dr. Schultze doubts the correctness of Biot's observations that the air-bladder in Fishes captured in the sea at a deep level contains great quantities of oxygen. Delaroche and Moreau, too, have proved the correctness of these observations to be beyond all doubt. Moreau (VI., pp. 79–84) has also shown that it is under the influence of the N. vagus that oxygen is received into the air-bladder. Dr. Bohr, Professor of Physiology at our university, has made experiments on the same subject, and has shown "that the formation of gas in the air-bladder is a true secretion of a highly oxygenated gaseous mixture, and that the secretion is so far under the control of the nervous system that it fails when the branches of the vagus which supply the air-bladder are cut" (*Jour. of Physiology*, vol. xv., No. 6, 1893, pp. 494–500).

lung in these animals is not respiratory. But what, in this case, we do not know, is the operation carried on in the capillaries in the recording section of the lung, whereas that which we do know is, that the lung in Snakes is an organ of respiration, *i.e.*, it is by turns receiving and expulsing the air. And we have not the slightest reason for doubting that the atmospheric air reaches the hinder part of the lung, if by no other means, at any rate by diffusion, by which means the air reaches into the bronchioli respiratorii and the alveoli pneumonales in the Mammalia. If we suppose that the hinder part of the lung in Snakes is not respiratory, we have simply turned things upside down; the conclusion being drawn from what was unknown to and against what is known: and yet it has given no offence. The very same method has been adopted with respect to the pseudobranchia and opercular gills in Fishes. We continually meet with the statement that when these organs receive venous blood their function is respiratory, but when they receive blood which has passed through the gill their function is not respiratory. And this theory is still maintained though there is not the slightest reason to suppose that, even in the latter case, the function of these organs should be another than that of the gills themselves with regard to the diosmotical relations. It is of some interest to see that none of the authors, who have supposed the air-bladder in some Fish to be a lung, has cared to get a notion of the mechanism necessary for the renewal of the air.

On account of the structure of the air-bladder and the relations of its arteries and veins, Sir Richard Owen judged the air-bladder in Protopterus to be an organ of respiration. Peters,[1] who has given the fullest account of the anatomy of this fish, gives a copious description of the ramification of the blood-vessels to the gills and the air-bladder, and says as follows (p. 17):—

"Im Ursprung unterscheidet sich die Lungenarterie, wie man sieht, nicht wesentlich von andern Körperarterien, die aus der Aorta Blut erhalten, und dies könnte es zweifelhaft machen, ob die Lungen des Lepidosiren [Protopterus] wirklich Lungen sind; diese Natur

[1] Peters: Ueber einen dem Lepidosiren annectens ähnlichen Fisch von Quellimane (*Archiv f. Anat. u. Physiol.*, Berlin, 1845, p. 1).

wird aber bewiesen theils durch die directe Einmündung der Lungenvene ins Herz, theils und noch bestimmter [dadurch] dass aus den Aesten des Truncus arteriosus schon Körperarterien entspringen, nähmlich die oben angezeigten." But afterwards, when he had an opportunity of dissecting the living Fish, he already began to doubt, and leaned to the opposite opinion, saying :[1]—" Dass die lungenähnliche Schwimmblase, ungeachtet des besonderen Eintritts ihrer Vene in das Atrium des Herzens, dennoch kaum als Lunge fungirt, scheint mir daraus hervorzugehen, dass ich an dem lebenden Thiere keinen Unterschied in der Färbung zwischen ihrem Blute und der der Körpervene bemerken konnte."

As to the habits of this animal, we know [2] that it becomes torpid during the dry season, like a fairly considerable number of tropical fishes, as in northern countries some fishes pass the cold season immersed in the mud, in a state of torpor; that it takes in atmospheric air ("occasionally, but at uncertain periods," according to Gray—" Anfangs . . . alle 4 bis 5 Minuten," according to MacDonnell). Where the air it has taken in is respirated, is unknown; the animal appears to swallow it and to expel it again through the mouth; and that it produces sounds : " Merkwürdig war mir," says MacDonnell, " dass es von dem Moment an, da ich es in das Wasser gesetzt hatte, aufhörte Töne von sich zu geben, selbst wenn man es aus dem Wasser herausnahm."

When Dr. Günther made known to the scientific world the Australian Ceratodus, which is so curious in many respects, nothing was known of its habits but what was stated in the passage quoted below. According to his observations, the airbladder—or, as it is called by Dr Günther, the "lung"—gives off blood to the heart, and one of its arteries can be injected from the arteria cœliaca, while no direct arterial connection exists between the air-bladder and the arcus aortæ, as in Lepidosiren. As to the manner in which Ceratodus is respiring, Dr. Günther proposed [3] the following hypothesis, exclusively based on the anatomical relations :—

---

[1] Peters : Reise nach Mozambique, IV. Berlin, 1868. *Flussfische*, p. 5.

[2] Peters, *l.c.*—Gray, J. E.: Observations on a living African Lepidosiren in the Crystal Palace (*Proc. of the Zool. Soc.*, London, 1856, p. 343-48).—MacDonnell, R.: Notiz über Lepidosiren annectens (*Zeitschr. f. wiss Zool.*, X., 1860, p. 409-11).—Duméril, A.: Observations sur des Lépidosiréniens (*Compt. rend. d. l'Acad. d. Sci.*, T. LXII. Paris, 1866, p. 97).

[3] Günther, Albert : Description of Ceratodus, . . . (*Phil. Trans. Roy. Soc.*, London, 1871, Pt. II., p. 511-571, p. 542).

"I think it much more probable that this animal rises now and then to the surface of the water in order to fill its lung with air, and then descends again until the air is so much deoxygenised as to render a renewal of it necessary. The Fish is said to make a grunting noise, which may be heard at night for some distance. This noise may be produced by the passage of the air through the œsophagus,[1] when it is expelled for the purpose of renewal. *From the perfect development of the gills we can hardly doubt that, when the fish is in water of normal composition, and sufficiently pure to yield the necessary supply of oxygen, these organs are sufficient for the purpose of breathing, that the respiratory function rests with them alone,* and that the lung receives arterial blood, returning venous blood, like all the other organs of the body. *But when the Fish is compelled to sojourn in thick, muddy water, charged with gases, which are the product of decomposing organic matter*[2] (and this must be the case very frequently during the droughts which annually exhaust the creeks of tropical Australia), *it commences to breathe air with its lung in the way indicated above.* Under this condition the pulmonary vein carries purely arterial blood to the heart, where it is mixed with venous blood and distributed to the various organs of the body. *If the medium in which the Fish happens to be is perfectly unfit for breathing, the gills cease to have any function;* if only in a less degree, the gills may still continue to assist in respiration. Ceratodus, in fact [in fact?!], can breathe by either gills or *lungs alone*,[3] or by both simultaneously."

What is the foundation of the suggestion propounded by Dr. Günther? It is true, he does not tell us so himself, but it is obvious that it is founded on two facts—that the air-bladder is cellular,[4] and that it sends its blood to the atrium of the heart;

---

[1] In these words the suggestion propounded by Joh. Müller (VII.) is seen to reappear.

[2] But if so, the Fish would, I think, nevertheless be lost, even if it were able to continue the respiration. While I stayed at Rio Paraguay a most extraordinary inundation took place; at that time it was quite a common view to see dead or dying Fishes near the riverside—in the midst of the stream I never saw any. That they had been poisoned by products of decomposition of putrefying organic matter, I could not doubt. Such matters are likely to be absorbed through the skin, a circumstance that Dr. Günther seems to have quite overlooked. That the lateral line serves to inform the Fish of the nature of the water, is no mere conjecture; that it is at least partly an organ of taste, is most certainly proved.

[3] This statement does not, however, agree too well with another passage of the author (p. 541): "the terminal branches of both arteries and veins [of the air-bladder] are rather wide, and can be injected with great facility."

[4] The remarkably well-drawn figure of the air-bladder in Dr. Günther's work (Pl. XXXVIII. fig. 2) shows me a structure which bears the greatest resemblance to air-bladders, that are sound-producing organs. But for the above-mentioned reason, I dare not—even as supposition—advance the opinion that this is the organ by which the animal produces the sounds which it is known to emit.

perhaps his suggestion that the air-bladder acts as an organ of respiration is right, perhaps it is wrong. I observe that Dr. Günther does not refer to the highly interesting and excellent investigations of Day (X*a*),[1] published three years before, on air-breathing fishes of India.

As to the biology of Ceratodus, we have, as far as I know, only a short report in a letter from E. Pierson Ramsay,[2] who kept some living specimens in the Australian museum at Sydney. And his report does not prove its breathing atmospheric air; for though he kept them "in a large tank," he does not say a word of their taking in air, but "when it rests on the bottom of the tank the pectorals are placed at nearly right angles to the body, the posterior fins lying parallel to the tail. If not disturbed they will remain in this position for hours, and only when stirred up think it necessary to use their fins and tail at all." For though it was "winter time and very cold," this behaviour offers a complete contrast to the proceeding of real air-breathing fishes, with regard to which Day says (X*a*, p. 275):— ". . . . . the purely water-breathers, if the term is admissible, can live without rising to the surface, unless under peculiar circumstances, whilst the compound breathers, as already mentioned, expire in a longer or shorter period if unable to reach the atmospheric air," whilst in another place (X*b*., p. 205)[3] he says:—"Of course under certain *abnormal* conditions, all species [of Fishes] rise to the surface, as I have already pointed out." I have lately seen in *Zoologischer Anzeiger* a notice on a work of R. Semon, "Zoologische Forschungsreise in Australien und dem Malayischen Archipel," the first volume of which work deals with Ceratodus. After the few words, added as a summary, the notice goes on to say:—"Das Tier atmet einmal in 30–40 Minuten." The work itself I have not seen as yet; it is to be hoped that the author, who has had an opportunity of studying the animal in its native habitat, has been able to give conclusive information whether the air taken in by the animal is respired in the air-

---

[1] X*a* : Day, Fr., "Observations on some of the Freshwater Fishes of India" (*Proc. of the Zool. Soc.*, London, 1868, pp. 274–288).

[2] *Proc. of the Zool. Soc.*, London, 1876, pp. 698–99.

[3] X*b* : Day, Fr., "On Amphibious and Migratory Fishes of Asia" (*Journ. of the Linn. Soc.*, London, vol. VIII., 1878, pp. 198–215).

bladder, or, as is the case in so many other Fishes, in the intestines, or in some other organ.

Posterior to Dr. Günther, a German author, Dr. Boas,[1] has found that the air-bladder in Ceratodus receives blood from the 4th gill-vein. Why the author considers the air-bladder to be a "lung" he does not tell us. But to him the circulation of blood in the organ cannot have been conclusive, as he regards the air-bladder as a lung, not only in Ceratodus and Protopterus, but also in Amia,[2] Lepidosteus,[3] and Polypterus,[4] although these fishes afford a striking contrast in that respect: for in some (in Ceratodus, according to Boas, and in Polypterus) the organ receives its blood from one pair, in others (Amia) from two pair of gill-veins; in others (Lepidosteus) from the aorta; and it returns it, now to the atrium of the heart (Ceratodus and Protopterus), now again to the kidneys (Lepidosteus), or the liver (Polypterus), or to the veins of the body (Amia, according to Franque). Accordingly, as it cannot be the relations of the blood-vessels that have determined the suggestion of this author, it must have been the circumstance that in all these forms the air-bladder is very cellular; thus his view is essentially older even than Joh. Müller.

Now, I am well aware that most readers will shake their heads at my presuming to question the accuracy of the generally adopted opinion that the "Dipnoi" respire by means of their air-bladder. But in order to point out how difficult it is to get a thorough knowledge of the function of an organ by mere

[1] Boas, J. E. V., Ueber Herz und Arterienbogen bei Ceratodus und Protopterus, *Morphol. Jahrbuch*, VI., 1880, p. 321–354.

[2] This animal is well known to take in atmospheric air (Wilder, B. G., in *Proc. Amer. Asso. Adv. Sci.*, 1875, Salem, 1876, p. 151; *ibid.*, 1877, p. 306–313); "Aereal Respiration in the Mud-Fish," in *Proc. Bost. Soc. Nat. Hist.*, XIX., 1878, p. 337; but we ignore by which organ it respires the air.

[3] Poey, F. (Memorias sobre la historia natural de isla de Cuba, T. II., 1856–58, p. 69), and B. G. Wilder (in his first quoted paper), have found that also this Fish takes in atmospheric air; but it is unknown in which organ it is respired. The observation of Poey is incorrectly referred by A. Duméril (*Hist. Nat. de Poissons*, T. II., 1870, p. 300).

[4] In a later paper (Ueber den conus arteriosus und die Arterienbogen der Amphibien. In *Morphol. Jahrbuch*, VII., 1882, pp. 488–572) Dr Boas has not included Polypterus among the Fishes which he supposes to breathe by means of a "lung." The reason (which the author has forgotten to tell his readers) is, that I myself informed him personally of the above-mentioned features of the habits of this Fish.

anatomical examinations, even if only mechanical principles are in question, I shall recall to the memory of my readers how the theory of Borelli as to the air-bladder as a hydrostatic apparatus was considered a fact beyond all contradiction, and yet it was crushed when put to the first[1] serious experimental test. Nay, even a thing so plain in appearance as the effect produced on an organ by the contraction of a muscle may be misinterpreted: not only Bridge and Haddon, but even a prominent author like Joh. Müller has been of opinion that the powerful extrinsic muscles of the air-bladder in the Pimelodina would by their contraction effect a compression of the organ, while the fact is that they effect an expansion, be it even momentary, of the said organ. And I do not doubt but that a reader who is not versed in the literature in question would, in perusing the physiological section of the work of Professors Bridge and Haddon, be filled with admiration at the acuteness and close reasoning displayed—and yet scarcely one passage is correct, because they have not examined the said organs in living specimens, and because they had but an insufficient knowledge of the literature in which are deposited the results of that kind of investigation.

It is worth remembering that in the air-breathing fishes, in which the matter is *well* known, it is not in the air-bladder that the respiration of the inhaled air takes place. Not to mention some of our European Cyprinoidæ, *e.g.*, *Carassius vulgaris*, which in the summer time, when the water is getting deficient in oxygen (if they live in smaller ponds), try to remedy this want by letting the atmospheric air pass, together with the water, over their gills, it is well known in the European species of the Cobitini that the respiration of atmospheric air, which frequently takes place when these fishes are chased out of the mud, where they use to live almost without stirring, though now and again, with long intervals however, they resort to the surface to take in atmospheric air, is brought about by the intestine.[2] The

---

[1] Valenciennes (XV., vol. xvi., 1842, pp. 14–16) had made experiments to empty the air-bladder of air in *Gobio fluviatilis* and he had seen it "très doucement" filled with air in the course of some hours, without the animal taking in atmospheric air. But Valenciennes did not see the bearing of his own observation.

[2] Erman, in his excellent paper, Untersuchungen über das Gas in der Schwimm-

same fact has been established, according to Jobert,[1] in Hypostomus sp. (XI*b*), Callichthys asper (XI*a*), Doras sp. (XI*b*), and Loricaria sp. (XI*a*), which Fishes cross the land to reach other waters, when their former dwelling-places are growing short of water. In his interesting investigations on several air-breathing Fishes of India, Day (X*a* and X*b*) has proved that the respiration of the atmospheric air is brought about in the labyrinthiform part of the gill-cavity in Polyacanthus, Osphromenus, Trichogaster, Ophiocephalus, and Rhyncobdella. The same author has pointed out that the air is respired in the accessory gill-cavity of Clarias,[2] Saccobranchus, and Amphipnous. The existence of an accessory "gill-snail" (Kiemensschnecke) has been proved by Hyrtl[3] in Heterotis Ehrenbergii, Chanos ("Lutodeira chanos"), Meletta thryssa, Chatoessus jacunda, Gonostoma Javanicum, Clupanodon aureus, Pellona Lechenaultii, and Hyodon claudalus, and an organ of a similar structure has been found by Kner[4] in Cœnotropus labyrinthicus, Curimatus vittatus, and C. cyprinoides. Most probably this organ is an air-breathing organ; however, before the year 1884, no physiological proof existed of this being the case.

When I said above that an air-bladder without a pneumatic duct, or with a long, narrow, and thin-walled pneumatic duct, may be determined as a sound-producing organ (according to our present knowledge), when furnished with extrinsic or intrinsic muscles, especially when its cavity is divided into inter-com-

blase der Fische, und über die Mitwirkung des Darmkanals zum Respirationsgeschäfte bei der Fischart Cobitis fossilis (*Gilbert Annalen der Physik.*, XXX., Halle, 1808, pp. 113–161).

[1] XI*a*: Jobert, Recherches pour servir à l'histoire de la respiration chez les Poissons (*Ann. d. Sci. Nat.*, 6 Sér., T. V., Paris, 1877, Art. No. 8).—XI*b*: Jobert, Recherches anatomiques et physiologiques pour servir à l'histoire de la respiration chez les Poissons (*ibid.*, 6 Sér., T. VII., Paris, 1878, Art. No. 5).

[2] I myself, who did not then know of the examinations of Day, have made the same at least almost evident with regard to Clarias macracanthus from the Nile (Om Aandedrœttet hos Clarias [On the Respiration in Cl.]. In *Naturhistorisk Tidskrift*, 3 R., Bd. XIII., Kjöbenhavn, 1883, p. 396–414).—What I there have said on the nature of the gill-rakers of Clarias is wrong.

[3] Hyrtl, J., Beitrag zur Anatomie von Heterotis Ehrenbergii (*Denkschr. d. k. Akad. d. Wiss. in Wien*, Bd. VIII., 1854, p. 74).—Hyrtl, J., Ueber besondere Eigenthümlichkeiten der Kiemen und des Skelette, und über das epigonale Kiemenorgan von Lutodeira (*ibid.*, Bd. XXI., 1862).

[4] Kner, R., Ueber Kiemen-Anhänge bei Characinen (*Ver. d. Zool. bot. Ges. in Wien*, 1861, p. 189).

municating chambers, I must point out that some naturalists have succeeded in proving, with regard to some Fishes, that the air-bladder serves to produce tones, even if it is *not* provided with special muscles. This fact has been proved by Dufossé, by means of vivisection of the animals, in *Peristedion cataphractum Trigla lyra, Hippocampus brevirostris, Sciæna aquila*, and *Umbrina cirrhosa*. In these Fishes the air-bladder sounds during the activity of the muscles, with the fascia of which the walls of the air-bladder are intimately connected. In these Fishes the conditions, at least according to the indications of Dufossé, are such that it would hardly have been possible to any one, by a mere anatomical examination, to recognise the air-bladder as a sound-producing organ. But, on the strength of Dufossé's physiological examinations, I am of opinion that the air-bladder has the same function in *Triacanthus brevirostris* and *Tr. biaculeatus*. In the following Fishes I have succeeded, by a mere anatomical examination, in recognising the air-bladder as a sound-producing organ, though it is not provided with special muscles: *Tetrodon fahaka, Balistes vetula, Monacanthus pardalis*, and *Holocentrum sogho*, in which Fishes the walls of the air-bladder are acted on in a somewhat different way by the muscles leading to the "coracoideum" (Cuv.). Finally, Professor Möbius [1] has recognised the air-bladder of *Balistes aculeatus* as a sound-producing organ, likewise employed in sounding under the action of muscles which lead to the same bone ("Postclaviculare"). And while I had arrived at the said result through anatomical examination alone with regard to the above-mentioned species of this genus, Professor Möbius had the opportunity to observe in *B. aculeatus*, in the living animal, "while the fish was drumming, a quick raising and sinking of a small spot of the skin," which spot proved by the following anatomical examination to be in immediate contact with part of the wall of the air-bladder.

[1] XII : Möbius, K., Balistes aculeatus, ein trommelnder Fisch. (*Sitzber. d. Akad. d. Wiss.*, Berlin, Bd. XLVI., 1889, p. 999–1006). Zacharias, Otto, "Trommelnde Fische" (I have but seen a copy of this short paper, four pages) contains no independent investigations, but is only a popular report of the paper of Möbius.

ARE THE EXTRINSIC MUSCLES OF THE AIR-BLADDER IN SOME SILUROIDÆ AND THE "ELASTIC SPRING" APPARATUS OF OTHERS SUBORDINATE TO THE VOLUNTARY PRODUCTION OF SOUNDS? WHAT IS, ACCORDING TO OUR PRESENT KNOWLEDGE, THE FUNCTION OF THE WEBERIAN OSSICLES? A Contribution to the Biology of Fishes. By William Sörensen, Copenhagen.

(*Continued from page* 139.)

I take the liberty here, as ten years ago, to point out that, by Moreau's statement of the manner in which the air-bladder of *Trigla hirundo* produces sounds, the reader may be led to suppose that Moreau attributes to the vibrations of the perforated internal transverse septum greater importance than is due to it, as far as I can judge; his words are apt to convey to the reader the impression that he thinks the fish owes to this diaphragm alone the production of sounds. In my opinion,[1] the air-bladder is capable of producing sound, but weaker ones it is true, even if the said diaphragm did not exist.

In *Sciæna aquila*, *Umbrina cirrhosa*, *Trigla lyra*, *Peristedion cataphractum*, and *Hippocampus brevirostris*, the air-bladder, according to Dufossé, produces sounds under the action of muscles with the fascia of which the air-bladder is intimately connected. While as regards the remaining fishes which he has examined, this author judges the air-bladder itself (with its gaseous contents) to produce the sound by the contraction of the extrinsic or intrinsic muscles, he is of opinion, with regard to the species here mentioned, that the air-bladder only serves to intensify the sound, which he considers to be produced by the vibration of the muscles while contracted; in other words, to be a muscular tone. Though I have not been fortunate enough to examine any living fish in which these conditions were

---

[1] As to this species, I have only examined it when dead. My opinion is not only based on my examination of living specimens of *Tr. gurnardus*, but on my whole knowledge of the air-bladders, which I have examined in living as well as in dead fishes.

present, I cannot subscribe to this opinion, but must consider the sound as being produced by vibrations of the air in the air-bladder and of the wall of the latter, when set in motion by the muscles with the fascia of which it is connected: firstly, because the sound, produced by the mere contraction of a muscle, is very faint in itself; and secondly, because, if his opinion were correct, it would be a mystery why sounds were produced by the contraction of some muscles and not of others, and that this is the case [1] has been proved by Dufossé (V*a*, p. 43): when he cut the nerve of the muscle in question on one side of the body the sound grew fainter, and quite ceased when the nerve on the other side was also divided. Further, Dufossé has found the height of the sound in *Peristedion cataphractum* and *Trigla lyra* to correspond with 517, nay, even with 870 vibrations in the course of a second; these very numbers appear to me too high by far to be reconciled with the sound being a muscular tone. And finally, it appears to me that the strength of the opinion of Dufossé is weakened by his own experiment (V*a*, p. 42, 43): when he extirpated the air-bladder of a *Trigla lyra* and replaced it by the foremost chamber of the air-bladder of a *Cyprinus carpio*, which chamber he then filled with air.

"Si j'ai opéré avec assez de promptitude et avec toutes les précautions que réclame cette expérience, le Poisson recommence à bruire, et les sons qu'il forme sont presque semblables à ceux qu'il émettait avant le commencement de la vivisection."

This experiment seems to me clearly to prove that the air-bladder (even if it be that of another fish) is rendered capable of emitting sounds when compressed by muscles capable of acting in a certain manner. It is here the place to add that, according to Dufossé, the musculi intracostales in question are extended between the os scapulare (Cuv.) and the 7th–10th vertebræ, and that they form two prominent masses of muscles, between which the air-bladder is imbedded.

Now, to return to the Siluroidæ, I have shown that the foremost strong, sometimes powerful, ray of the pectoral fins, and,

---

[1] This is still more confirmed, as far as I can see, when we consider the production of sounds (by means of the os præoperculare) in Cottus; but it would lead me too far here to explain this fact more fully.

in a manner, the strong ray of the dorsal fin, are sound-producing organs. This fact is explained in the following manner:—The animal has in its power to render these more or less powerful

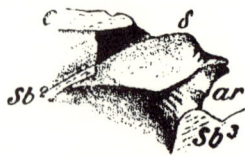

FIG. 4.—*Doras maculatus*, Cuv. et Val., of a specimen measuring 50 cm. of length; natural size. The hinder part of the helmet (*C*) and the muscular crests of the interspinous bones (Sb^2, Sb^3) are persected. Laid open to the view are δ, the left side of the roof-like keel of the second interspinous bone ; *ar*, the articular cavity for the median part of the articular face of the strong ray.

weapons as it were immovably fixed by means of a mechanism, which, by the way in which it operates, bears the greatest resemblance to the brake of a wheel,—for instance, that of a railway carriage. In the first ray of the pectoral fins this brake consists in a crest-like arched process, springing from the upper

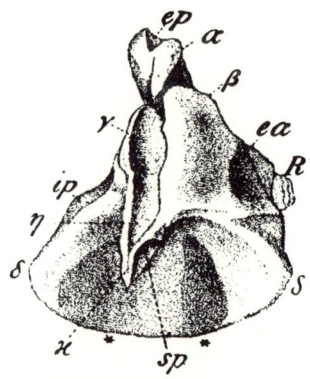

FIG. 5.—*Doras maculatus*, Cuv. et Val. The first ray of the right pectoral fin, seen from the base and a little obliquely from above; natural size. Of a specimen measuring 69 cm. of length.

side of the base of the ray, and passing forwards and backwards in an excavation of the "humerus" (Cuv.) during the movements of the ray, and pressed against the walls of the excavation when the ray is to be fixed.[1] In the dorsal fin this is effected by

[1] Professor Gegenbaur, in his well-known work "Schultergürtel der Wirbelthiere, Brustflosse der Fische" (*Untersuchungen zur vergleichenden Anatomie der Wirbelthiere*, Heft 2, Leipzig, 1865), among the Teleosteans, has made the study of the Siluroidæ his particular task. This work being generally considered as

means of the preceding "rudimentary" ray, which ray is wanting when the large ray is flexible all through, and accordingly no weapon. The posterior surface of this "rudimentary" ray is excavated, and may be pressed against a roof-like process

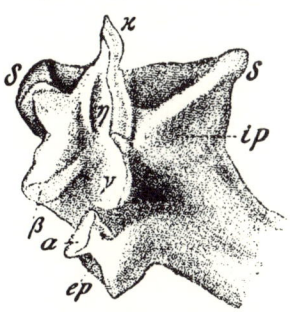

Fig. 6.—*Doras maculatus*, Cuv. et Val. The same object as in Fig. 5, seen obliquely from the base, from behind and from below.

(fig. 4) which springs from the interspinous bone.[1] Now, as a sound is produced when the brake of a railway carriage in motion is put in action, in the same manner here. When the pectoral fins are moved forwards or backwards, and when the dorsal fin is moved backwards—for the latter can never be fixed when moved forwards—a discordant sound is produced when the brake begins operating.[2]

one of the most important works on these subjects, I must point out that he is mistaken in mentioning (pp. 116–117 and 153–155) this process as part of the diarthrosis of the ray. Firstly, on functional grounds, it cannot be said to form part of the diarthrosis, as it does not serve to facilitate the movement, but to prevent it. Nor can it, from a morphological point of view, be called part of the diarthrosis: the rays of the pectoral fins do not in any fish articulate with the bone called by Cuvier humerus, and by Geoffroy Saint-Hilaire clavicula. On this point Cuvier et Valenciennes (XV., T. XIV. p. 318) were already well informed.

[1] Professor Haddon ("On the Stridulating Apparatus of *Callomystax gagata*," *Jour. of Anat. and Phys.*, T. XV., 1881, pp. 322–326) has arrived at the result that a sound is produced in *Callomystax gagata* when the channeled faces of the first and second interspinous bones are rubbed against other likewise channeled faces of the confluent spinous processes of the 4th and 5th vertebræ. When I referred to this result in my book (II*b*, p. 120) I expressed a doubt whether this suggestion was not based on imperfect observation. This doubt I now find still more confirmed. For though I do not know the anatomy of this fish, fig. 68 (Pl. XVI.) of the lately-published work of Professors Bridge and Haddon (I*b*) quite clearly shows not only the "rudimentary" ray, but also the well-known roof-like process behind it, on the interspinous bone.

[2] Starting from the Siluroidæ, I have pointed out similar conditions in a series of several Fishes:—At the dorsal fin in Balistes, Monacanthus, Acanthurus, Capros,

As, however, the large ray of the dorsal fin and the first ray of the pectoral fins in the great majority of the Siluroidæ are developed as a weapon, I was, with regard to this family, prevented from using one of the methods I generally employed to enlarge my knowledge of the air-bladder as a sound-producing organ,—that

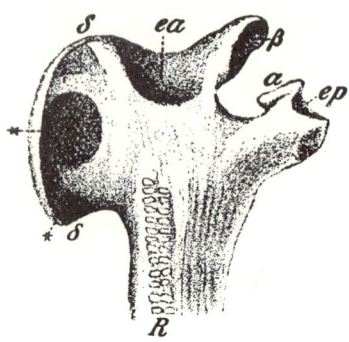

Fig. 7.—*Doras maculatus*, Cuv. et Val. The same object as in figs. 5 and 6, seen in front, somewhat obliquely.—In Figs. 5-7 the different parts, designed by letters, are: R, the foremost denticulated edge of the ray; δ, δ, the process, acting as a brake; *, *, the scouring faces of this process; γ, η, the vertical part of the diarthrosis; a, β, the two remaining parts of the diarthrosis, each on its process; ea, the place of insertion of the two muscles, leading forwards the ray; ip, the place of insertion of the muscle, leading backwards the ray; ep, the place of insertion of the muscle, which permits the ray to move backwards, but prevents it from moving forwards; sp, the place of insertion of the muscle, which permits the ray to move forwards, but prevents it from moving backwards. (This muscle, which does only exist in Doras, is a specialised portion of one of those muscles, which move forwards the ray.) κ, the process, carrying "sp."

is, the examination of such fishes as were said, in the literature, to produce sounds. *Malapterurus electricus* being provided, according to Joh. Müller, with an "elastic spring" apparatus, it was not unlikely that the air-bladder was capable of producing sounds. And the first ray of the pectoral fins being quite flexible, while the dorsal is altogether wanting,—the only remainder is one interspinous bone,—most likely this fish, *if* it were able to produce sounds, would have to do so by means of the air-bladder. But though this fish, on account of its electric organ, has often been the object of examinations, I did not succeed in finding any indication as to its capability to produce

Triacanthus, Centriscus, Gasterosteus, and Anarrhicas; at the ventral fins in Triacanthus, Capros, and Gasterosteus; at the os hyomandibulare in Dactylopterus; at the os præoperculare in Cottus (II*b*, pp. 5-82, where the details are given).

sounds. I therefore asked my sister, Mrs Ida Leschly, who was then staying in Mansourah on the Nile, to procure a living specimen of this species, that she might learn whether it were able to produce sounds or no. Though the fishermen were not aware of the fact, she was fortunate enough to hear, and very distinctly, sounds repeatedly produced by one of the three specimens she got, and which had been deposited in a jar of water immediately after being captured. When the fish was

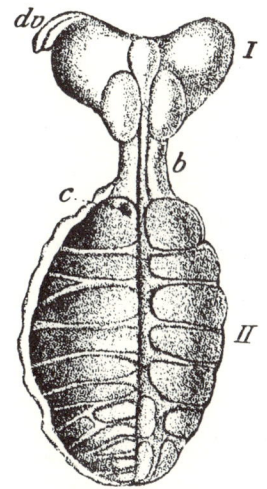

Fig. 8.—*Malapterurus electricus* L. The air-bladder seen from below. The second portion is opened along with the longitudinal septum; to the right the wall is moved a little aside, so as to show the low transverse septa. b, the constricted part between the anterior (I) and posterior (II) portions of the airbladder; c, the hindmost opening of the right channel, uniting the anterior and the posterior portions; dv, one of the two circular plates at the fore-end of the air-bladder; nothing but the edge is visible, being a little removed from the air-bladder.

brought to her, it had recently been taken in the river; it was all alive, frisky, and unapproachable on account of the vigorous shocks which it imparted, in spite of its inconsiderable size. She described the sound as not unlike the hissing of a cat. This appeared somewhat strange to me; but, after I had dissected the fish, I could easily account for it by the long, narrow ducts through which the air has to pass from the anterior to the posterior chamber of the air-bladder. My sister was able to state that the sound was not accompanied by any gasping for air or any beating with the tail.

It is a matter of course that, on account of this statement, it was so much easier to me to consider the air-bladder as a sound-producing organ when the "elastic spring" apparatus does exist.

Starting from the suggestion that this apparatus does exist in all species of the genera Doras, Oxydoras, Rhinodoras,[1] Auchenipterus, Euanemus, Malapterurus, and Synodontis, and that the extrinsic muscles have the same function in all species of this family as in the two genera I had observed in living specimens, it may *now* be stated that the air-bladder serves to produce sounds in sixty-eight species [2] of this family, or in more than the tenth part of the species belonging to this family,—unquestionably, a result differing somewhat from that at which Bridge and Haddon have arrived; according to these authors (*Ib*, p. 270), "doubtful examples of the association of the air-bladder with phonation in a few Siluridæ and Cyprinidæ." But I may be allowed to add that there is sufficient reason to *presume* the air-bladder to be a sound-producing organ in more species than the

---

[1] Kner, R. ("Ueber einige Sexualunterschiede bei der Gattung Callichthys und die Schwimmblase bei Doras," in *Sitzungsberichte d. k. Akad. d. Wiss. Wien*, Vol. XI., 1855, pp. 138-146; and "Ichthyologische Beiträge," *ibid.*, Vol. XVII., 1855, pp. 92-162), has found this apparatus in 9 species of the genus Doras, in 7 species of the genus Oxydoras, and in 2 of the genus Rhinodoras. From his memoir it appears that he has found this apparatus in all the species examined.

[2] Günther (XVI., vol. v.) enumerates 13 species of the genus Doras (besides 3 species mentioned), 7 species of the genus Oxydoras, 3 species of the genus Rhinodoras, 9 species of the genus Auchenipterus, 1 Euanemus, 3 species of the genus Malapterurus, 12 species of the genus Synodontis. According to Bridge and Haddon (*Ib*), an "elastic spring" apparatus does also exist in 4 species of the genus Pangasius. Extrinsic muscles exist in *Centromochlus megalops, Trachelyopterus tæniatus,* and *Sorubim lima* (according to Kner, "Ichthyologische Beiträge II." in *Sitzungsberichte d. k. Akad. d. Wiss. in Wien*, vol. 26, 1857, pp. 373-448); *Platystoma fasciatum* (acc. to Joh. Müller); *Pl. sturio, Pl. Vaillantii,* and *Pl. platyrhynchus* (acc. to Kner); *Pl. tigrinum* (acc. to Bridge and Haddon); *Pl. Orbignyanum*? (acc. to me); *Pseudaroides clarias* (acc. to me); *Piramutina piramuta* (acc. to Bridge and Haddon); *Piratinga filamentosa* (acc. to Joh. Müller); *Pimelodus gracilis* (acc. to Kner); *P. ornatus* (acc. to Kner, Bridge and Haddon); *P. ranina* (acc. to Cuvier and Valenciennes); *P. maculatus* (acc. to Cuv. and Val., Bridge and Haddon). Bridge and Haddon (*Ib*, p. 246) having declared not to have found extrinsic muscles in any fish of this family except in the Pimelodina, we had perhaps better, at the present, doubt the existence of such muscles in *Rita pavimentata, Ælurichthys Gronovii,* and *Æl. marinus* (acc. to Cuv. and Val.); *Amiurus cauda-furcatus* (acc. to Kner); *Am. catus* and *Arius Milbertii* (acc. to Cuv. and Val.). In *Arius cœlatus* these authors state that they exist, but Bridge and Haddon deny their existence in this species.

above-mentioned. At anyrate, there is some reason to *presume* that the division of the air-bladder, effected either by external diverticula or by transverse septa, or in some other way, which by different authors is said to exist in several Siluroidæ, where extrinsic muscles or "elastic spring" apparatus are not present, may have some relation to the production of sounds. I here take the liberty to direct the attention of other naturalists to this subject.

I ought further to add, that in *Silurus glanis* it was impossible for me, at the present stage of the knowledge of the air-bladder as a sound-producing organ, to see that this organ had any function in this respect. And in *Clarias macracanthus* and *Plecostomus Villarsii* it does not appear possible to me that this organ could have that function.

As may be seen from what I have quoted above (pp. 122–124) from my observations during the vivisection of the mentioned animals, intricate or difficult physiological examinations are out of the question. On the contrary, the fishes that I have been vivisecting being animals of the length of 25 cm. up to 1 metre, and their air-bladder, as well as the muscles which cause the production of sounds, being organs of considerable size, the examinations are easy enough, provided that a sufficient material be at hand (but the Rio Paraguay is very rich in these fishes) for, of course, life is not long in being extinguished in fishes on shore and with opened bellies.

If it has not been sufficiently proved by what has already been said, I must direct the reader's attention to the fact that the extrinsic muscles of the air-bladder in Platystoma and Pseudaroides do not produce the effect attributed to them by Bridge and Haddon (Ib, p. 115, on *Platystoma tigrinum*): "As the contraction of these muscles must necessarily lead to the forcible compression of the anterior chamber of the bladder, . . . we shall in future mention them as the 'compressor muscles.'" They do not, by any means, compress the air-bladder: the immediate effect—which lasts only a moment—of their contraction consists in moving forwards the anterior part of the wall of the air-bladder. But I feel bound to add that the authors are not to be blamed for having drawn this conclusion from the purely anatomical data, for they only followed Joh. Müller (VIII.).

With regard to the morphological interpretation of the muscles of the "elastic springs," the authors are of opinion (I*b*, p. 230) that they are specialised portions of the dorso-lateral musculature. I cannot indorse this opinion. This may *perhaps* be the case in Synodontis, where these muscles have quite a different position[1] from that of the other genera; but in these genera they appear to me, according to their whole position, to be quite homologous with the extrinsic muscles of the air-bladder in the Pimelodina.

Bridge and Haddon have made a discovery which appears to me most interesting, viz., a pair of small muscles, which are said always to exist when "compressor" muscles are present. They "arise from the ex-occipitals, and are inserted into the anterior wall of the anterior chamber of the air-bladder." I have overlooked these muscles, most probably because, as I am led to presume by the figures of Bridge and Haddon, they are concealed by the pronephros; and, living in the wild forests of that country, I was only able to dissect the animals the very same day they were taken; the next day they were rotten, on account of the hot climate. What is the function of these muscles I cannot, therefore, say for certain; however, they serve, I suppose, to support the big extrinsic muscles. But what I firmly believe is, that they do not fulfil the function attributed to them by Bridge and Haddon, viz., that "of limiting the violent excursions of the tripodes [mallei] which might otherwise take place when the anterior chamber is forcibly compressed by the contraction of its compressor muscles."

I should think that whosoever has perused my "Observations on the producing of the sound" quoted above, even if he should judge the opinion set forth there about the *manner* in which the sounds are produced to be quite inaccurate, will agree with me that the production of sound is brought about by the said muscles, so that this opinion of mine is not "an ingenious theory," but a plain fact. And when Professors Bridge and Haddon are of opinion—and as to this opinion I quite agree with them—that the theories which they have propounded on the function of

---

[1] See my book (II*b*, p. 125, fig. 48), o Joh. Müller (VIII*c*., Tab. III. fig. 4).

these muscles and the "elastic spring" mechanism are incompatible with my "theory," then their suggestions must be considered, for that very reason, as null and void.

And if they had known my paper, published in 1884, and that of Dufossé (V*b*), published in 1874, they might probably have been able to multiply, even to a considerable degree perhaps, the number of the Siluroidæ in which the air-bladder is a sound-producing organ. As for me, I can only say, according to the text and figures of the authors, that beyond what has hitherto been known, that is the case in *Pangasius Buchanani, P. djambal, P. macronema, P. juaro; Platystoma tigrinum* and *Piramutina piramuta*. For, though it is most probable that several others, besides those which I have mentioned here, might, if subjected to an anatomical examination with this aim in view, be recognised as being furnished with an air-bladder, serving as a sound-producing organ, I am not able, with no other basis than their examinations, to state it as a fact with regard to any more forms.

I would add that the part of the memoir of Professors Bridge and Haddon which deals with the function of the air-bladder and the Weberian ossicles, would have presented quite a different aspect if they had known my paper on sound-producing organs in fishes.[1] This part of their work would also have looked quite differently if they had thoroughly studied the physiological investigations made in modern times on the subject of the functions of the air-bladder.

More than 200 years ago, Borelli[2] (Lib. I. p. 332 *seq*.) set

---

[1] In a "Note on the Production of Sounds by the Air-bladder of Certain Siluroid Fishes," communicated to the Royal Society of London, April 26, 1894, and which appeared in the part of the *Proceedings* issued 24th July 1894, some weeks after this part of the manuscript was sent to the Editors of the *Journal*, Professors Bridge and Haddon "draw the attention of those interested in the subject to Dr. Sörensen's researches," and "express our regret at the injustice we have unintentionally done him." I cannot refrain from observing that I have not attracted the attention of the authors to this fact; the Note has been published against my wish; the two passages of my paper which the authors have translated in their note, are in mutual contradiction because of their having translated, in the passage on Doras, p. 89, the Danish word "utvivlsomt" by "very doubtful" instead of "without doubt." (See quotation from my paper in the October number of this *Journal*, p. 123, where these two words are printed in italics.)

[2] Borelli, *De motu animalium*, Romæ, 1680.

forth the hypothesis that, by compressing the air-bladder, the fishes lowered themselves in the water, and that, by distending it, they rose again, as if their body were the stiff bulk of a vessel incapable of changing direction in the same manner as other Vertebrata by bending their bodies. Though this hypothesis cannot be said ever to have been borne out, and in spite of the weighty arguments urged against it by Delaroche[1] (pp. 249–250) at the beginning of this century, this opinion has been prevalent until the last quarter of our century. No wonder, then, that Joh. Müller,[2] when he discovered the remarkable "elastic spring" apparatus in the mentioned genera of the Siluroidæ, set forth as his opinion that this mechanism served to help the fish to sink and rise in the water, according to the muscles being relaxed or contracted to a smaller or greater extent; nay, that more probably ("vielmehr") they served to give the fish an oblique direction, downwards, when the muscles of the springs were relaxed, upwards when they were contracted (VIIIc).

Delaroche, in opposition to the hypothesis of Borelli, put forth another, which for certain reasons I take the liberty to quote. He says as follows (XIV. p. 261):—

"Il résulte de ce que je viens de dire, que la vessie n'a pas d'autre usage bien constaté que celui de mettre la pesanteur spécifique des poissons en équilibre avec celle du milieu ambiant."

If he had confined himself to these words, it would have been so much the better, for that is correct, and even he was not able to support this statement with as weighty arguments as did Moreau about seventy years later,—the reasons he gave were of great weight. But less fortunately he adds (*ibid.*, pp. 262–264):—

"Les muscles propres qui sont fixés à ses parois dans un grand nombre d'espèces, ont probablement pour usage de comprimer plus ou moins fortement le gaz qu'elle renferme, non comme le supposent ceux qui ont adopté l'hypothèse de Borelli pour changer la pesanteur spécifique du poisson, mais au contraire pour le maintenir toujours au même

---

[1] XIV: Delaroche, F., "Observations sur la vessie aërienne des Poissons" (*Annales du Mus. d'Hist. Nat.*, T. XIV., Paris, 1809, pp. 184–217; pp. 245–289).

[2] VIIIa: Müller, Joh., "Beobachtungen über die Schwimmblase der Fische, mit Bezug auf einige neue Fischgattungen (*Archiv f. Anat. u. Physiol.*, Berlin, 1842, pp. 307–329).—VIIIb: Müller, Joh., The same matter (*Monatsber. d. k. Akad. d. Wiss.*, Berlin, 1842, p. 202).

point. . . . Il faut donc, pour que leur pesanteur spécifique ne varie pas, qu'il y ait une cause toujours agissante qui empêche cette condensation et cette dilatation. Telle paroît être la fonction des muscles propres de la vessie. . . . On peut raisonnablement supposer que les muscles abdominaux remplacent ces muscles propres chez les poissons qui en sont privés." Though this is certainly incorrect, yet it appears to me an improvement of the Borellian hypothesis.

As I am here going to mention several[1] of the theories put forth in the course of time touching the functions of the air-bladder, I feel bound not to pass the suggestions of Cuvier and Valenciennes. With regard to the form and structure of the air-bladder in the Sciænoidei, these authors say[2] (T. V., 1836, p. 3):—

". . . et bien que ces vessies natatoires ne paraissent pas avoir de communication avec l'extérieur, comme presque toutes les sciénoïdes font entendre des bruits, des grognements, encore plus marqués que ceux des trigles, il est difficile de croire que la disposition de ces organes n'ait pas quelque rapport avec cette propriété." And as to *Batrachus grunniens* (*ibid.*, T. XII., 1837, p. 471):—"Cet appareil musculaire de la vessie doit contribuer, comme dans les autres poissons grondeurs, au bruit que les batrachoïdes font entendre." Nay, Valenciennes afterwards, when mentioning the sounds produced by the Siluroid genus Synodontis, feels so sure about this theory, that, with much more weight than he was entitled to, he declared (*ibid.*, T. XV. p. 251):— "Tous ceux, d'ailleurs, qui connaissent l'histoire naturelle des poissons, savent que les sons que ces animaux font entendre sont dus au mouvement qu'ils peuvent donner à l'air de leur vessie natatoire, en exerçant sur cet organe une compression plus ou moins forte quand il est pourvu de muscles constricteurs . . ."

This was, maybe—in the real meaning of the words—an "ingenious theory," but nothing more. For, with the exception of Aristotle, whose words[3] I have placed at the head of the first

[1] Such theories as must be said to have been mere vagaries or not to have been generally adopted, I entirely pass.

[2] XV: Cuvier, G. et Valenciennes, A.: *Histoire naturelle des Poissons*, Paris, 1828–45.

[3] I may perhaps be allowed, though no linguist, to point out that the philologists have, quite naturally, misunderstood this sentence. At anyrate, all the interpreters (4–5) of Aristotle whom I know of have understood it as if the meaning were: "others (of these Fishes) produce the sound by means of their bowels in the vicinity of their stomach. Each of these bowels contains air." But ἕκαστον does not refer to τοῖς ἐντὸς περὶ τὴν κοιλίαν but to πάντα δὲ ταῦτα (o: all these Fishes).

And now, this sentence being no longer obscure, it is clear that it is to be translated thus: "But all these (Fishes) produce something like a voice, some

section of this paper, which are perhaps—and most likely too—based on real investigations, all theories touching the manner in which fishes are producing sounds, propounded before Holbrook, Dufossé, and Moreau, are mere theories, or, properly speaking, mere hypotheses.

In 1876 Moreau, in his work (VI.), which made an epoch in this point of natural history, leaning upon physiological experiments, proved that the air-bladder—to express myself in a few words—serves to equilibrate the body of the fish with the water at a certain level after the lapse of some time, the capillaries of its walls either absorbing air from the air-bladder or secerning air into it. And with regard to the only fish among those furnished with muscles in the air-bladder which he had examined, viz., Trigla, he proved, as did also Dufossé, that they served to make the air-bladder sound.

A thorough examination of the function attributed by Bridge and Haddon to the elastic spring mechanism (Ib, pp. 298–300), as to the extrinsic muscles (Ib, pp. 300–301), on which subject I have above quoted the principal remarks of these authors, will show that they are of opinion that these organs serve, in a slightly varying manner, as a " pressure adjustment," *i.e.*, they adjust the volume of the air-bladder to the pressure of water to which the fish is exposed, and perhaps also by expelling some of the air through the pneumatic duct. In other words, apart from this last phrase, their theory is the very same as that which Delaroche had put forth eighty-four years before, not, it is true, as to the Siluroidæ, but as to fishes in general. Though Bridge and Haddon have presumably[1] known the theory of Delaroche which I have referred to, they have not been aware that their theory is quite identical with that of Delaroche. But the long interval makes a great difference: when the hypothesis was put forth by Delaroche, no facts were known with which it was incompatible, and it was in fact an advance upon the then prevalent theory of Borelli. But in 1893, when it was repeated by

---

by rubbing the gill-arches [one against the other], these organs being furnished with teeth [literally thorns]; others by means of the air-bladder [literally : the bowels around the cavity]. Each of these (Fishes) contains air, by the rubbing and moving of which the sound is produced."

[1] The theory of Delaroche is amply reported by Moreau.

Bridge and Haddon, it has been crushed, as may be seen, several years ago, not only by my examinations (unknown to them) of the very Siluroidæ, but also by authors with whom they have been acquainted, viz., Moreau and Charbonnel-Salle,[1] which French authors they obviously have read. How they have been able to bring this theory of theirs to accord with the experimental investigations made about the air-bladder by Moreau and Charbonnel-Salle is a mystery to me. The latter having examined both physoclyst (Perca) and physostome fishes (Esox and several species of Cyprinoidei), says as follows :—

"Quand le Poisson nage tranquillement, sans provocation extérieure, le tracé [by means of an apparatus analogous with the sphygmoscope by Chauveau et Marey] de la vessie est identique et parallèle à celui de l'ampoule hydrostatique; aucune inflexion brusque ne signale une contraction de muscles [the air-bladder of these fishes has neither extrinsic nor intrinsic muscles] agissant sur la vessie, soit pour la comprimer (hypothèse de Borelli), soit pour la dilater (Geoffroy Saint-Hilaire), *soit enfin pour rétablir après chaque déplacement le poids spécifique modifié par la pression variable de l'eau (Delaroche).* La sensibilité de l'appareil permet d'affirmer que, dans ces conditions, des actes musculaires, même très faibles, ne passeraient pas inaperçus. Lorsque, au contraire, par des foulées énergiques de la nageoire caudale, le Poisson fuit avec vitesse, le parallélisme général des deux courbes est conservé, mais un élément se surajoute au tracé de la vessie natatoire : de véritables secousses des muscles latéraux hérissent le tracé et témoignent d'une brusque augmentation de la tension intérieure, tension qui retombe au zéro, au moment où la nageoire caudale, après s'être incurvée à droite, se recourbe à gauche en repassant par l'axe du corps. Il importe de remarquer que toute augmentation notable de tension est liée à l'incurvation du tronc; en dehors de cette condition, des secousses musculaires ont une action très faible sur la vessie. Or cette incurvation est exceptionelle dans la locomotion ordinaire du Poisson. *En outre, la brève diminution de volume ainsi produite a lieu aussi bien quand l'animal fait effort pour monter que lorsqu'il tend vers la profondeur.* Ce fait suffirait à prouver que l'augmentation de poids spécifique résultant de cette contraction ne joue aucun rôle dans la locomotion ; car en admettant qu'elle favorise la descente, il faudrait admettre qu'elle entrave l'ascension. .—. . chez des poissons de 80 gr à 100 gr, l'augmentation du poids spécifique n'atteint, dans aucun cas, 0 gr, 50 et que cette force minime est appliquée au centre de gravité de l'animal pendant 5 à 7 centièmes de seconde, durée moyenne de la période de raccourcissement de la fibre musculaire. . . . En résumé, la vessie natatoire peut être comprimée par les muscles du tronc au

---

[1] Charbonnel-Salle : "Sur les fonctions hydrostatiques de la vessie natatoire" (*Compt. rend. d. l'Acad. d. Sci.*, T. CIV., Paris, 1887, pp. 1330–33).

même titre que les autres organes contenus dans la cavité abdominale. *Les changements de volume qu'elle subit, n'ont aucune signification fonctionelle,* ils n'aident nullement le Poisson dans ses changements de niveau ou dans ses changements de direction. Les deux théories classiques résumées ci-dessus doivent être définitivement abandonnées."

It is well worth noticing what Professors Bridge and Haddon consider to be the result of the physiological investigations of the two French authors. They say (I*b*, p. 279): "From the conclusions established by Moreau and Charbonnel-Salle, it becomes obvious that the varying degrees of tension of the gaseous contents of the air-bladder due to variations in the height of the superincumbent column of water, constitute an important factor in the physiology of locomotion in Fishes."[1] Moreau, in the first page of his memoir, says: "Mais l'expérience II. . . . nous oblige de reconnaître que le rôle de la vessie natatoire admis pour la locomotion est imaginaire." And in another place[2] he says: "La vessie natatoire n'est point un organe de locomotion, mais elle est un organe d'équilibration." And (VI. p. 53) he says: "Delaroche accepte une manière de voir que les faits ne justifient pas." And Charbonnel-Salle distinctly declares that his experiments disprove the theory of Delaroche. Nay, to show whether Professors Bridge and Haddon have well understood the result to which Moreau and Charbonnel-Salle have been led by their experiments on the air-bladder as a hydrostatic apparatus, I confine myself to quote in addition the following passage of these authors, in a footnote on the page (I*b*, p. 278) where they refer to the results of the above-mentioned French physiologists: "A Fish in equipoise in the water resembles the philosophical toy known as the 'Carthusian Diver,' and the slightest exertion of its fins will readily cause motion in the vertical direction," along with some words of Professor Charbonnel-Salle in another paper,[3] where, on the first page (p. 305), he mentions how the naturalists interpreted a Fish according to the ancient theories, he uses the following terms: "Il serait un ludion[4] portant en lui-même la cause active de ses déplacements," and (*ibid.*, p. 319) he says: "Ce résultat constant des expériences paraîtra peut-être suffisant pour juger la théorie du Poisson-ludion . . ."

[1] When these authors continue, "and hence, in the absence of any other tenable hypothesis as to its function, there is a strong *a priori* probability that the object of the Weberian mechanism is to acquaint the Fish with the varying degrees of tension to which the air-bladder may be subjected," here already I must declare this conclusion to be quite invalid, starting as it does from erroneous premises.

[2] Moreau, F. A., *Mémoires de Physiologie*, Paris, 1877, p. 179.

[3] Charbonnel-Salle, L.: "Recherches expérimentales sur les fonctions hydrostatiques de la vessie natatoire" (*Annal. d. sci. nat.*, 7 sér. T. II. Paris, 1887, pp. 305-331).

[4] The great dictionary of Littré: "Ludion, Terme de physique. Petite figure qui flotte dans une bouteille de verre pleine d'eau, et qui est construite de manière qu'on peut, à volonté, sans y toucher, la faire monter ou descendre par l'effet de la pression de l'air."

Though Moreau and Charbonnel-Salle are of opinion that they have crushed the Carthusian-Diver Fish, according to Bridge and Haddon, however, he is still diving.

I confess that I can only agree with three of the physiological views suggested by Professors Bridge and Haddon: firstly, that the experiments of Professor Jobert tending to bear out that Erythrinus respires atmospheric air in the air-bladder are not decisive [1]; secondly, that the theory of Sagemehl on the function of the Weberian ossicles is untenable; thirdly, that the sound produced by Clarias does not arise from the expulsion of air from the air-bladder. On this fish I have written eleven years ago [2] (p. 406):

"It emits two kinds of sounds, of which one, produced simultaneously with a movement of the pectoral fins, was discordant ... while the other sound was dull but distinct, and not unlike an eructation. The latter of these sounds has been observed several times as well in as out of water, the Fish at the same time opening its mouth without moving the pectoral fins." And I can add that the sounds which this Fish emits *under water when disturbed* are caused by the movements of the pectoral fins.

I said above that the curious air-bladders represented in the "Introduction" of Dr. Günther (with the exception of one) are those of different genera belonging to the *Sciænoidei*, in which family the air-bladder is well known to be a sound-producing organ. This, however, is not the impression derived from the perusal of Dr. Günther's book. The statement made about the manner in which fishes of this family produce sounds runs as follows (XIII. pp. 427-428):—" To this fish, (*P.*[*ogonias*] *chromis*) more especially is given the name of 'Drum,' from the extraordinary sounds which are produced by it and other allied Sciænoids. These sounds are better expressed by the word drumming than by any other, and are frequently noticed by persons in vessels lying at anchor on the coasts of the United States, where those fishes abound. It is still a matter of uncertainty by what means the 'Drum' produces the sounds. Some naturalists believe that it

---

[1] The reasons why I could not adopt the suggestion of this author, *i.e.* that the air-bladder is an organ of respiration, I have stated in my book, (II*b*, pp. 187-190).

[2] *Naturhistorisk Tidsskrift*, 3 R. Bd. xiii., Kjöbenhavn, 1883.

is caused by the clapping together of the pharyngeal teeth,[1] which are very large molar teeth. However, if it be true that the sounds are accompanied by a tremulous motion of the vessel, it seems more probable that they are produced by the fishes beating their tails against the bottom of the vessel in order to get rid of the parasites with which that part of their body is infested."

I take the liberty here to quote, in a somewhat different order, what I wrote ten years ago, where I had gathered all the elucidations published until 1884.[2] In doing so, I am able to show the reader, I suppose, that it is not a rarity whatever that Fishes produce sounds by means of the air-bladder.

With the exception of the Siluroidæ, the Triglidæ, and the Pristipomatidæ, none of the great families of fishes count so many species known as sound-producing. And the air-bladder in these fishes appears in the most varied and at the same time in the most complicated forms, now furnished with, now devoid of muscles, its cœcal diverticula extending into the musculature of the body and the tail, conditions which, by way of induction, led Cuvier and Valenciennes—at a time when there did not exist one investigation as to the mode in which sounds were produced in fishes—to the conclusion quoted above, that the air-bladder—and more particularly in this family—had something to do with the production of sounds. The following species are known to be sound-producing:—*Umbrina cirrhosa*, L.; *Sciæna aquila*, Lac.; *Sc. adusta*, Ag. (according to Ihering [3]); *Otolithus regalis*, Bl. (*Labrus squeteague*, Mitchill,[4] p. 398); and *Pogonias chromis*, L. (*Sciæna fusca*, Mitch., *l.c.*, p. 411). The fact is less certain with regard to *Pogonias fasciatus*, Lac. (*Labrus grunniens*, Mitch., *l.c.*, p. 405); *Corvina ronchus*, Cuv. et Val. (T. V., p. 107);

---

[1] That several Fishes of this family are able to produce sounds by gnashing their teeth is quite probable: nearly all Fishes may do the like. I have seen a *Gadus morrhua* in its agony gnashing its pharyngeal teeth, though they do not appear to be adapted for this purpose. But at anyrate this is a matter quite apart from the drumming sounds for which the Sciænoidei are so renowned.

[2] Since 1884 I have not followed the literature on this subject. The elucidations of Ihering, however, are of more recent date.

[3] Ihering, H. v., "Die Lagoa Dos Patos" (*Deutsche Geographische Blätter*, T. VIII. fasc. 2, p. 185).

[4] Mitchill, S. L., "On the Fishes of New York" (*Trans. o. t. lit. a. phil. soc. New York*, T. I., 1815, pp. 355-492).

Umbrina ronchus, Val.,[1] and *Umbrina Canariensis*, Val. (*ibid.*, p. 24); *Sciæna ocellata*, L. (*Sc. imberbis*, Mitch., *l.c.*, p. 411); *Larimus dentex*, Cuv. et Val. (T. V., p. 139); and *Micropogon undulatus*, L. (Cuv. et Val., T. V., p. 217 and p. 221; *Bodianus costatus*, Mitch., *l.c.*, p. 417).

Of all these Fishes I have only succeeded in examining (a dead specimen of)—

Micropogon undulatus.—The form of the air-bladder needs no other description than that given in the figure 9. On the middle[2] a pair of slender, round, hollow horns project, which

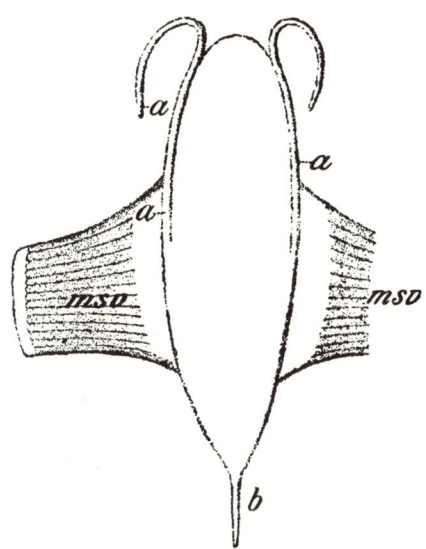

Fig. 9.—*Micropogon undulatus*, Linn. The air-bladder seen from below; natural size. (To the left the muscles are only partially shown.) *a,a,a,* the lateral horns of the air-bladder; *b*, the hinder point of the air-bladder; *msv*, the muscles, which must make the air-bladder act as a sound-producing organ.

extend forwards along the lateral wall of the air-bladder, at whose fore-end their final parts again turn backwards. The outer membrane is fibrous, stiff, of a silvery lustre; the inner one is pellucid, the "red corpus," formed like a horse-shoe, placed at the ventral side. On the dorsal side, the air-bladder is intimately attached to the vertebral column.

Musculature.—Below, on the ventral-lateral wall of the body,

[1] Valenciennes, in Webb et Berthelot, *Histoire naturelle des Iles Canariens*, T. II., pt. 2 (p. 24).

[2] Much farther behind in the specimen examined by Cuvier et Valenciennes (T. V., pl. 138).

near the lower end of this wall and that of the ribs, issue a pair of broad aponeuroses which ascend obliquely, inclosing the air-bladder, with which it is interwoven on the dorsal side a little before the base of the horns. The hindmost part of these aponeuroses is provided with a layer of muscles,[1] whose fibres turn obliquely upwards and forwards. The muscles, as well as the aponeuroses, have their greatest length (vertically) near their posterior end, and gradually grow shorter in front.

According to its structure, the air-bladder of this fish must be a sound-producing organ. Most probably the contraction of the muscles will, for a moment, compress the air-bladder and strain its dorsal wall, each of which operations must separately be able to bring the air-bladder to produce sound.

In *Otolithus regalis*, Bl. (Cuv. et Val., T. V., Pl. 138), *O. Cayennensis*, Lac. (*O. toe-roe*, Cuv. et Val., *ibid.*, Pl. 138), *O. guatucupa*, Cuv. et Val., *Ancylodon jaculidens*, Cuv. et Val., and *A. parvipinnis*, Cuv. et Val., the air-bladder has a similar form, according to Cuvier et Valenciennes, but the horns are considerably wider. These authors do not, however, mention whether it is provided with muscles or no.[2]

According to the description of the same authors (T. V., p. 200, Pl.[3] 139) *Pogonias chromis*, L., has in its air-bladder a sound-producing organ, no doubt the most powerful to be found in any fish (perhaps with the exceptions of Ganoids and "Dipnoi"). It is oval, broad behind, narrower in front, where laterally it expands into great wide cavities, which are in their turn lobed and incised. Its outer membrane is "gélatineuse et fibreuse," and is nearly half an inch thick[4] (in a specimen of

---

[1] When I had only been able to examine dead specimens of a species, I have always made sure, by means of the microscope, that what I called muscles were really so.

[2] *L.c.*, p. 200, they say: "J'ai déjà fait remarquer que la plupart des Sciénoïdes les plus remarquables par cette faculté [to produce sounds] ont de grandes vessies natatoires, très-épaisses, munies de muscles très-forts." The presence of muscles, however, is only mentioned in Pogonias and Micropogon. Generally it is not noted whether the ends of the diverticula of the air-bladder extend into the lateral musculature of the body. Here is a wide field open to anatomical investigations.

[3] The figure in question is copied in the "Introduction" of Dr. Günther.

[4] Bridge and Haddon (I*b*, pp. 273-274): ". . . it may be objected that in many Siluridæ the walls of that organ are too thick to admit of their vibrating synchronously with rapidly recurring sound waves."

three feet and a half in length). The hinder half, which is not furnished with appendages, is on either side covered with a very thick layer of muscles with transverse fibres; besides, the extremities of the ramifications of the lateral cavities partly penetrate between the ribs into the lateral muscles of the body. The "drum" is said to attain to a weight of a hundred pounds, and the tones which it produces seem to be of an enormous strength.[1] On this fish, Holbrook[2] has made the following observations:—" Frequent examinations of the structure and arrangement of the air-bladder, as well as observations on the living animal just taken from the water, when the sound is at intervals still continued, have satisfied me that it is made in the air-bladder itself; that the vibrations are produced by the air being forced by strong muscular contractions through a large opening, from one large cavity, that of the air-bladder, to another, that of the cavity of the lateral horn; and if the hands are placed on the sides of the animal, vibrations will be felt in the lateral horn, corresponding with each sound." This passage, then, is, as far as I know, the first statement in modern times of observations of the fact that the air-bladder is capable of acting as a sound-producing organ. Though the investigation itself has evidently been rather superficial, the correctness of the main fact is open to no doubt, when we bear in mind that this statement, the first in modern times, no longer stands alone But I very much doubt the correctness of the opinion of Holbrook as to the *manner* in which the sounds are produced. With a view to what I am going to state further on as to the importance of the production of sounds in the Siluroidæ, I take the liberty to quote the following passage from the same author: —" At this time [April] . . . begins its drumming noise; this season [the spawning-season] passed, the sound is no longer heard, and the fish is then rarely taken." The elucidations given by

---

[1] *Vide* Cuvier and Valenciennes, T. V., p. 198.

[2] Holbrook, J. E., *Ichthyology of South Carolina*, T. I., 1860, p. 118. This work is faunistic; of anatomical details it does not contain many, and the statement quoted above is the only piece of physiological information, as far as I know. It therefore appears to me that it would have been more natural if it were I who was unacquainted with this work of Holbrook, as I am by no means a systematical ichthyologist, and do not read English with any greater facility than I do the other languages of Gothic and Latin root.

Ihering (*loc. cit.*) on the habits of this fish also afford some interest: they "sollen nie vormittags, sondern von Mittag bis in die Nacht herein 'trommeln,' weshalb denn auch nur nachmittags und abends ihnen nachgestellt wird."

In *Pogonias fasciatus*, Lac., the air-bladder appears to be of the same structure. It would be most desirable if the air-bladder in one of these species were submitted to a renewed examination, which, even if it were merely anatomical, would undoubtedly bring to light interesting facts.

Sciæna aquila, Lac. In this fish whose musical performances probably have given rise to the Homeric fable of the song of the Sirens, the structure of the air-bladder is, according to Dufossé (V*b*, p. 5–22), briefly this: It is large (one-third of the total length of the fish), shaped like a long bag, and pointed behind; on either side it is furnished with 34–42 short ramified diverticula, among which the 6th–10th are the biggest. A great number of them, especially the larger ones, are intimately connected with the "aponeuroses" of the nearest among the lateral muscles of the body,—nay, the very largest penetrate through these muscles, even to an extent of 4 cm. Besides, the air-bladder is fused ("se soude") with the centra of the foremost vertebræ, and intimately attached to some of the processus transversi of the succeeding vertebræ and to the foremost ribs by means of very strong "aponeuroses"; and on its dorsal side it is intimately connected with the aponeuroses in the roof of the abdominal cavity. The wall is very thick;[1] the outer membrane fibrous and very compact; the inner membrane, which is thicker than is generally the case, forms a horizontal septum with a large oval aperture towards the fore-end, and in large specimens it extends above the orifice of the diverticula, so that the compartments of these diverticula are separated from the main compartments of the air-bladder. When solitary, the fishes of this species, according to Dufossé, but rarely emit sounds; and if so, these sounds are faint, dull, and of short duration; but when in flocks, and, above all, in shoals during the spawning season, they incessantly emit tones of such a power and duration, that all

---

[1] 7 mm.—1 cm. There is no indication as to the length of the specimens examined. The species attains to a length of 2 metres, and to a weight of 25–30 kilos.

their strength would seem to be quite exhausted by these exertions. The sound is so intense that specimens—large ones, I suppose—keeping at a distance of 18 metres under the surface of the water, are distinctly to be heard by a listener whose ear is 2 metres above the water. Previous authors—Bonaparte, Duhamel—state them to be still audible at a distance of 20 fathoms. By means of vivisections Dufossé has proved that tones can be produced by the activity of most of the muscles,[1] which, coated with "aponeuroses," are in immediate contact with the diverticula of the air-bladder, but that the most frequent and most intense tones are produced by the activity of those muscles, which, "completely naked," are placed around the long branches of the largest diverticula. The tones may be of different pitch, in perfect accordance with their being formed in different places (and under the influence of different muscles).

To judge from the structure of the air-bladder as represented by Cuvier and Valenciennes it must also be considered a sound-producing organ in the following Sciænoidei: *Sciæna ocellata*, L., where the diverticula penetrate between the ribs; *Sc. diacanthus*, Lac. (*Corvina catalea*, Cuv. et Val.), with 20 pair of diverticula, of which 18 are ramified in a high degree (T. V., Pl. 139); *Corvina coitor*, Buch., with 10 pair of diverticula, of which 9 are ramified; *C. Belengerii*, Cuv. et Val., where the air-bladder is, moreover, constricted near the fore-end; *C. lobata*, Cuv. et Val. (*ibid.*, Pl. 139), where it does not differ from that of the preceding species, but is very wide before the constricted place. It is less obvious in *Sciæna obliqua*, Mitch. (*Leiostomus humeralis*, Cuv. et Val.), where, as in *Sc. xanthurus*, Lac., it is very much like that of *Otolithus regalis*; *Corvina nigrita*, Cuv. et Val. (*ibid.*, Pl. 138), where from the fore-end issue two long horns, divided into several branches; *C. acoupa*, Cuv. et Val., whose horns are wider, but only bi-branchiate; *C. furcræa*, Lac., where the structure is the same; and *C. biloba*, Cuv. et Val., *C. axillaris*, Cuv. et Val., and *Nebris microps*, Cuv., with a pair of long unramified horns.—(In *Corvina*

---

[1] Dufossé points out that these muscles are of a more intensely red colour than the other parts of the lateral musculature of the body. In the Siluroidæ, Characini, and Trigla, which I have examined when alive, the muscles which make the air-bladder able to act as a sound-producing organ are also more intensely red than the rest of the muscles.

ronchus, Cuv. et Val., and *C. Senegalla*, Cuv. et Val., the air-bladder is like that of *C. nigra*, L., without any outer division or appendage whatever.)—In *Collichthys pama* (*ibid.*, Pl. 138) the form of the air-bladder is essentially as in Micropogon, but the horns are divided in front into several rather short branches. In *C. lucida*, Rich., it appears in one of the most remarkable forms to be met with; but, apart from this feature and from its stiff thick wall, nothing in the detailed description of it given by Günther[1] (T. II., p. 312) goes to prove its being an organ of sounds. The same must be said about the description by the same author (*ibid.*, p. 317) of the air-bladder of *Lonchurus depressus*, Bl.; "a process arising from the third vertebra," which process is connected with the air-bladder, is, however, mentioned in such a manner that we are led to suppose the existence of an ossification of the wall of the air-bladder. A renewed examination of this organ in these two species would therefore be most desirable.

Umbrina cirrhosa, L.—As to the relative size of the fusiform air-bladder, the manner in which it is attached to the skeleton and to the aponeuroses of the abdominal cavity, and as to the thickness of its wall, this species is, according to Dufossé (V*b*, p. 22–28), quite analogous to *Sciæna aquila*; but instead of the ramified diverticula, it presents only three pair of lateral bulgings, bipartite by a transverse fold, and connected with (". . . rapport de contiguité . . .") the parts of the deepest (innermost) layers of the lateral muscles of the body, which fill up the intervals between the 3rd–5th ribs. Inside it is found, as in *Sciæna aquila*, a thin horizontal septum with a large, oblong hole on the middle. When the fish is out of the water, the sound can only be heard, at most, at a distance of 2 metres and a half, and is produced (at the greater part at least) by the activity of the said muscle portions, which are in contact with the air-bladder. Dufossé also relates the interesting circumstance that in young not yet pubescent specimens of a length of 2–3 dcm. (the fish attains a length of more than 2 feet), he observed the same vibratory movements ("frémissements") of the muscles, which in grown-up specimens produce sounds, but which had now no such effect. The vibratory movement was

[1] XVI: Günther, A., *Catalogue of the Fishes in the British Museum*, London, 1859–70.

easily felt at the surface of the body, but nearly imperceptible on the inside of the abdominal wall. On this occasion Dufossé makes the following remarks:—

"La répartition de l'intensité de ces mouvements explique pourquoi ils ne seraient pas bruyants, lors même que la vessie aurait acquis un degré de développement plus avancé que celui auquel elle est parvenue à l'âge des individus dont il est ici question. Toujours est = il que la multiplicité de ces mouvements de frémissement démontre l'aptitude que possèdent les fibres musculaires de ces animaux à engendrer des mouvements de cette espèce, aptitude qui s'étend alors à un grand nombre de muscles, et qui, par cela même, est un peu confuse chez les jeunes sujets, mais qui se concentrera plus tard dans les muscles de la couche profonde des grands latéraux, à l'âge où l'appareil de renforcement [the air-bladder], étant suffisament développé, sera convenablement disposé pour recevoir ces petits mouvements et pour leur donner la force nécessaire à ébranler le milieu ambiant."

But this suggestion, based as it is on the supposition that the sound arises from the vibration of the muscles themselves, is untenable. The difference must be owing to another cause, whether it be that the air-bladder is not connected with the abdominal wall in the same manner as at a later stage, or of some other reason. To a zoologist, residing on the shores of the Mediterranean, it would be an interesting task to study this problem.

It has been established already, in 1864, by investigations of Moreau,[1] and afterwards by those of Dufossé (published in 1874), who has examined no less than eight species of the genus Trigla, that the sounds emitted by these animals are produced when the walls of the air-bladder and its gaseous contents are thrown into vibrations. But though these examinations have been carried on with living animals, and though the two French authors agree on the main points,[2] yet the learned ichthyologist, Dr. Günther, tells us, in 1880 (XIII., p. 479):—"The grunting noise made by Gurnards when taken out of water [as well as in the water] is caused by the escape of gas from the air-bladder through the open pneumatic duct." Unfortunately there exists no pneumatic duct at all[3] in these fishes. From whom, then,

---

[1] Moreau, A.: "Sur le voix des Poissons" (*Compt. rend. d. l'Acad. d. sci.*, T. LIX., Paris, 1864, pp. 436–37). Also in VI., p. 67.

[2] As, by the way, they do with my examinations of living specimens of *Tr. gurnardus*.

[3] *Trigla hirundo* certainly is, from a morphological point of view, furnished with a pneumatic duct, but this "duct" is a perfectly solid string.

does this error arise? From Sir Richard Owen, I think, who says[1] as follows:—"In a few genera (Trigla, Pogonias) the air-bladder and its duct are subservient to the production of sounds." But though it may be understood how this eminent anatomist could forget that this genus belongs to the physoclyst fishes, it is indeed a little strange that the great systematic Ichthyologist, Dr. Günther, could commit the same error. And whence, then, arises the error of Sir Richard Owen? I cannot with certainty answer this question, and I do not think it sufficiently important to trace it in literature. But as far as I have been able to judge, this error arises from misinterpretations, at second hand, of the remarks of Cuvier and Valenciennes, quoted above.

I have quoted the statements of Dr. Günther, in order to show justice to Professors Bridge and Haddon. I cannot fail to see that, so generally consulted as is the "Introduction" of Dr. Günther in England, in the Scandinavian countries and Germany, this book is not without fault in leading Professors Bridge and Haddon *a priori* to judge the investigations recorded in my *Lydorganer hos Fiske* to be a mere "theory."

To conclude, I feel bound to point out a deficiency in my observations of the air-bladder as a sound-producing organ. According to the structure of the air-bladder in *Gadus morrhua* and *G. æglefinus*, I had to consider this organ as a sound-producing organ, but a very weak one, because of the slight size of the extrinsic muscles. In both species I irritated, by means of an electric stream, the muscles, as well as the nerves leading to these muscles. But the muscular contractions produced by these operations had only this effect, that the wall of the air-bladder bulged a little inwards and outwards on the place where it was covered by the muscles. But no sound whatever was emitted. The negative result at which I arrived may perhaps be due to the circumstance that I am not accustomed to that kind of experiments. It would, therefore, be most desirable if such an examination were undertaken by an experimental physiologist *ex professo*.

[1] Lectures on the Comp. Anat. and Physiol. of the Vertebrate Animals, Pt. I., p. 278.—On the Anatomy of Vertebrates, Vol. I., p. 497. In the latter book the word "Pogonias" is omitted.

Reprinted from *Proc. Roy. Soc. London*, **55**, 439–441 (1894)

VIII. "Note on the Production of Sounds by the Air-bladder of certain Siluroid Fishes." By Professors T. W. BRIDGE and A. C. HADDON. Communicated by Professor A. NEWTON, F.R.S. Received April 17, 1894.

Dr. William Sörensen, of Copenhagen, has drawn our attention to the fact that, in our memoir on "The Air-bladder and Weberian Ossicles in the Siluroid Fishes," published in the 'Philosophical Transactions' last year (vol. 184, pp. 65—333), we failed to do justice to the results of certain investigations which are embodied in his paper, entitled "Om Lydorganer hos Fiske: en physiologisk og comparativ-anatomisk Undersögelse," and published at Copenhagen in 1884. In this paper Dr. Sörensen treats of the various methods of sound production in Fishes in general, and in the case of the Siluroid Fishes, describes the production of sounds by means of certain stridulating mechanisms (friction of the dorsal and pectoral spines), the "elastic spring" apparatus, and the paired extrinsic muscles of the air-bladder in the *Pimelodinæ*. We do not here wish to criticise his morphological conclusions, but to point out that, contrary to the assumption on pp. 270 and 301 of our paper, Dr. Sörensen did make some experiments on living Fish. After describing the nature of the "elastic spring" and the disposition of its muscles in the South American Siluroid, *Doras maculatus*, on p. 88 of his paper, he says, concerning this Fish:—

"*Observations on the Production of Sounds.*—When one opens the abdomen of a recently caught fish and quickly extracts the intestines with everything that is attached to them so that the swim-bladder is exposed, one can very easily perceive that the swim-bladder is in a convulsive vibratory motion at the same time that the sound is produced. It is a very deep murmuring note, which is so strong that it can be distinctly heard at a distance of 100 ft. when the animal is out of the water. Unlike the sounds produced by the movements of the pectoral fin, the tones produced by the swim-bladder are not grating, and therefore not disagreeable to the ear. As far as I am able to judge, the swim-bladder commands only one note, but this can be stronger or weaker according to the will of the fish. If one moves the fingers backwards and forwards over the swim-bladder, one will soon perceive that the vibrating motion, beginning at the same time as the sound, is strongest in the front, especially at the "muscle-springs," and also that the muscles passing to these contract at the same time that the sound is produced. If the muscles are cut through, the sound is no longer produced. If one makes a little hole in the swim-bladder, the sound will not be very much weaker; but if a larger opening is made in it, the sound will considerably diminish in strength. If one takes out the swim-bladder, the note will become very weak, but may still be heard; it is then produced only by the vibrations of the springs. By ordinary observations I have not been able to prove that those bars or cross-walls (transverse septa) which project into the lumen of the chief compartment of the swim-bladder, or its external diverticula, assist in the production of sound; but if one compares this with what I state later on in Pseudaroides, I believe it would prove very doubtful, on account of their incomplete partition walls, that the diverticula of the swim-bladder even to a great degree serve to strengthen the sound by the air passing to and fro over them. By looking more particularly one will observe that the anterior cutaneous plate at the side of the body also vibrates when the sound is produced. I suppose that the action of the ligament, which connects it with the circular plate of the muscular spring, besides transferring the sound vibrations of the swim-bladder to the water, consists in preventing a too violent recoil of the spring when the muscle is relaxed."

After describing the air bladder, and the arrangement of its paired extrinsic muscles in *Platystoma orbignyanum*, and *Pseudaroides clarias*, Dr. Sörensen continues (p. 93):—" When the swim-bladder is laid open in the living animal, it is very easy to perceive that the contractions of the previously mentioned muscles [extrinsic muscles] occur at the same time as the production of a strong, deep, murmuring sound, whilst the wall of the swim-bladder is put into strong vibratory motion. The majority of the specimens I have examined of Pseudaroides had at the most a total length of 25—35 cm. The walls of the swim-bladder were, therefore, not so thick, but I was able to distinguish the internal transverse septa as darker transverse lines; I could therefore see very distinctly that when the sounds were produced, the septa were in a state of rapid vibration forwards and backwards. This is sufficient to prove that they play a very important part in tending to increase the sounds by the fact that the air vibrates over their free edges, from one chamber to the other. If one makes a small hole in the swim-bladder of Platystoma, the strength of the sounds will not be very much diminished. If an even smaller incision is made the sound becomes fainter and fainter, and at length dies away, even though the muscles are functional."

"So far as I have been able to see, only one muscular contraction takes place, as in Doras, for every time sound is produced. This always lasts a certain period, is fainter at the end, but ceases suddenly. About the nature of the sound, the same can be said as I have stated about Doras. The sound a Platystoma produces can be heard at a distance of more than 20 feet, when the animal is on the land."

From these observations it is also clear that the sounds produced by these fishes are not caused by the expulsion of air through the ductus pneumaticus, as erroneously assumed by us on pp. 298 and p. 301 of our memoir, but are caused by the vibration of the air within the air-bladder, which is set in motion either by the "elastic spring" apparatus, or by the extrinsic muscles.

The investigations of Dr. Sörensen seem to show that, under certain conditions, the "elastic spring" mechanism, and the paired extrinsic muscles of the *Pimelodinæ*, are structures subordinate to sound production, and are not, as we suggested, related to any method of adjustment to varying hydrostatic pressures. Had we appreciated this fact earlier, we should have modified certain of the tentative conclusions suggested on pp. 298—301 of our memoir. On the present occasion we wish to draw the attention of those interested in the subject to Dr. Sörensen's researches, and at the same time to express our regret at the injustice we have unintentionally done him.

The Society adjourned over Ascension Day to Thursday, May 10.

THE PRODUCTION OF SOUND IN THE DRUMFISHES, THE SEA-ROBIN AND THE TOADFISH.[1]

By R. W. Tower.

CONTENTS.

Introduction.
Anatomy of the swim-bladder.
    The drumfishes:
        Bearded drum (*Pogonias cromis*),
        Squeteague (*Cynoscion regalis*),
        Croaker (*Micropogon undulatus*),
        Other drumfishes examined.
    The sea-robin and the toadfish:
        Sea-robin (*Prionotus carolinus*),
        Toadfish (*Opsanus tau*).
Sound production in the drumfishes:
    Recorded observations and theories,
    Experiments to determine cause of sound,
    Experiments to determine character of muscular contraction,
    Experiments to determine pressure of gas in swim-bladder.
Sound production in the sea-robin and the toadfish:
    Experiments to determine cause of sound and character of mechanism.
Conclusions.
Literature cited.

Introduction.

The production of sounds by certain fishes has long been an interesting subject of investigation. Some species, as *Scomber brachyurus*, by rubbing together the pharyngeal teeth make a noise resembling a harsh grunt; some, as the puffer, or swellfish, make a similar sound by rubbing together the incisor teeth of the upper and lower jaw. In other cases stridulation has been recorded, and sounds are also said to be produced by the forcing of air through the pneumatic duct in those fishes in which the air-bladder

[1] Read by title at the meeting of the Academy on 13 April, 1908.

communicates with the exterior. Besides these kinds of sound production, which are of no special interest in this discussion, there are two others. One is the drumming of the squeteague, croaker and other drumfishes (Sciænidæ); the other is the so-called grunt of the sea-robin (*Prionotus*) and the common toadfish (*Opsanus*). With the difference in the kind of sound made by the drumming and the grunting fishes there will be found to be a distinct difference in the structure of the swim-bladder, which is the organ chiefly involved in the production of sound by these species.

Anatomy of the Swim-Bladder.

THE DRUMFISHES.

Bearded drum (*Pogonias cromis*).— The swim-bladder of the drum is characterized by its large size and the enormous number of its diverticula. The bladder occupies, as is the case in nearly all of the sciænoid fishes, the entire length of the abdominal cavity. The diverticula are finger-like processes which arise laterally from the bladder and open into its large cavity. These tube-like appendages in the adult ramify through the connective tissue, and in many cases adhere firmly to the aponeuroses of the neighboring muscles. The air-bladder itself lies free in the abdominal cavity, attached on the dorsal side to the body of the fourth vertebra and covered on the ventral side by the peritoneum, which is continued from the parietal walls. When examined carefully, the air-bladder is seen to be made up of three layers: the outside is of a hyaline character and is composed of extremely tough fibrous tissue; the middle layer, which is separated from the outer layer only with great difficulty, is connective tissue containing elastic fibres; the inner layer is a very delicate connecting tissue, lined with pavement epithelium. Jäger (1903) has recently discovered that this inner layer does not cover the entire bladder-lumen, but on the dorsal surface there is an oval space in which the inner layer disappears, with the exception of the pavement epithelium. This space he calls the "oval," and maintains that it can be increased or diminished by the action of small muscles. In the middle layer ramify all the blood-vessels, which break into small branches and then enter the inner layer, where, in the region of the "oval," they form an anastomosing capillary net-work almost as complete as is found in the "red-body." This net-work is thus separated from the lumen of the air-bladder only by the single layer of pavement epithelium. The function of the "oval," according to Jäger, is the absorption of oxygen and the diminution of the amount of gas in the bladders of fishes having no pneumatic duct. Thus the "oval" and the pneumatic ducts serve the same physiological function. Adhering to a portion of the dorsal surface of the air-bladder, just posterior to the point of attachment to the vertebræ in the male, is the central tendon of the two red drumming muscles.[1] Upon opening the bladder of the drum, there is found on the inside, running almost the entire length, the red vascular body which has been described as the blood gland, or "red body."

Squeteague (Cynoscion regalis).— In the squeteague the swim bladder (fig. 1) is a long carrot-shaped organ, tapering to a point at the posterior end, and sending out from the broad anterior end three diverticula— two lateral horns and a central rounded "head." The dorsal surface of the "head" is attached by its outer or fibrous tunic to the sides of the body of the fourth vertebra, which broadens out to receive it. The lateral appendages of the swim-bladder of the drum are wanting in the air-bladder of the squeteague, which has nothing to mar its smooth even contour except the two lateral horns already described, which arise from the most anterior part of the organ. On the inside of the air-bladder is found the characteristic "red-body," or "blood-gland," which is present in the drum. The drumming muscles are present in the male squeteague only. Their insertion is lateral in the common fascia of the *rectus abdominis* muscle, about a half-inch from the mid-ventral line. The muscles, one on either side, are bilaterally symmetrical and originate from a central tendon, which lies free in the mid-dorsal line just above the swim-bladder and between it and the kidney. The anterior extremity of this central tendon is inserted by its middle third into the dorsal surface of the neck of the swim-bladder, while the right and left thirds merge into the fascia that support the peritoneum. Posteriorly, in the region of the anus, the tendon narrows down to a cup-shaped extremity that receives the tip of the swim-bladder, and then gradually tapers to a point, which is inserted into the base of the first anal fin-ray. A closed cavity is thus formed, bounded laterally by the two drumming muscles, ventrally by the confluent abdominal muscles, and dorsally by the central tendon. This closed bag or cavity contains the viscera

---

[1] Dufossé (Annales des Sciences Naturelles, ser. V, vol. XIX. 1874, p. 39) has described in *Trigla lyra* two red muscles which he called intra-costal muscles. From his description I am unable to identify them with the "drumming muscle" just mentioned. In no case has it been possible to find these "drumming muscles" in any of the Triglidæ.

Dufossé attributed to these intra-costal muscles the function of motor agents of the skeleton. "Considérés uniquement comme agents moteurs du squelette, ces muscles intra-costaux ont evidemment pour functions: d'une part, de fléchir latéralement ou de maintenir l'épine dorsale dans sa rectitude ordinaire, suivant qu'un seul muscle se contracte, ou bien que la contraction de ces deux muscles est simultanée, quand les os scapulaires leur servent de point fixe; d'autre part, d'attirer en dedans ces derniers os, et par suite les scapulaires et les humeraux, (Cuvier), lorsque la colonne vertébrale est préalablement fixée." It is evidently from this supposed function that Dufossé gave the name of intra-costals to these muscles.

and swim-bladder, the latter having the "central tendon" directly applied to its dorsal surface.

The blood-vessels and nerves supplying the drumming muscles are only

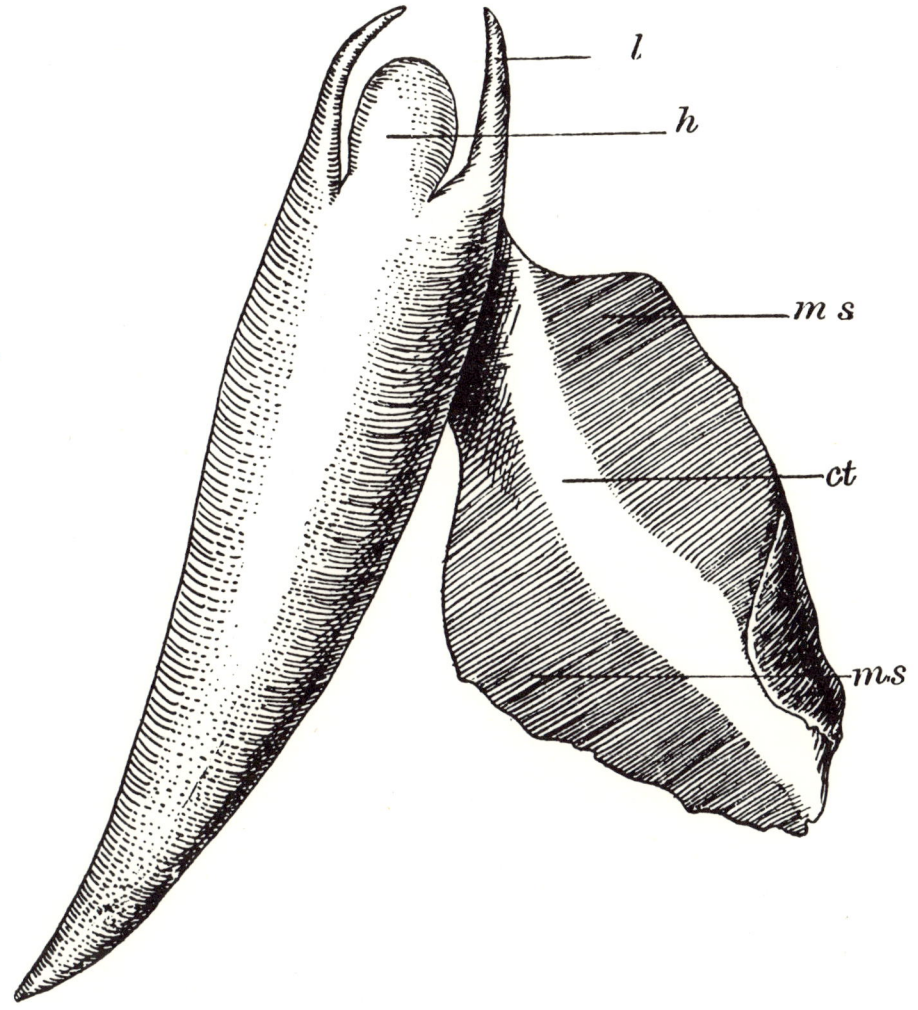

Fig. 1. Swim-bladder of Cynoscion regalis.

The two m. sonifici (m s) are shown laterally displaced. h, head; l, lateral horn; c t, central tendon of m. sonifici.

accessory branches from the arteries and nerves of the abdominal muscles. An embryological study of the origin of these accessory blood-vessels and nerves and their relation to the muscle at the time when it is first laid down would be instructive. For the drumming muscles the name *musculus sonificus* has been suggested and is used in the following discussion.[1]

In young squeteague two inches long, it is impossible to distinguish macroscopically a differentiation of this muscle. But if a piece of the peritoneum with the underlying fascia is removed and examined under the microscope there are seen striations typical of voluntary muscles. The muscle fibres run in the direction of the short diameter of the fish, *i. e.*, circularly around the air-bladder. These young squeteague have been heard to "drum," and the contractions of the *m. sonificus* can be easily felt when the fish is held firmly in the hand. In the young this muscle has not acquired the deep red color that so characterizes it in the adult.

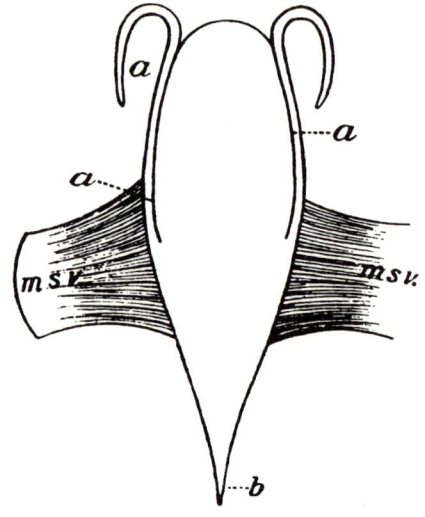

a. a. a. Lateral horns of air-bladder.
b. Hinder point of air-bladder.
msv. The muscles which must make the air-bladder act as a sound producing organ.

Fig. 2. Swim-bladder of Micropogon undulatus after Sörensen.

Croaker (Micropogon undulatus).— Another sciænoid, known in American waters as the croaker, is of interest from an anatomical standpoint. The difference between the bladder of this fish and that of the squeteague, except for its being considerably smaller, is that the central head is not present, and the two lateral horns are reduced to two very small tubes. It is therefore an even more simple organ than that of the squeteague. The two bilateral *sonifici* have the same arrangement in both animals, and the description of the muscles of one applies equally well to those of the other. Sörensen states that "the form of the air-bladder needs no other description than that given in figure 9" (a copy of which is here appended as fig. 2).

---

[1] Dr. Hugh M. Smith and Dr. Theodore Gill suggested several anatomical names, from which *musculus sonificus* was selected as being the most appropriate. The author is greatly indebted to these two well-known ichthyologists for their assistance.

Sörensen made his dissection on one specimen preserved in alcohol, consequently the diagram is somewhat misleading, as can be seen by comparing it with the bladder and muscle taken from a fresh specimen (Pl. VI. fig. 1). Bridge uses this figure from Sörensen to verify the following statement: "In other fishes, the air-bladder, without possessing special muscles of its own, may, nevertheless, be partially invested by tendinous or partly muscular and partly tendinous, extensions from the muscles of the body wall." This muscle (*m. sonificus*) cannot be considered an extension of the muscles of the body wall but a unique, specific muscle which has been developed for the purpose of sound production. The muscles with the aponeuroses are united with the swim-bladder by means only of a tendon on the dorsal side immediately anterior to the base of the horns, and in no way attach themselves directly to the bladder, which is completely surrounded by the muscles and tendons. Sörensen states: "According to its structure, the air-bladder of this fish must be a sound-producing organ. Most probably the contractions of the muscles will, for a moment, compress the air-bladder and strain its dorsal wall, each of which operations must separately be able to bring the air-bladder to produce sound." Sörensen did not make any physiological experiments and based his conclusions entirely upon anatomical data. In the light of experiments soon to be described it is evident that he did not understand the "drumming" mechanism.

Other drumfishes examined.— Through the courtesy of Dr. Hugh M. Smith (1905) of the Bureau of Fisheries, it has been possible for me to examine specimens of the southern squeteague (*Cynoscion nebulosum*, Pl. VI, fig. 2), the yellow-tail (*Bairdiella chrysura*, Pl. VII, fig. 1), and the spot (*Leiostomus xanthurus*, Pl. VII, fig. 2).[1] The anatomical relations of the air-bladder and the *m. sonificus* are so similar to those noted above that no further description is necessary. In the spot the peritoneum is so pigmented with black that the *m. sonificus* is somewhat hidden.

The Sea-Robin and the Toadfish.

In these fishes there is found a swim-bladder which is so radically different in its outward appearances from that of the sciænoid fishes, and at the same time is so characteristic, that attention is immediately attracted to this organ. The sound produced, described as a grunt, differs markedly in character from the drumming of the Sciænidæ.

---

[1] These drawings were made from dissections completed by T. E. B. Pope of the Bureau of Fisheries.

Dufossé (1874) in his memoir on "Sons Expressifs Produits par les Poissons d'Europe" has given an accurate and complete anatomical description of the air-bladders of the European *Zeus faber*, *Dactylopterus volitans* and various Triglidæ. Inasmuch as the air-bladders of the Triglidæ of the North American waters differ in some respects from those described by Dufossé, I will here state briefly the structure in the species under examination, *Prionotus carolinus*, or the red-winged sea-robin, as well as of *Opsanus tau*, or the common toad-fish.

Sea-robin (Prionotus carolinus).— The air bladder of *Prionotus* (Fig. 3)

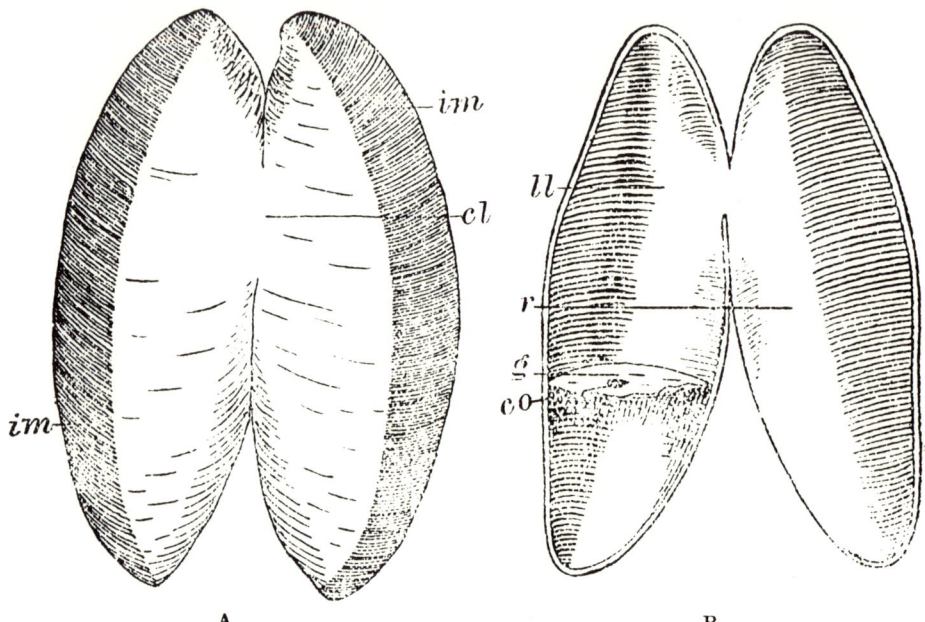

Fig. 3. SWIM-BLADDER OF PRIONOTUS CAROLINUS.

A. Viewed externally. i m, intrinsic muscle;. c l, connecting lumen.
B. Longitudinally bisected. r, right lobe; g, internal septum; c o, central opening of septum.

is a deeply bi-lobed organ, occupying about two-thirds the space of the abdominal cavity. The two lobes are connected near the anterior end by a rather small tube. Along the outside portions of the respective lobes is found a muscle, red in color, and running from the anterior end of the lobe to the posterior end. The muscles adhere strongly to the underlying coat of the air-bladder, and can be separated from it only with difficulty. The muscle-fibres run in the plane of the short axis of the bladder. These muscles correspond to the "intrinsic muscles" of Dufossé. The bladder is not connected with the exterior by a pneumatic duct, Günther (1880) to the contrary notwithstanding, for the entire bladder has been removed from the abdominal cavity without losing any of the contained gas, an operation which would be impossible if there were any means of communication between it and the exterior.

The air-bladder itself consists of three layers — an external, a middle and an internal — together with the pair of muscles just described. The outer and middle layers are composed of thick, compact tissue, containing both elastic and non-elastic fibres. The inner membrane is a mucous

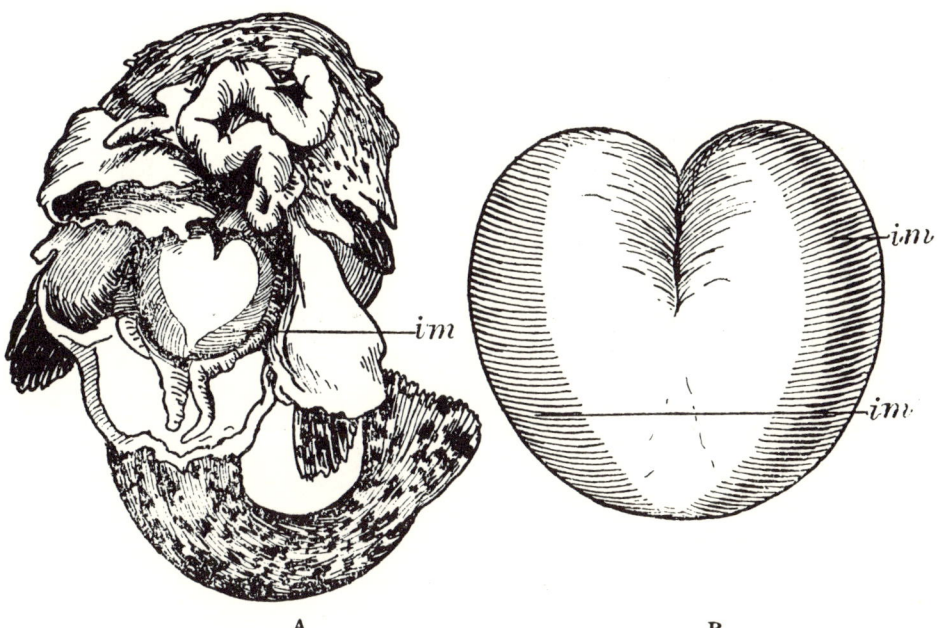

Fig. 4. Swim-bladder of Opsanus tau.

A. Viewed in situ. i m, intrinsic muscle.
B. Viewed externally.

tissue provided with numerous blood vessels. Lying in this tissue are found also the blood glands or red-bodies which were described in the bladders of the sciænoids.

The left lobe (fig. 3 B) in all specimens of *Prionotus* is divided into two parts by a partition formed of the internal tunic or membrane. In the centre of this partition is a small opening, a little larger than the head of a pin. The right lobe is never divided. This perforated partition was present in all specimens examined of both sexes. The embryological history of this partition has never been investigated. There is no difference in the structure of the swim-bladder in the male and female, the intrinsic muscles being present in both. It is evident that we have here anatomically a very different structure from that in the swim-bladder of the Sciænidæ, a fact which will play a very important part in the interpretation of the physiological experiments soon to be described.

Toadfish (Opsanus tau).—The swim-bladder of *Opsanus* (fig. 4) is relatively a much smaller organ than in *Prionotus*. When examined externally (fig. 4 B), it seems to be deeply bi-lobed on the anterior half; but when viewed in longitudinal section (fig. 5), it is seen that less than one half of the organ is actually divided. The swim-bladder is supplied with

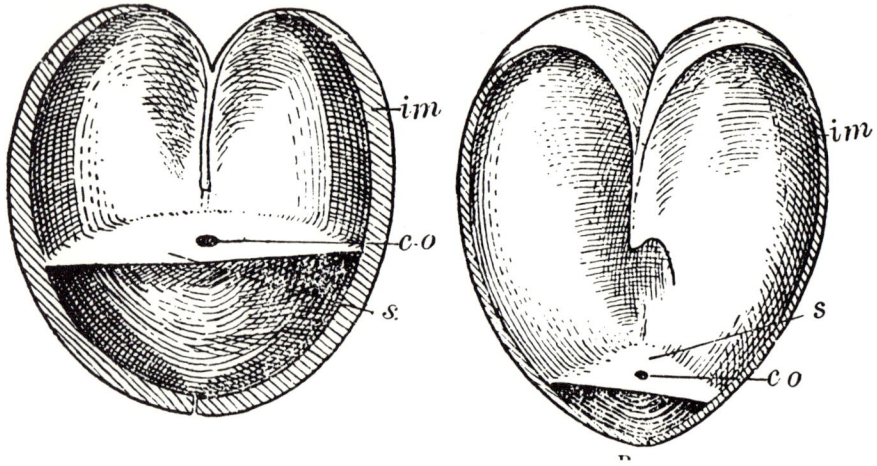

Fig. 5. SWIM-BLADDER OF OPSANUS TAU.

A. A specimen longitudinally bisected showing the position of the internal septum (s). c o, central opening of septum; i m, intrinsic muscle.
B. Another specimen longitudinally bisected showing the variation of the internal septum (s).

the same intrinsic muscles as that of the sea-robin. The muscles arise at the most anterior part of the right and left lobes respectively, and are separated posteriorly by only a small tendon. The muscular tissue is thick and strong, the fibres running transversely to the long diameter of the swim-bladder. An internal septum divides the bladder into two parts, an anterior and a posterior. The septum is perforated by a small opening, which forms the only means of communication between the two cavities. In the posterior cavity alone is found the blood gland.

Most interesting is the great variation found in the position of the transverse septum in different specimens. In some cases, the partition is fully one-third of the distance from the posterior end (fig. 5 A) while in others it is less than one-sixth of the distance (fig. 5 B). Indeed, in the large number of specimens examined, no two were found to be alike. Whether the variation is accompanied by any change in function, I was unable to determine.

Sound Production in the Drumfishes.

RECORDED OBSERVATIONS AND THEORIES.

It has been noticed by many fishermen that the common squeteague at times makes a very plain and unmistakable drumming noise. As to how this noise is produced they can give no explanation nor is there any account of it in scientific literature, with the possible exception of Dufossé's memoirs, which seem to be too little known at the present time. Their observations do not tell us whether it is the male[1], female, or both that produce this characteristic noise. In the anatomical discussion, it was found that only the male was supplied with the red drumming muscle which, from its relation to the air-bladder, was considered to be connected functionally with the latter organ. Further observations demonstrated that the drumming occurred only in those animals in which this red muscle was present — that is, in the male squeteague. In some other species, as *Micropogon undulatus*, drumming occurs in both male and female, and likewise the *m. sonifici* are present in both sexes.

In the rather limited amount of study that has been given to the noises produced by these fishes, some of the conclusions are mere deductions from anatomical data, without any experimental or physiological proof. In other instances, the authors confuse the sounds produced by fishes of entirely different orders, and which have swim-bladders both anatomically and physiologically different. For this reason it is very difficult to deduce correct conclusions from their writings.

As noted in a previous paper, Aristotle spoke of fishes that produce sound by some mechanism involving their air-bladder. The fact was thus known to fishermen and scientists very early; but no scientific explanations were offered nor were any experiments made which would account for these noises. Cuvier (1834) writes that "these fishes [sciænoids] swim in a troop and send forth a bellowing louder than that of the gurnards, and

---

[1] Since the above was written Dr. H. M. Smith has published the results of some observations, which show that both male and female of *Micropogon* make the drumming sound and that the male only in *Pogonias, Sciænops, Cynoscion, Leiostomus* and *Bairdiella* produce the drumming sound. (Science, vol. 22, 1905, p. 376.)

it has occurred that the fishermen, guided by their noises alone, have taken twenty *sciænæ* at a single throw of the net." The fishermen assure us that the noise of the *sciænæ* is sufficiently loud to be heard through twenty fathoms (120 feet) of water, and that they are careful from time to time to place their ears over the edges of the boat, that they may be directed by the noise. Some say that it is a dull humming sound; others that it is a rather sharp hissing. Some fishermen contend that the males alone make this noise in spawning time, and that it is possible to take them by imitating it and without employing any bait. That these fishes do produce noises that can be heard long distances is an undisputed fact.

The fish of this family best known to us is the "weak-fish," described by Dr. Mitchill under the name of *Labrus squeteague*. It was known by the Narragansett Indians as the squeteague; and by the French of New Orleans as the trout. The fishermen of Cuvier's time "attributed to it certain dull sounds similar to that of a drum, which are heard sometimes under the water and only in the season when it is abundant."

Another sound-producing fish of American waters which is described by Cuvier is the drum (*Pogonias cromis*). Cuvier states that "various accounts are given concerning the nature of the noise of these drums." According to Dr. Mitchill, it is when they are taken out of the water that they send forth this noise, but Schoepf says that "it is under the water that this noise is dull and hollow; that several individuals assemble around the keel of ships at anchor, and that then their noise is most sensible and continuous." This agrees with the report made by Lieut. John White, U. S. N., in 1824, in which he describes how his crew and himself, while at the mouth of the river Cambodia, were astonished by some extraordinary sounds which were heard around the bottom of the boat. It was like a mixture of the bass of the organ, the sound of bells, the guttural cries of a large frog and the tones which imagination might attribute to an enormous harp; one might have said that the vessel trembled with it. These noises increased, and finally formed a universal chorus over the entire length of the vessel and the two sides. The natives told Lieutenant White that the noises were produced by a troop of fishes. M. Humboldt describes a similar phenomenon in the South Sea on February 20th, 1803. Towards seven in the morning, he says, the whole crew were awakened by this extraordinary noise, which resembled drums beating the air. It was afterwards learned that the noise was produced by one of these sciænoids. Cuvier, in speaking of the same species, states that, "it would be an object of curious research to find out the organs in these fishes which seem to produce such strong and such continuous sounds, and that at the bottom of the water and without any communication with the external air. Most of the sciænoids have a large natatory bladder, very thick, provided with very strong muscles, but the bladder has no communication either with the intestinal canal, or with the exterior generally." This represents all that was actually known up to the time of Cuvier concerning the mechanism of the sound-producing organs. It was evidently thought by Cuvier that the air-bladder and attending muscles were of some importance in producing this phenomenon.

Somewhat later (1860), Holbrook stated that "frequent examinations of the structure and the arrangement of the air-bladders, as well as observations on the living animal just taken from the water, when the sound is at intervals still continued, have satisfied me that it is made in the air-bladder itself; that the vibrations are produced by the air being forced by strong muscular contraction through a large opening, from one large cavity, that of the air-bladder, to another, that of the cavity of the lateral horn; and if the hands are placed on the sides of the animal, vibrations will be felt in the lateral horn, corresponding with each sound."

It was not until Dufossé published his memoir on the sounds and noises produced by the fishes of Europe, in 1874, that we had any physiological explanation of the phenomenon in the sciænoid fishes which is based on actual experiments. In 1864, Moreau published the results of his experiments on the "grunting" mechanism in the Triglidæ, a process, however, which is entirely different from the "drumming" of the Sciænidæ, and should not be confounded with it. With regard to the sounds produced by certain muscles Dufossé says,

> Le phénomène physiologique connu généralement sous le nom de *trépidation* ou *trémulation* musculaire, et que Wollaston a assimilé, avec raison, à un mouvement de vibration, n'a pour ainsi dire été observé que chez l'Homme, et n'a jamais été le sujet d'une étude approfondie, soit au point de vue biologique, soit au point de vue la physique proprement dite, quelques physiologistes pensent même encore que ce mouvement assez rapide pour produire un léger bruit, désigné sous le nom de *bruit de rotation* par Laënnec et sous celui de *bruit de contraction des muscles* par d'autres auteurs, est trop faible par lui-même et trop peu important par ses effets pour devenir jamais d'un certain intérêt en physiologie générale.

This is a concise statement of what was known concerning the physiology of this noise at the time when Dufossé wrote his memoir. Dufossé divided his work into two propositions, viz:

1. Quelques muscles de certains poissons bruyants deviennent en se contractant susceptibles d'un mouvement de vibration:
2. Ce mouvement est le principe des sons que font entendre ces animaux.

To prove these propositions, Dufossé made two physiological experiments. In the first, he inserted his finger into the stomach of a lyre capable of producing an intense noise. During the production of the noise, he noticed an intense vibration which coincided exactly in duration with the sounds heard by his ear. He then punctured the wall of the air-bladder and drew out all the gas. The sound ceased, but the vibrations could still be felt. He then removed the entire swim-bladder, and applied his finger successively to the muscles and aponeuroses which lie alongside the vertebral column, and he found that all the organs were in repose except the intra-costal muscle, which vibrated and gave to his finger the same sensation as when the air-bladder was in its natural position.

In the second experiment, Dufossé opened the abdomen of a lyre just in front of the anus, and extirpated the swim-bladder entirely ("j'extirpe la vessie pneumatique tout entière"). He then inserted an artificial bladder ("poche membraneuse") and inflated it. The fish commenced again to produce a noise similar to that made before the operation. In another he cut the nerve supplying the intra-costal muscles, first the right and then the left. After both were cut, the noise ceased and could not be again renewed.

From these data, Dufossé argues that there are two factors in the producing of the noise, viz: the contraction of the intra-costal muscle, which is the primary cause of the sound, and secondarily, the reënforcement of these vibrations of the swim-bladder. The producing of the noise is voluntary. Dufossé recognizes many difficulties in this explanation, however, because the facts do not agree with those of theoretical physics, as can be seen from the following quotation:—

"Le mécanisme de la production des sons chez ces poissons a pour complément la transmission des vibrations sonores des muscles à la vessie qui est en contract avec eux. Les parois de cet organe communiquent ces vibrations au gaz qu'elle renferme, et ceux-ci vibrent de telle façon comme le prouvent surabondamment mes deux premières expériences, que l'intensité de ces vibrations est incomparablement augmentée. D' après ce résultat et en considérant que la vessie est une cavité close à parois membraneuses et souples se moulant si exactement sur la surface des organes qui les environment qu'elles ne peuvent vibrer que comme elles le feraient si elles étaient réellement adhérentes par tous les points de leur superficie à la masse de ces organes on ne peut expliquer, conformément aux principes de la physique, le renforcement des vibrations sonores qu'en admettant que le volume des gaz contenus dans vessie, ou, ce qui est le même chose, que la capacité de cet organe a naturellement des rapports exacts de grandeur avec celle des nombres de vibrations sonores que lui sont transmises. L'exactitude des rapports que suppose cette explication ne s'accordant pas avec plusieurs faits ichthyologiques, entre autres avec les incessants changements de volume que submit nécessairement la vessie pneumatique quand le poisson vient du fond de l'eau à la surface ou s'enfonce dans la profondeur des mers, cette explication n'est acceptable qu'en admettant que si ces rapports existent réellement ils doivent pouvoir varier d'une certaine quantite sans que le degré de renforcement des sons soit grandement modifié."

In speaking of the Sciænidæ, Dufossé says that the sound is produced for the most part by muscles; but a little later, in speaking of *Pseudosciæna aquila*, he says

"Le mécanisme de la production des sons chez les individus de l'espèce *Sciæna aquila* est plus compliqué que celui des poissons dont j'ai parlé jusqu' à présent (*lyre*). Je n'ai nullement la pretention de donner la théorie de ce mecanisme."

Just why Dufossé makes this statement is not intelligible, for all the drumming fishes of America that have been examined have the same mechanism, and it is very evident that the sound is produced in exactly the same way. As Dufossé examined only European forms, however, he may have observed a difference in structure that is not present in the American species.

The fact that Dufossé stated that he could not explain the mechanism in *Sciæna*, while for the Gurnard lyre his explanation was not in accordance with physical phenomenon, has possibly led more recent ichthyologists to ignore his work. Thus we find that Günther (1880) in speaking of the American drum (*Pogonias cromis*) says,

"It is still a matter of uncertainty by what means the "Drum" produces sounds. Some naturalists believe that it is caused by the clapping together of the pharyngeal teeth, which are very large molar teeth. However, if it be true that the sounds are accompanied by a tremulous motion of the vessel, it seems more probable that they are produced by the fishes beating their tails against the bottom of the vessel in order to get rid of the parasites with which that part of their body is infested."

That these explanations of Günther are unwarranted will be seen from experiments soon to be described.

Sörensen (1895) disagrees with Dufossé's statement that it is the vibration of the muscles while contracted which produce the sound and that the air-bladder only intensifies the sound. Sörensen considers the sound as being produced by vibrations of the air in the air-bladder and of the walls of the latter when set in motion by the muscles with the fascia of which it is connected.

Jordan and Evermann (1902) say that "most of the species make a peculiar noise, variously called croaking, grunting, drumming, or snoring, supposed to be produced by forcing air from the air-bladder into one of the lateral horns."

We have presented to us then, four distinct mechanisms in the Sciænoids:
1. Muscular tone; the vibrating muscle producing a sound which is intensified by the air-bladder (Dufossé).

2. Clapping together of pharyngeal teeth (Günther).
3. Vibrations of air in air-bladder and of the walls of the latter when set in motion by certain muscles (Sörensen).
4. Forcing of air from air-bladder into one of lateral horns (Holbrook, Jordan and Evermann).

Experiments to Determine Cause of Sound.

That the explanation given by Günther is wrong can be very easily seen from the following experiments, in all of which the animals were kept alive by artificial respiration, i. e. by irrigating the gills with a stream of fresh water.

Experiment I. The air was drawn from the swim-bladder of a squeteague by means of a trochar and the drumming immediately ceased.

a. The stomach was then filled with water. The drumming returned but not as loud as normal.

Experiment. II.—An incision one inch long was made in the mid-ventral line. Through this a portion of the air-bladder was pulled out but the drumming continued.

a. The bladder was now amputated and the drumming ceased.

b. A collapsed rubber balloon was then inserted into the abdominal cavity, and, as soon as it was inflated, the drumming returned with apparently normal intensity and pitch.

c. The rubber balloon was filled with salt water instead of air. The drumming continued until the water was allowed to escape; then it ceased. The tone is low and apparently changed but little under the different conditions.

Experiment III.—The air-bladder of a male squeteague was removed. The drumming ceased. The air-bladder from a female squeteague, which can produce no noise, was inserted into the abdominal cavity of the male, and the drumming immediately returned.

These three experiments show conclusively that the "clapping together of the pharyngeal teeth" has nothing to do with the production of the drumming noise in the squeteague. It is also shown in experiment III that there is no difference between the function of the bladder in the male, which drums, and that of the female, which does not drum, as far as the noise-production is concerned.

Experiment IV.—The entire viscera (intestines, spleen, liver, reproductive organs and air-bladder) were removed from a male squeteague. The drumming stopped, and the *sonificus* contracted as usual, but there was no noise. A rubber balloon filled with air was then inserted into the abdominal cavity. The drumming again returned, but was not of normal character.

a. The balloon was filled with water and the drumming continued, but was weaker and of apparently different pitch.

This experiment shows that an inflated rubber balloon can take the place of all the abdominal organs in respect to noise-production; although it does not prove that these organs may not play some part in the normal mechanism.

To determine whether there is any experimental basis for the view held by Jordan and Evermann, viz: that the drumming is produced by forcing air from the air-bladder into one of the lateral horns, the following experiments were undertaken.

Experiment V.— An incision about one inch long was made in the mid-ventral line of a male squeteague. The air-bladder was ligatured in the middle, thus separating the organ into two chambers,— the anterior containing the lateral horns, and the posterior remaining a simple closed cavity. Drumming, however, went on as in normal animals.

a. The part of the bladder posterior to the ligature was punctured. The drumming continued only in the region of the anterior part of the bladder, which remained inflated.

b. Another animal was prepared in the same way, and the part anterior to the ligation was punctured. The drumming continued only in the region of the posterior portion of the bladder, which remained inflated. In this part of the bladder there are no lateral horns.

c. The posterior end of the bladder was then folded into the anterior part of the abdomen. The drumming noise then came from the anterior part of the abdomen in the region of the inflated half-bladder.

d. The anterior half of the bladder was amputated, leaving the posterior part still inflated. This was inserted at different places and the drumming noise occurred wherever this part of the bladder was placed.

Experiment VI.— An incision was made in the mid-ventral line of a large "drummer," about half way between pectoral and anal fins. At the right angles to this incision, longitudinal incisions were made on both sides, extending nearly to the region of the kidney. These incisions were made through the drumming muscles. The air-bladder and viscera were lifted up with forceps, and the remaining part of the drumming muscle and central tendon was cut. This separated the entire muscle, tendon and insertion into halves — an anterior and a posterior part. The drumming still continued on both sides of the bisection. In order to show this still more completely, the anterior half of the abdomen was raised by inserting two fingers, which prevented the drumming in this (anterior) part, while the posterior gave the same characteristic noise. Next, the posterior half of the abdomen was raised in the same manner, and the drumming stopped in the posterior part but continued in the anterior. Upon removing the fingers, the noise continued as in normal animals.

a. The air-bladder was ligatured in the same place as the bisection of the muscle. The drumming occurred as before. Again the anterior and the posterior parts were in turn raised, and the drumming was made to occur in either part at will.

The two experiments V and VI, as well as the previous ones, prove conclusively that the lateral horns have nothing to do with producing the drumming noise; and the forcing of air into the lateral horns, if such takes place, is not the true explanation.

It remains now to consider the views of Dufossé and of Sörensen. Is this drumming a muscular tone, i. e., a sound produced by the vibrating muscle and intensified by the air-bladder (Dufossé), or do certain muscles set into vibration the air in the air-bladder and the walls of the latter (Sörensen)?

Experiment VII.— The entire abdominal viscera except the air-bladder were removed. Contractions of the muscles occurred, but no noise. The rubber balloon was inserted into the abdominal cavity and inflated (the air-bladder being intact and inflated). The drumming returned. When the balloon was allowed to collapse, the noise ceased.

a. The abdominal cavity was packed tight with cloth (the air-bladder being intact and full of air). The drumming was loud, and when the cut edges were drawn together, it increased to a normal drum. When the cloth packing was removed, the muscle still contracted, but no noise was heard. When the cavity was packed a second time with cloth, the drumming became again audible.

Experiment VIII.— The entire viscera, including the air-bladder, were next removed. Notwithstanding the large hemorrhage that occurred, the *sonifici* still contracted. The rubber balloon was inserted and inflated with air. The drumming noise returned of apparently the same pitch but not so loud as normal.

a. The central tendon was then cut longitudinally into two parts. The muscles on either side contracted rhythmically, as could be seen from the vibrations of the cut ends of the tendon. There were, however, no vibrations of the abdominal muscles, such as are seen in normal animals. This was as might be expected, because after cutting the central tendon the two drumming muscles have nothing to work against. The inflated rubber balloon now produced no sound. This seems to show that the air-bladder does not act as an intensifier of muscular tone. The experiment suggests that the air-bladder functions either in maintaining the tension inside of the abdominal cavity, or as a vibrating organ or both.

Experiment IX.— Incisions were made on both sides of the median line of the abdomen. After this operation the drumming remained perfectly normal. The *m. sonifici* were then cut from their origin on the abdominal muscles. One side was amputated first, and the drumming still continued. While the one on the opposite side was being cut, the drumming died away gradually until the drumming muscle was severed its entire length, when the noise ceased. Yet at this time the muscle contracted, as could be easily felt by touching it with the finger.

If now the air-bladder served as an intensifier of the muscular vibrations, we might ask why it suddenly ceases to fulfill that function in the above experiment. Also, in experiment II *c*, drumming occurs when the air-bladder is replaced by a rubber balloon filled with water. This water-bladder cannot be looked upon as an intensifier of sound or a resonator.

The foregoing experiments show that any part of the muscle can produce the drumming when conditions are suitable. We have also seen that by lifting a part of the abdominal muscles, the drumming over that part immediately ceases. That the most ventral parts of the abdomen are active in drumming is evident from the vibrations of this part of the body of a squeteague. The whole mid-ventral area, from pectoral fins to anus, pulsates in a strong rhythmical manner, which corresponds to the contraction of the *m. sonificus* as can be readily seen from the appended kymograph tracings.

Experiment X.— An incision one inch long was made about half way between the pectoral fins and the anus and at right angles to the long axis of the body. Great care was taken in order not to injure the drumming muscles. Between the ventral muscles and air-bladder was inserted a piece of sheet cork about two and one-half inches long and two inches wide. This stretched severely the mid-ventral part of the abdominal muscle and held it rigid, so that it could not be pulled in when the *sonificus* contracted. No noise was produced, yet the muscle apparently contracted in a perfectly normal manner. This would again show that the drumming is not a muscular tone intensified by the air-bladder.

The drumming is undoubtedly a sexual character, for in the squeteague the male only makes this noise. The female not having developed any drumming muscles is not able to produce this sound. In some other sciænoids, as the croaker, both male and female produce the drumming, but the former is said to produce a much more intense noise than the female. I have often observed that the drumming muscles in the male croaker are much thicker and heavier than in the female.

The conclusion is that by each contraction of the *m. sonificus* a sudden blow is dealt which throws into vibration the abdominal walls and organs. The physics of this phenomenon is very complex, as undoubtedly all of the abdominal parts play a rôle. But the organ that chiefly participates in the vibration is the swim-bladder with its walls made tense by the pressure of the contained gas. It is well known that in man the chest walls and abdominal walls can be set into irregular vibration by being percussed and that there is here a resonance effect produced by a resonance cavity or semifluid material which is selectively set in resonance vibration. The gas pressure in the air-bladder as well as the character of the muscular contractions which will be immediately described indicate the same conclusion.

In all of the above experiments the pitch of the drumming sound was not determined with scientific accuracy. Undoubtedly if the tone could have been determined by physical apparatus the pitch, which to the ear was apparently the same, would have been found to be different in the various experiments.

EXPERIMENTS TO DETERMINE CHARACTER OF THE MUSCULAR CONTRACTION.

The character of the contraction of the red drumming muscles has never been studied, nor has the relation of the contractions to the pitch of the drumming been accurately recorded. Dufossé has given the pitch of the drumming of the meagre as well as he could determine it by the ear alone. The following experiments were performed in carrying on the present study:

Experiment XI.— The first experiment was made so as to record the number of vibrations produced by the abdominal tissue in the mid-ventral line during the process of drumming. To accomplish this, a light wooden lever was made, with a piece of sheet cork two inches long and one half inch wide attached at the bottom, and a fine wire inserted in the top at right angles to the lever. The cork was held in place on the abdomen of the squeteague by two rubber bands going around the fish and over each end of the cork strip. The revolving drum of the kymograph was then placed so that the wire point would trace on the smoked paper of the drum. Thus when the animal commenced to drum, the vibration of the part of the abdomen under the lever would be traced by the writing point on the smoked paper. The drum of the kymograph revolved once in 4.848 seconds, and its circumference was 48.5 cm. The tracings are given on Pl. VIII, fig. 1. The number of vibrations per second, as determined by comparison with the tracings of a tuning fork vibrated 100 times per second, is 24.

Experiment XII.— A control experiment was made the next day on another squeteague, but with the drum of the kymograph revolving only once in 20.202 seconds. The number of vibrations should agree or at least be within the limits of experimental error. The tracings are given on Pl. VIII, fig. 2. The number of vibrations is again 24 per second.

In both of the above experiments the lever was placed on the mid-ventral line just posterior to the pectoral fins.

Experiment XIII.— The next experiment was to determine whether the anterior and posterior ends of the abdomen vibrated synchronously, or whether the vibration passed over the abdomen like a wave, from anterior to posterior, or vice-versa. Mere observation as well as the resting of the fingers on the anterior and posterior parts at the same time detected that all the muscle-fibres contracted synchronously. To determine this more accurately, two levers were arranged — one being placed just posterior to the pectoral fins, and the other just anterior to the anus — so that they should write under each other on the smoked paper. The traces indicated that the entire abdominal mechanism vibrated synchronously; hence all the fibres of the two drumming muscles contract at the same time under stimuli controlled by the central nervous system of the animal.

Experiment XIV.— In a fresh male squeteague an incision one inch long was made on one side through the thick, white abdominal muscles until the red *m. sonificus* was exposed. The cork base of the lever was inserted through this opening until it rested on the red muscle within. With the lever in this position and the animal on its side, the writing point should not move up and down in a perpendicular plane, but should move horizontally, back and forward. This, then, should give in the kymograph reading a series of dots, representing the apex of each curve. Such a tracing was recorded on the drum of a kymograph. The vibrations were 24 per second. From this experiment it is seen that the muscle itself and the abdominal tissue vibrate at exactly the same rate.

Experiment XV.— Another experiment was made to show the effect of substituting an inflated rubber balloon for the air-bladder. The number of vibrations was 24 per second. It is thus evident that the vibrations produced in the presence of the rubber balloon are the same as in the normal condition of the animal.

The five preceding experiments agree in the rate of vibrations of the abdominal part which is in immediate relation to the drumming muscles, and which is directly connected with sound production according to our present views. It was next necessary to record the contractions of the muscle itself, and for this purpose the following two experiments were performed.

Experiment XVI.— An incision two inches long was made in the mid-ventral line just posterior to the pectoral fins. Through this opening was inserted a slender wire with a sharp hook on one end and an eye on the other. The hook was fastened directly into the fibers of the *m. sonificus*. The eye was attached to an ordinary muscle-lever which was supplied with a writing point. The kymograph was then placed so that it would receive the tracings made by the writing point.

In this experiment none of the viscera were disturbed and the noise produced differed in no way from that of the normal animal. To measure the time, a tuning fork was used, whose double vibrations of 100 per second were registered on the revolving drum. The rate of the contractions is 24 per second, which is identical with the experiments made on the abdominal walls. As is shown in Pl. VIII, fig. 3, the amplitude of the contractions is much more than in the experiments made on the abdominal walls. This is undoubtedly due to the release in tension caused by the separation of the right and left abdominal portions to which these muscles are attached, together with some resistance caused by the rubber bands. To determine this point another experiment was made in which the vibrations of the abdominal wall were registerd by a wire hook attached at one end to the *rectus abdominis* and the other to the muscle-lever. With this method the amplitude of vibration is nearly the same as that of the muscles. It was noticed, too, that when the amplitude was the greatest the loudest sound was produced in both the experiments on the abdominal walls and on the drumming muscle.

Experiment XVII.— Experiment XVI was repeated, except that the air-bladder was punctured. The drumming noise stopped. The contractions of the drumming muscles, registered as in the preceding experiment, are given on Pl. VIII, fig. 4. The number of contractions computed from those of a tuning fork is 24 per second.

It is very evident that there is no difference between the contractions when the swim-bladder is full of air and when it is collapsed, and that this organ has no effect upon the contractions of the drumming muscle. This is especially well demonstrated in the tracings where the register of the muscular contractions in an animal with the bladder intact is placed directly over these from an animal in which the bladder is collapsed (Pl. VIII, fig. 4).

Experiment XVIII.— The viscera, including the swim-bladder, were removed from a squeteague after an incision had been made in the mid-ventral line from the pectoral fins to the anus. The wire hook of the registering apparatus was inserted into the middle of the central tendon. No noise was produced. The number of contractions was 24 per second. The amplitude of vibration was less than some registered by the muscle and more than others. The experiment revealed no new factor.

In the experiments just described each contraction of the muscle, represented in the tracing by the apex of the curves, is simultaneous with the sound produced, and thus the rapid series of contractions institute the roll or "drumming."

EXPERIMENTS TO DETERMINE THE PRESSURE OF THE GAS IN THE SWIM-BLADDER.

Experiments were made to discover, if possible, the pressure exerted on the air-bladder by the contraction of the drumming muscles.

Experiment XIX.— The pressure of gas in the air-bladder of a female squeteague (which has no drumming muscles and can not drum) was determined by making an incision one inch long in the mid-ventral line two inches anterior to the anal fin. The posterior end of the swim-bladder was ligatured and then amputated just back of the ligature. The open end of a small mercurial manometer was inserted and tied by another ligature. The first ligature was then removed and the mercury rose to a height of 4 mm., which was produced by the normal pressure of the gas in the air-bladder. The animal was kept alive by artificial respiration.

  a. The same experiment was then tried on the swim bladder of a male squeteague, both while it was quiet and while it was drumming. In the quiet animal, the pressure rose to 4 mm. and remained there until drumming occurred, when it rose to 6 mm. In other words, the increased pressure brought about by the contraction of the drumming muscles equalled 2 mm. of mercury. During the drumming the meniscus of the mercury could be seen to oscillate between 4 mm. and 6 mm., as the muscles successively contracted and then relaxed.

One interesting feature is that in all the animals examined the normal pressure in the bladder was 4 mm. in the male and female — the large and small animals alike. The gas pressure within the swim-bladder maintains a tension on the elastic walls, while the increased density of the gas due to the pressure tends to produce a louder sound than would otherwise occur. These experiments show that

1. The chief cause of the drumming is the contraction of the drumming muscles.

2. As the myogram distinctly shows, the contraction of the drumming muscles is of the nature of a series of simple contractions.

3. These muscles contract at a definite rate, viz.: 24 vibrations per second.

4. By the force of each contraction the abdominal organs are set into vibration, especially the walls of the air-bladder.

5. The elastic walls of the air-bladder are always tense, because of the pressure of the contained gas. This pressure is increased each time the drumming muscles contract.

6. The vibration of the tense walls of the air-bladder and the contained gas are sufficient to produce the drumming noise.

7. The sound produced is low. The actual number of sound vibrations was not determined.

Sound production in the Sea-Robin and the Toadfish.

EXPERIMENTS TO DETERMINE CAUSE OF SOUND AND CHARACTER OF MECHANISM.

If a sea-robin is examined under artificial respiration, the single twitch of the abdomen when a grunt is made can be very easily observed. If the animal is opened along the mid-ventral line, both the contraction of the intrinsic muscles and the single twitch of the swim-bladder can be observed. The noise is of the same pitch and loudness after the abdomen has been opened as before. The removal of all the viscera except the air-bladder has no effect on the noise produced. It is noticed that the two muscles contract simultaneously.

Experiment XX.— An animal under artificial respiration was opened, and various parts of the bladder were stimulated by a current from an induction coil, viz.:

a. One of the two nerves supplying the bladder was stimulated. A perfectly normal grunt was produced.

b. The fibrous part of the bladder was then stimulated. A normal grunt was not produced.

c. The muscle itself was stimulated directly. Again a perfectly normal sound was produced.

These experiments show only that artificial stimulation of either nerve or muscle will cause a normal sound to be produced.

Experiment XXI.— The swim-bladder was removed from a fresh specimen and laid upon the operating table. The nerves and the muscles of the bladder were then stimulated successively as in experiment XX. In each case there was a grunt of the same pitch and intensity as is produced by the normal animal.

This shows very clearly that the sound-producing mechanism of the sea-robin is entirely within the bladder and its intrinsic muscles. This mechanism, then, stands in direct contrast to that of the drumfishes, just discussed.

Experiment XXII.— The swim-bladder was removed from a sea-robin. The muscle was stimulated and an audible grunt was produced. The bladder was then placed on an improvised registering apparatus, so arranged that the bladder was connected with a muscle lever and writing point. The muscle was then stimulated. An audible grunt resulted. The vibration of the bladder was registered on the drum of the kymograph. The grunt is produced by one single sharp contraction of the intrinsic muscle (Pl. VIII, fig. 6). This was repeated each time that the muscle was stimulated.

a. One of the lobes of the bladder was now punctured. Both lobes collapsed. Through the opening was inserted the rubber balloon (collapsed). This was inflated; the muscle was thus superimposed over the inflated rubber balloon. The muscle was then stimulated as before. It contracted and produced a grunt the same as in the isolated bladder full of air. Moreau (1876) concluded that it was the vibration of the perforated internal septum which was the direct cause of phonation. That this septum vibrates is true, but from the foregoing experiment it would seem that the walls of the air-bladder are the chief vibrating organ. In the sea-robin the left lobe only possessed the internal septum, but it made no difference with the sound produced whether the right lobe or the left lobe was used for the experiment.

b. The uninjured lobe was filled with salt water and closed by a ligature. The muscle was then stimulated by a current from an induction coil. A grunt occurred as when the swim-bladder was filled with air, although not so loud. These contractions were recorded by means of a kymograph and are given on Pl. VIII, fig. 7. On comparing the record with those given on Pl. VIII, fig. 6, it is evident that the curves have about the same amplitude, but are not so well sustained.

Experiment XXIV.— The swim-bladder was removed from a sea-robin as quickly as possible. The muscle was stimulated by a current from an induction coil. An audible grunt resulted. This sound was more intense when the bladder rested on the table. It is interesting to note that this particular animal did not produce any noise while alive. The isolated bladder was then placed on the registering apparatus, and records were obtained under single stimulations and also by stimulations continued for several seconds. The records are given on Pl. VIII, fig. 8. The character of the curve is changed by the continued stimulation, the muscles going into incomplete and then complete tetanus. Tetanic contraction does not appear to be the normal procedure, but is produced by artificial stimulation. And as far as could be determined, the sound was produced at the beginning of the tetanus, i. e. at the first up-stroke of the lever, and died out during the remainder of the contraction. The loudest grunts were produced at single full contractions of the intrinsic muscles. The sound produced starts with a grunt, which gradually dies out. It does not resemble drumming.

a. The bladder was then punctured and all of the air expelled from both lobes. The muscle was again stimulated, but there was no sound, although the muscle contracted as usual. The collapsed rubber balloon was inserted into one lobe of the bladder, and then inflated. Upon stimulation a grunt was produced. The bladder was now inflated still more, and upon stimulation a grunt of higher pitch was produced. When the bladder was inflated still more, the pitch became yet higher.

b. The rubber balloon was now filled with sea-water and the muscles stimulated. A grunt was produced, although the pitch was apparently changed.

It is very evident then, that in the sea-robin and the toadfish the swim-bladder with its intrinsic muscle is an organ for the production of sound. By the contraction of the intrinsic muscle the tense walls of the air-bladder are made to vibrate, thus producing the sound.

These grunts can be imitated very closely by drawing the forefinger and thumb towards each other over the surface of an inflated rubber balloon, especially if the rubber is dry or has been resined.

Conclusions.

I. The sciænoid fishes that make a drumming noise have specific sound-producing muscles which are only superficially attached to the swim-bladder. For this drumming muscle the name *musculus sonificus* has been proposed and adopted.

II. The chief cause of the drumming noise is the contraction of the *m. sonificus*, which produces a vibration of the abdominal walls and organs, especially the swim-bladder.

III. The sea-robin and the toadfish, which make a grunting noise, have muscles which are intrinsically connected with the swim-bladder and are known as *intrinsic* muscles.

IV. The cause of the grunting noise is the contraction of the intrinsic muscles which produce a vibration in the walls of the air-bladder.

V. The mechanism in the Sciænoidæ is adapted to the production of rapidly repeated sounds or rolls.

VI. The mechanism in the sea-robin and the toadfish is adapted to the production of sounds repeated at more or less long intervals.

Literature Cited.

Cuvier, Georges. The animal kingdom arranged in conformity with its organization, by the Baron Cuvier, with supplementary additions to each order, by Edward Griffith and others. Vol. X, Class Pisces, with supplementary additions, by Edward Griffith and Charles Hamilton Smith. London, 1834.

Bridge, T. W. Cambridge Natural History, vol. VII, 1904, Fishes, p. 359.

Dufossé. Recherches sur les bruits et les sons expressifs que font entendre les poissons d'Europe et sur les organes producteurs de ces phénomènes acoustiques ainsi que sur les appareils de l'audition de plusieurs de ces animaux. Annales des Sciences Naturelles, 5me ser., t. XIX, 1874, art. 5, 53 p., pl. 16, 17, 18, 19, and t. XX, art. 3, 134 p.

Günther, A. An introduction to the study of fishes. 1880.

Holbrook, J. E. Ichthyology of South Carolina. 1860. [2nd ed.]

Jäger, A. Die Physiologie und Morphologie der Schwimmblase der Fische. Archiv für die gesammte Physiologie, bd. XCIV, 1903, p. 65–138.

Jordan, D. S., and Evermann, B. W. American food and game fishes. 1902.

Lichtenfelt, H. Literatur zur Fischkunde, p. 68, 1906.

Moreau, A. Sur la voix des poissons. Comptes Rendus de l'Académie des Sciences, Paris, 1864, Aug. 29, p. 436.

—— Recherches experimentales sur les fonctions de la vessie natatoire. Annales des Sciences Naturelles, 6me ser., t. IV, 1876, art. 8, 85 p.

Smith, H. M. The drumming of the drumfishes (Sciænidæ). Science, n. s. vol. 22, 1905, p. 376.

Sörensen, W. Are the extrinsic muscles of the air-bladder in some Siluroidæ and the "elastic spring" apparatus of others subordinate to the voluntary production of sounds? What is, according to our present knowledge, the function of the Weberian ossicles? Journal of Anatomy and Physiology, vol. XXIX, n. s. vol. IX, 1895, p. 109–139, 205–229, 399–423, 518–552.

—— Om Lydorganerhos Fiske. Kjöbenhaven, 1884.

White, J. Voyage to the seas of China. 1824.

PLATE VI.

Fig. 1. SWIM-BLADDER OF MICROPOGON UNDULATUS.
    In normal position resting on the central tendon which joins the m. sonificus of either side.
    The two lateral horns extend back over the bladder three-fourths of its entire length.

Fig. 2. SWIM-BLADDER OF CYNOSCION NEBULOSUM.
    In normal position resting on the central tendon which joins the m. sonificus of either side.

(176)

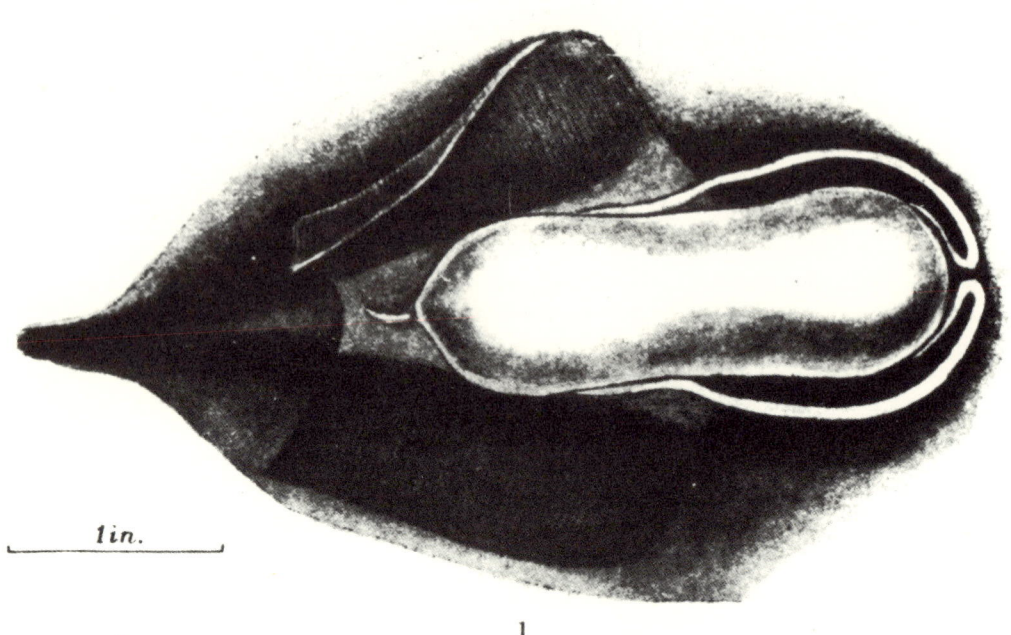

1

2

PLATE VII.

Fig. 1. SWIM-BLADDER OF BAIRDIELLA CHRYSURA.
    In normal position resting on the central tendon which joins the m. sonificus of either side.

Fig. 2. SWIM-BLADDER OF LEIOSTOMUS XANTHURUS.
    In normal position resting on the central tendon which joins the m. sonificus of either side.
    The undissected portion at the anterior of the bladder shows how the two m. sonifici completely inclose the swim-bladder.

(178)

1

2

PLATE VIII.

KYMOGRAPH RECORDS OF SOUND-PRODUCING SWIM-BLADDERS.

Fig. 1. Squetague. Normal swim-bladder. Twenty-four vibrations per second (rapidly revolving drum).
Fig. 2. Squeteague. Normal swim-bladder. Twenty-four vibrations per second (slowly revolving drum).
Fig. 3. Squeteague. Myogram of m. sonificus. Twenty-four vibrations per second.
Fig. 4. Squeteague. Myogram of m. sonificus.
    a. Swim-bladder normal. b. Swim-bladder collapsed.
Fig. 5. Tuning fork having one-hundred double vibrations per second. Kymograph drum revolving at same rate as for Figs. 1, 3, and 4.
Fig. 6. Prionotus. Swim-bladder removed.
Fig. 7. Prionotus. Swim-bladder removed and filled with sea-water.
Fig. 8. Prionotus. Swim-bladder removed. Prolonged stimulation.

(180)

ANNALS N.Y. ACAD. SCI. VOL. XVIII, PLATE VIII.

Fig. 1.

Fig. 2.

Fig. 3.

Fig. 4. a
b

Fig. 5.

Fig. 6.

Fig. 7.

Fig. 8.

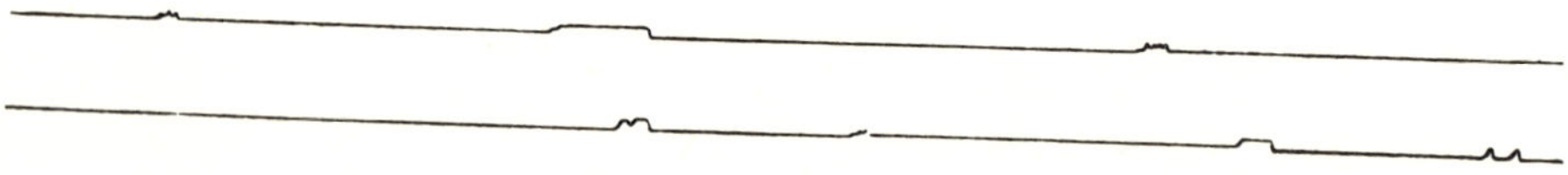

Sound Production in the Haemulidae

By Martin D. Burkenroad

SMITH (1907) says that the "grunt" of haemulids is produced "by the air-bladder." Gudger (1929) makes the same statement. While at the Carnegie Laboratory at Tortugas in the summer of 1929, the writer had the opportunity of making several observations on grunts, and will here attempt to give a more detailed account of sound production in these fishes than is supplied by the earlier workers. The writer wishes to express his indebtedness to Dr. W. H. Longley, Mr. C. M. Breder, Jr., and Dr. E. S. Hathaway, for their aid.

The "grunt" of *Haemulon plumieri* is a loud rasping croak, given when the fish is in difficulty (possibly also at other times), and produced both under water and in air. Noise-making is not confined to adults, in *H. sciurus* at least, since fishes of this species, down to 30 mm. and possibly smaller, grunt; however, the sound in small fish is very faint.

When an adult of *H. plumieri* is held with the mouth open, the upper and lower pharyngeal teeth can be seen to grate against one another as the "grunt" is produced. On dissection, the upper pharyngeal teeth are seen to be in a broader patch than the lower ones, and to be more freely movable than the latter. On the dorsal surface of the lower part of the fourth pair of gill arches are hard teeth of the same character as those of the pharyngeal patches, so that the total width of the lower tooth-bearing area is equal to the width of the upper pharyngeal patch. In a dissected specimen, when the teeth are in the "grunting position," *i.e.*, closely approximated, the posterior bony edge of the lower pharyngeals is seen to abut against a swelling formed by the anterior end of the swim bladder, which presses against the dorsal wall of the oesophagus. The anterior end of the swim bladder is held in place by several strong tendinous insertions.

If the swim bladder is deflated, the pharyngeals placed in "grunting position," and either patch scraped with a metallic instrument, a dry, grating sound is heard. If the swim bladder is now inflated, and the scratching repeated, the upper pharyngeals continue to produce a dry rasp, but the lower ones produce a deep noise approximating the normal "grunt."

If a cloth is passed between the teeth of a living grunt, so that they cannot scrape together, the fish can make no sound, although it makes the attempt, since the same vibration of the body which accompanies normal "grunting" can be seen. A living *H. plumieri* with the swim bladder deflated by puncture "grunts," but the sound is a dry rasp, and not the normal deeper note.

Hence the "grunt" of *H. plumieri* seems to be produced by the scraping of the upper, more movable pharyngeal teeth against the teeth on the lower pharyngeals and fourth gill arches, while the swim bladder acts as resonator.

The sound-making apparatus in very small *H. sciurus*, in adult *H.*

plumieri, and in the pig-fish (*Orthopristis chrysopterus*) of the Louisiana coast, seems, on dissection, to be identical, so that it is probable that the grunt is produced in the same way in all haemulids.

LITERATURE CITED

SMITH, H. M.
    1907 Fishes of North Carolina. *N. C. Geol. and Econ. Surv.*, 2: 209.

GUDGER, E. W.
    1929 Teleostean fishes of Tortugas. *Pap. Tortugas Lab. Carn. Inst. Wash.*, 26: 180.

TULANE UNIVERSITY, NEW ORLEANS, LOUISIANA.

Notes on the Sound-Producing Marine Fishes of Louisiana

By Martin D. Burkenroad

I

DURING the winter and summer of 1929-30, the writer, working in the field with the fishermen of the Louisiana coast, was afforded considerable opportunity to collect the fishes of the coastal waters. A number of the species were found to be capable of producing sound. Since aquarium facilities were lacking, notes were made on the vocal species as they were taken alive from the hook, seine or otter-trawl; a few of the fishes were kept under observation for a time in water-buckets or other small containers. These observations were supplemented by the examination of preserved material. At no time did the writer hear any vocal sounds attributable to fishes uncaptured and undisturbed. He is therefore unable to offer any evidence as to the function of vocalization, except to point out that, whatever other conditions call forth the production of sound by fishes, those stimuli attendant on capture and handling seem to excite noise-making in fishes capable of it. The major exceptions to this rule which he has noted are among the batrachoidids: not a single individual, among nine of *Porichthys porossissimus* and among several of *Opsanus tau* observed, made the swim-bladder noise of which these fishes are reported to be capable (Tower, 1908; Hubbs, 1920; Greene, 1924). It must be noted, however, that a certain number of the individuals of each of the species which usually produce sound on capture did not do so. The sounds produced on capture by vocal fishes, i.e., those possessing organs apparently specially adapted to the end of sound-production, did not seem to the writer to be produced involuntarily during the convulsive undirected contractions of all the muscles, as has been the impression of some observers (Sørensen, 1894). The fishes, in the intervals between flurries, often remained perfectly passive except for respiratory and sound-producing movements; fishes held firmly in the hand, making no struggle, usually became exceedingly vocal. It seems probable that such fishes are excited to produce sound in response to stimuli in their natural environment, which are comparable with those supplied by capture by man, and that this response, only in the special case of capture by man, becomes of no apparent adaptive value.

II

Several fishes were observed which do not seem to have been previously described as sound-producing.

1. The first of these, *Hyporamphus unifasciatus,* belongs to an order, the Synentognathi, of which no members have been previously recorded as vocal. Only one individual was observed. This animal produced a fairly

loud, cricket-like stridulatory sound whenever it was lifted from the water of its container, apparently by the scraping together of the rapidly vibrated pharyngeal patches. These patches are opposable, and are covered with small, hard, firmly set teeth. The vibration of the pharyngeal region was synchronous with the production of the sound.

2. The second of these undescribed forms, *Chaetodipterus faber*, is referred to the family Ephippidae, which is grouped with the Chaetodontidae in the suborder Squamipinnes. A chaetodont fish, *Holacanthus*, has been described on morphological grounds, as probably being capable of producing sound (Sørensen, 1894). The swim-bladder of individuals of *Chaetodipterus* of all sexes and sizes is provided with a thin red patch of muscle. A faint drumming noise is produced by fishes of this species, apparently through the agency of the swim-bladder and its intrinsic muscles. The thin-walled sack extends anteriorly almost to the base of the skull, ending in two short caeca. At its posterior end are two long, tubular caeca, which extend backward for some distance behind the visceral cavity. The swim-bladder, which lacks internal partitions, is loosely attached to the dorsal body wall. The deeply colored muscle is compact, flat and broad. It is very firmly attached to the ventral surface of the posterior half of the swim-bladder, and extends dorsally a short distance on the sides, and posteriorly a short distance out on the caeca. This patch of muscle can be observed to vibrate as the sound is produced, in a fish whose belly has been opened. If the swim-bladder is cut open, sound production ceases, although the muscle may continue to vibrate. In addition, a few specimens of *Chaetodipterus* produced a grating croak by the scraping together of the upper and lower patches of pharyngeal teeth.

3. *Syngnathus louisianae* was observed to produce a click quite similar to that of an elaterid beetle, by repeatedly snapping the head very sharply up.

4. *Lagodon rhomboides* produced a scraping clash by the sliding on each other of the upper and lower incisor teeth. The noise is made during a peculiar sneeze-like violent gasp interpolated occasionally among the usual respiratory movements of the fish kept out of water. It is not at all certain that the sounds produced by *Syngnathus* and *Lagodon* are anything but the incidental and involuntary accompaniments of actions directed to some function other than the vocal.

III

A number of species were observed, which are closely related to other, previously described sound-producers, and which agree with them in their method of vocalization.

1.—A number of carangids produce a stridulatory sound, a harsh, almost continuous croak, by scraping the upper and lower pharyngeal patches together. *Chloroscombrus chrysurus* and *Vomer setapinnis* may be added to this list. *Caranx hippos* makes a similar noise, as has been noted by Bridge (1904). Individuals of this species whose swim-bladders were deflated produced a fainter, dryer noise than did normal ones; it is therefore probable that in this carangid, the swim-bladder acts as a resonator for the stridulatory sound, in the same way that it does in the haemulids (Burkenroad, 1930).

2.—*Spheroides nephelus,* down to very small sizes, and *Chilomycterus spinosus,* in agreement with other gymnodonts (Tower, 1908), produce a stridulatory sound by the grating of the incisor teeth. The sound, in the individuals observed by the writer, was a high-pitched, nasal, whining scrape, produced during and after inflation.

3.—In common with other sciaenid species previously reported as sound-producers, the males of *Cynoscion arenarius, Stellifer lanceolatus* and *Larimus fasciatus* make a drumming noise by means of the air-bladder and its extrinsic muscles.

4.—The triglids *Prionotus tribulus* and *P. punctatus* (?) produce the grunt described for other species of the family, by the action of the intrinsic muscles of the swim-bladder.

5.—The batrachoidid *Porichthys porossissimus,* which, as noted above, was never observed by the writer to produce any sound, possesses a swim-bladder—intrinsic muscle apparatus similar to that described for the species *notatus* of the Pacific coast by Hubbs and Greene, and is no doubt similarly capable of sound production.

6.—*Galeichthys milberti* seems to agree with other siluroids (Dufossé, 1874; Sørensen, 1884, 1894) in its manner of sound production. The writer has observed this species to make a dull, low-pitched grunting noise, which seems to come from the air-bladder. Synchronously with the grunt, a small area of skin just back of the cranial bones on each side of the dorsal median line can be seen to be in very rapid vibration. On dissection, the swim-bladder is found to be heart-shaped, with the apex posterior. It is divided into an anterior and a posterior chamber by a transverse partition which is incomplete laterally, so that the anterior and posterior chambers communicate by two narrow openings. The posterior chamber is further subdivided by a median longitudinal, and by two pairs of incomplete sagittal, partitions. The slender pneumatic duct opens on the middle of the ventral wall of the anterior chamber. The dorsal wall of the swim-bladder is firmly attached, in the region of the juncture of the anterior transverse and the longitudinal partitions, to the vertebrae. The middle of the anterior face of the swim-bladder is attached to a rounded, peg-like vertebral projection. In close contact with the wall of the air-bladder, and loosely attached to it, across its convex anterior face, are two broad tendinous bands. The median ends of these are attached to the peg-like projection, while their outer ends are attached on each side to a springy, curved, slender, bone, the base of which is a flat, thin sheet of bone expanded along the dorsal surface of the body cavity. On the dorsal surface of these bony plates are inserted short muscle fibers, the axis of which is dorso-ventral, and which originate on the posterior cranial bones and the thick dorsal skin. It is the vibrations of these muscles which may be seen as the grunting noise is produced. This apparatus would appear to be an "elastic spring" mechanism essentially similar to that described for other siluroids by Sørensen (1884), and to it, probably, can the "grunt" of this fish be ascribed; however, this conclusion has not been checked experimentally.

In addition, *Galeichthys* makes a second noise, often concurrently with the first; a whining mew apparently produced by stridulation at the articulation of the bony ray of the pectoral fin. Sørensen, in connection with this phenomenon as observed by him in other siluroid species, states that the pectoral spine is fixed in position as a defensive weapon by the friction of an arched crest of its base against the adjacent "scouring faces" of the articulation. He believes that the "brake-like" action as the spine is moved to various defensive positions incidentally causes the sounds, which, however, he thinks may have a secondary function in frightening the assailant. Thilo (1896), with whom Sørensen (1896) disagrees, while not mentioning sound production, believes that the arched crest (designated by Sørensen as δ) is a portion of the diarthrosis of the joint. Thilo believes that the essential portion of the defensive locking mechanism is a prop-like projection (*Hemmfortsatz*) of the base of the spine; Sørensen, on the contrary, believes this peg, which he designates as β, to be a part of the diarthrosis. Dufossé (1874), has described the base of the pectoral spine of a siluroid both as a sound-producing and as a locking mechanism. The structure and action of the pectoral spine of *Galeichthys milberti* seems to be similar to that of other siluroids described by previous workers.

The writer wishes to point out the bearing of the defensive behavior of *Galeichthys milberti* on the controversy outlined above. In this species, the pectoral spine appears to have but a single defensive position: When the fish is roughly handled, the spine is swung forward as far as it will go, and there, standing at right angles to the body-axis, locked into place. At other times, captured catfish kept the pectoral spines in rapid to-and-fro motion, at which time the mewing sound was produced. A strong stimulus promptly caused the locking reaction; the spine was always fixed into place at the extreme forward end of its range of movement. The noise ceased with the cessation of the movement of the spine. The fixing of the spine in the defensive position seems to be accomplished in this way: If the spine is rotated counter-clockwise, that is, in such a way that its anterior edge is pulled ventrally, the peg (*Hemmfortsatz*, β) fits into a socket in the articulating bone in such a way that it acts as a prop. The spine can now be moved forward, but not backward. The crest (δ) is essential to this process in that it acts as a guide, and prevents the peg from lifting out of its socket when pressure is applied to the spine. If, on the other hand, the rotation of the spine is reversed, the propping peg is lifted from contact with its socket, and the spine can be moved backward, but can not be moved forward very freely. This is because of the friction caused by the scraping of the crest against the walls of its runway-like socket. In what may be called the neutral position of the spine, with regard to rotation, both crest and peg seem to act as part of the diarthrosis, and the spine can move freely in either direction. Thus, the only way in which the spine can be immovably erected is by bringing it forward to the end of its range, at the same time rotating it counter-clockwise. It is thus stopped from moving forward by striking the bony anterior wall of the articulation-cavity, and from moving backward by the propping action of the peg. This is, as the writer has pointed out, the defensive position of the spine of living fish. It therefore is probable that the friction of the crest against the articulating surface, when the spine is rotated clockwise, is not adapted to defensive fixation of the spine, but solely to sound production. Along the lateral contact surface of the crest are a number of fine vertical striations which are thus probably to be considered as stridulatory ridges. For this species, at least, it would seem that Sørensen's belief that sound is produced incidentally to the defensive fixation of the spine, does not hold. *Felichthys felis* appears to be similar to *Galeichthys* in its sound-producing apparatus.

IV

Two groups of fishes, while previously reported as sound-producing, appear to make sounds by means other than those previously ascribed to them.

1.—*Monacanthus hispidus*[1] was observed to produce a sharp, whining, scraping noise by sliding the biting edges of the lower incisors upward over the sloping posterior surfaces of the upper incisors during the rapidly repeated closing of the mouth. The posterior surfaces of the two median pairs of upper incisors of the inner series are striated with a number of fine transverse ridges, which are not present on any of the other tooth surfaces. These are certainly to be considered stridulatory ridges. The possession of such specially modified stridulatory surfaces on the teeth does not appear to have been previously noted in fishes.

2.—Certain sciaenids, which also produce sound by means of the swim-bladder and drumming muscles[2] produce a second kind of noise, often concurrently with the first, by pharyngeal patch stridulation. Individuals of *Micropogon undulatus,* of both sexes, make a loud drumming noise by means of the swim-bladder apparatus. Individuals of this species, in addition, produce a croak like that of a haemulid by the friction of the patches of pharyngeal teeth. The two parts of the upper patch are turned inward toward each other, and drawn backward over the lower patch. No differences were observed between the sexes in the production of this sound. The males of *Stellifer lanceolatus* make a faint drumming noise with the swim-bladder apparatus. Both sexes produce a croak similar to, but fainter than that of *Micropogon,* by pharyngeal stridulation. It seemed to the writer that the females of this species, voiceless as far as "drumming" was concerned, produced the "croak" more readily and persistently than did the males. A number of individuals of *Bairdiella chrysura,* especially the small ones, croaked with the phryngeal teeth. The male of this species drums with the swim-bladder and its muscles.

None of the other sciaenids observed ever croaked with the pharyngeals. These other species are *Cynoscion nebulosus, C. nothus* and *C. arenarius, Leiostomus xanthurus, Larimus fasciatus, Sciaenops ocellatus* and *Pogonias chromis.* In all of these the male drums with the swim-bladder apparatus. *Menticirrhus americanus* appears to be completely voiceless. The pharyngeal patches of *Micropogon, Stellifer* and

---

[1] Fishes of this genus have been previously described as producing sound by means of the swim-bladder, and by dorsal spine stridulation, neither of which methods was used by this species.
[2] For a description of this phenomenon, see Smith (1905) and Tower (1908).

Bairdiella are thickly set with short, hard, firmly set teeth, presenting an even surface which, when scraped with a hard object, makes a noisy, scratchy sound. The pharyngeal teeth of the species of *Cynoscion* are very similar, except that a few teeth are longer than the rest. The pharyngeal patches of *Larimus* present a softer surface which does not scrape very noisily. The pharyngeal patches, or parts of them, of *Leiostomus* and *Pogonias* are set with flattened, pavement-like teeth. Those of *Sciaenops* and *Menticirrhus* are sparsely set with long, strong teeth. The upper and lower patches of all of these fishes are opposable. The pharyngeal teeth of all of the sciaenids which do not use them for sound production, except, perhaps, the last two mentioned, certainly appear as if they might be so used; in fact the sound made by *Pogonias* was formerly attributed to the friction of its pharyngeal teeth. Nevertheless, these fishes do not make any movement of their pharyngeal patches, under conditions which cause the first three stridulating species listed above to move the patches over each other. The writer, therefore, is inclined to believe that the movement of the pharyngeal teeth of *Micropogon, Stellifer,* and *Bairdiella* represents a modification of the more usual scaenid behavior, the end of which is the production of sound.

Sørensen (1894) makes the statement that it is probable that nearly all fishes are able to make a noise by "gnashing their teeth," and seems not to attach any significance to this as a method of sound production. The writer wishes to point out that tooth-stridulation is no more universal among the fishes he has observed than is vocalization by any other means. While the teeth of many fishes which employ them as sound-producing organs do not seem to be specially modified to this end, they are so modified in some forms, as *Monacanthus*. Also, the modification of behavior attendant on sound-production by "gnashing of the teeth" must certainly, as is pointed out by the differences in this respect among the sciaenids, be regarded as being as significant as is a morphological adaptation.

The writer wishes to call attention to the fact that one sciaenid of those observed by him, *Pogonias,* is exceptional in that it has drumming muscles which are completely attached to, and confined to, the swim-bladder. Tower (1908) seems to have overlooked this fact, since he states that the drumming muscles of sound-producing sciaenids are only superficially attached to the swim-bladder. Tower concludes that this type of apparatus is adapted to producing a rapidly repeated series of grunts. He contrasts with the sciaenid apparatus that of the sea-robin and the toad-fish, with the swim-bladder of which the sound-producing muscles are intrinsically connected, and concludes that this latter type of apparatus is adapted to producing single grunts, not rapidly repeated. Since the muscle of *Pogonias* is intrinsic to the swim-bladder, while the fish nevertheless produces a rapidly repeated series of grunts, the writer believes that no general validity may be accredited to Tower's conclusions.

One Louisiana fish, *Orthopristis chrysopterus,* has been mentioned by the writer (1930) in a previous paper in Copeia. This haemulid makes a croaking noise by the grating together of its pharyngeal teeth; the swim-bladder probably acts as a resonator, as in other haemulids.

If the twenty-five sound-producing fishes discussed in this paper are arranged in taxonomic order, it can be seen that the manner of sound production shows no very clear relation to taxonomic grouping, although the list is perhaps suggestive. Each family has its own typical method of vocalization, which may accord with that of related families.

1. *Nematognathi.* Two vocal species. Swim-bladder—elastic spring apparatus, and pectoral spine stridulation.
2. *Synentognathi.* One species. Pharyngeal stridulation.
3. *Acanthopteri.*
   A. *Scombroidei.*
      a. *Carangidae.* Three species. Pharyngeal stridulation.
   B. *Percoidea.*
      a. *Haemulidae.* One species. Pharyngeal stridulation.
      b. *Sciaenidae.* Ten species. Swim-bladder—ex- or intrinsic muscle apparatus, and pharyngeal stridulation.
   C. *Squamipinnes.*
      a. *Ephippidae.* One species. Swim-bladder—intrinsic muscle apparatus, and pharyngeal stridulation.
   D. *Sclerodermi.*
      a. *Monacanthidae.* One species. Incisor tooth stridulation.
   E. *Gymnodontes.*
      a. *Tetraodontidae.* One species. Incisor tooth stridulation.
      b. *Diodontidae.* One species. Incisor tooth stridulation.
   F. *Craniomi.*
      a. *Triglidae.* Two species. Swim-bladder—intrinsic muscle apparatus.
   G. *Haplodoci.*
      a. *Batrachoididae.* Two species. Swim-bladder—intrinsic muscle apparatus.

In only one group, the Sciaenidae, is there any sexual differentiation of sound-production. In three groups—the Sciaenidae, the Ephippidae, and the Siluridae—we find the same fish producing sound by more than one means.

Summary

1. Capture by man of sound-producing fishes is probably to be regarded as a stimulus to sound production similar in effect to some stimulus provided by the normal environment, to which the response of the fish is adaptive. Sound production in many fishes probably has a defensive function.

2. Sound production in *Hyporamphus unifasciatus* and *Chaetodipterus faber* is described.

3. Sound production is described in a number of fishes closely related to other, previously described, sound-producers. It is pointed out that the sound produced by the movement of the pectoral spine of *Galeichthys milberti* is not to be regarded as merely incidental to the defensive fixation of the spine.

4. Methods of sound-production not previously described for these forms are noted in *Monacanthus hispidus* and certain sciaenids.

5. The importance of the teeth as sound-producing organs of fishes is emphasized.

6. It is pointed out that the drumming muscles of one sciaenid, *Pogonias*, are intrinsic to the swim-bladder.

7. Twenty-five sound-producing fishes of Louisiana are classified. The list suggests that sound production has arisen independently in the different groups.

ACKNOWLEDGMENTS

Mr. Isaac Ginsburg, of the U. S. Bureau of Fisheries, to whom I am also indebted for the means of collecting a quantity of the material, has kindly identified a number of fishes.

Mr. C. M. Breder, Jr., of the New York Aquarium, who suggested a study of the pharyngeal teeth of sciaenids as sound-producing organs, has been kind enough to read and criticize the manuscript, as has also Dr. J. E. Lynch of Tulane University and Dr. Carl L. Hubbs of Michigan. Some of the work was done while the writer was a guest of Dr. E. H. Behre, directress of the Louisiana State University Summer Laboratory, to whom he wishes to express his appreciation. The writer wishes to thank Dr. E. S. Hathaway, of Tulane University, and the personnel of the Department of Conservation of Louisiana for the facilities extended by them.

LITERATURE CITED OR CONSULTED

AGASSIZ, J. L.
    1850 Manner of producing sounds in catfish and drumfish. *Proc. Am. Acad. Arts and Sci.*, 2: 238.

BRIDGE, T. W.
    1904 *Cambridge Nat. Hist.*, 7: 355.

BURKENROAD, M. D.
    1930 Sound Production in the Haemulidae. COPEIA, 1930 (1): 17.

DUFOSSÉ, M.
    1874 Recherches sur les sons expresifs que font entendre les Poissons d'Europe. *Ann. Sci. Nat., Zool.* (5), 9 (5) and 20 (3).

GILL, T. N.
    1881 Record of recent scientific progress in zoology—Fishes, etc. *Ann. Rept. Smiths. Inst.*, 1881: 366.

GREENE, C. W.
    1924 Sound production in Porichthys notatus. *Am. Jour. Physiol.*, 70.

HUBBS, CARL L.
    1920 The bionomics of *Porichthys notatus* Girard. *Am. Nat.*, 54: 380-384.

SMITH, H. M.
    1905 Drumming of sciaenids. *Science*, Sept. 22: 376-378.

SØRENSEN, W.
    1884 Om Lydorganer hos Fiske. Copenhagen. *Dissertation.*
    1894 Are the Extrinsic Muscles of the air bladder in some Siluroidae and the "elastic spring" apparatus of others subordinate to the voluntary production of sounds, etc. *Jour. Anat. and Physiol.*, 29: 110-139, 205-229, 518-552.
    1896 Some remarks on Dr. Thilo's memoir on "Die Umbildungen an den Gliedmassen der Fische." *Morph. Jahrb.*, 25: 170-189.

THILO, O.
  1896 Die Umbildung an den Gliedmassen der Fische. *Morph. Jahrb.*, 25.
  1898 Ergänzungen zu meiner Abhandlung "Die Umbildungen an den Gliedmassen der Fische." *Morph. Jahrb.*, 25 : 81-90.

TOWER, R. W.
  1908 Production of Sound in drumfish, toadfish, and searobins. *Ann N. Y. Acad. Sci.*. 18 : 149-180.

WEISS, O.
  1908 Erzeugung von Geräuschen und Tonen. In Winterstein's *Handbuch Vergl. Physiol.*, 3 (1) : 249-318.

DIVISION OF RESEARCH, DEPARTMENT OF CONSERVATION OF LOUISIANA, NEW ORLEANS, LOUISIANA.

Part III

THE POST-WAR PERIOD

Editor's Comments
on Papers 8 Through 11

8 DOBRIN
Measurements of Underwater Noise Produced by Marine Life

9 KNUDSEN, ALFORD, and EMLING
Underwater Ambient Noise

10 FISH, KELSEY, and MOWBRAY
Studies on the Production of Underwater Sound by North Atlantic Coastal Fishes

11 MARSHALL
The Biology of Sound-producing Fishes

The creation of this interdisciplinary area of science was made possible by the otherwise traumatic stimulus of World War II. Prior to the war, interest in underwater sounds of marine organisms was the province of a few keen-eared fishermen and even fewer ivory-tower biologists. With the Battle of the Atlantic came a sort of Merrimac-Monitor combat on a global scale, and the unique acoustic characteristics of water suddenly became crucial to both sides. From these roots sprung ASDIC and SONAR. Simultaneously came the realization that all was not quiet and the sea was far from silent. Choruses of snapping shrimp and buzzing dolphins cluttered the high audio frequencies, and spawning croakers, drumfish, catfish, and many others created a din in the estuaries. Toadfish, seeking a shelter, were found to sidle up against acoustic mines, and commit involuntary suicide upon uttering their first "boat-whistle." Understandably, much of this information was classified and not released for general public or even scientific information until after the war. One of the first such reports was that of M. B. Dobrin. Almost simultaneously, several other articles appeared in which the contribution of marine life to ambient sea noise was described, evaluated, and, in some instances, identified to species. An important example of such reports, in addition to that of Dobrin, is included in this collection primarily because of its emphasis on the biological sound sources. This is the paper by Knudsen, Alford, and Emling. V. O. Knudsen was a well-known oceanographer who today is best known for the graphs he published on the relationship of ambient sea noise to surface sea state, and these "Knudsen curves" are included in the reprinted article. In later publications, however, these graphs have been considerably augmented and extended to cover a greater variety of sonic sources.[1]

Following the above revelations, the phase of collecting and identifying biological sound sources began. A notable pioneer in this effort was Marie Poland Fish, who tirelessly "auditioned" every species of marine organism she could possibly collect, and whose benchmark paper is reprinted here.[2] From these studies, it became clear that a large number of fish species were at least capable of producing sounds. Unfortunately, such catalogs did not turn out to be as useful as had been hoped. For one thing, the "signature" or species-characteristic sounds were not unique enough in spectral or other ways. For another, many of the sounds were evoked from animals in abnormal captive conditions, often by prodding or electric shock. Furthermore, the anomalous acoustic conditions in the tanks created artifacts in the acoustic analyses.

On a morphological basis alone, it became evident that many species of fishes, particularly among the benthic and abyssal forms, should be capable of sound production. The British ichthyologist N. B. Marshall predicted that perhaps a large percentage of the twenty to thirty thousand known species of teleosts may be capable of sound emission, and his major paper on this is reprinted here.

[1]Wenz, G. M. (1964). Curious noises and the sonic environment in the ocean. In *Marine Bio-Acoustics*, W. N. Tavolga, ed., pp. 101–119, Pergamon Press: Oxford.
[2]A more detailed version of this paper was: Fish, M. P. (1954). The character and significance of sound production among fishes of the western North Atlantic. *Bull. Bingham Oceanographic Collection*, **14**(3), 1–109. A more recent compilation, accompanied by tape recordings, is: Fish, M. P., and Mowbray, W. H. (1970) *Sounds of Western North Atlantic Fishes*. Johns Hopkins Press: Baltimore.

Copyright © 1947 by the American Association for the Advancement of Science
Reprinted from Science, 105, 19–23 (1947)

MEASUREMENTS OF UNDERWATER NOISE PRODUCED BY MARINE LIFE[1]

M. B. Dobrin
Naval Ordnance Laboratory, Washington, D. C.

That certain fish species make noise under water has been common knowledge among fishermen since ancient times. For at least a century, observations on this phenomenon have been published by naturalists and zoologists. Until recently, however, all observations upon noise produced in this way were incidental and qualitative. No physical measurements of its frequency distribution or intensity are reported anywhere in the biological literature. It was not until the recent war that a need was felt for exact quantitative data on biological water noise. The introduction during the war of underwater acoustic equipment, such as listening devices, submarine detecting gear, acoustic mine mechanisms, and homing torpedoes, raised questions as to the interference that might be expected from natural background noises in the water. For this reason, information was required on the nature and magnitude of the water noise to be expected at various localities and under various conditions. Since no data of the type needed were available in the general literature, it was necessary for war research agencies working in underwater sound to make their own measurements. A large body of data was accumulated in this way which should considerably augment previously available knowledge of natural water noise and its production.

Although waves, wind, and tidal currents give rise to a measurable amount of water noise, this is seldom of a higher order of magnitude than 1 dyne/cm.2 in an octave band and is usually much lower. Biological sources, on the other hand, can be responsible for sustained noises with an octave pressure of several hundred dynes per square centimeter.

MEASUREMENT OF NATURAL WATER NOISE

The Naval Ordnance Laboratory carried on background measurements at several field locations where biological noises were particularly intense, and it has recorded the highest natural water-noise levels that have been observed anywhere. At the same time, a systematic effort was made to identify the species giving rise to the different kinds of fish noises recorded. This involved elaborate tests on segregated fish species, both in aquaria and in experimental ponds.

---

[1] The field measurements and data reductions upon which this report is based were carried on by the following staff members of the Naval Ordnance Laboratory: L. G. Swart, G. E. Brown, L. C. Bell, D. L. Bobroff, R. F. Grunwald, G. R. Irish, W. E. Loomis, and the author. The consulting biologist for much of the work was Cdr. Charles J. Fish, USNR, of the Mine Warfare Operational Research Group. Walter H. Chute, director of the John G. Shedd Aquarium, gave the Laboratory substantial cooperation in its measurements there, and H. F. Prytherch, director of the U. S. Fishery Biological Laboratory, Beaufort, North Carolina, generously granted use of facilities for the field measurements and segregation tests reported from that area.

Procedure. Underwater background noises were picked up at all locations by underwater hydrophones and recorded on discs which were later played through an octave analyzer into a series of Esterline-Angus tape recorders, each octave over the range 50–3,200 c.p.s. being recorded on a separate tape. Calibration was by 1-kc. signal corresponding to a known sound pressure injected into the hydrophone circuit. Spectra were calculated and plotted from the octave tapes. At the same time, the disc recordings were available for listening and identification.

The hydrophones were of the Brush C-21 rubber-covered crystal type or the RCA 2A condenser type with a Monel diaphragm. Each kind had preamplifier inside the case. Discs were made with RCA cutting heads operated through standard Brush or RCA recording amplifiers. Filters in the analyzing system were of the ERPI octave type. The recording system is designed to give mean rectified sound pressure instead of root mean square or peak pressure. For the case of impulsive noises, such as the drumming and grinding produced by fish, the peak noise would be some 40 per cent higher than the mean rectified signal.

DIRECT FIELD DATA

Disc recordings of underwater background noise in open water have been made by the Naval Ordnance Laboratory at various points along the East Coast of the United States from Florida to Cape May. During the course of these surveys, biological noise of many kinds has been heard and recorded. The measurements have been under a wide variety of circumstances as regards oceanographic conditions, season, time of day, etc.

Frequency Spectra. The large variation that can be expected in the background at a single location is illustrated in Fig. 1 by spectra obtained at typical locations along the East Coast. The ordinates represent the sound pressure observed within an octave band. These are plotted at the mid-band frequency of the octaves extending from 50 to 3,200 c.p.s., the points being connected by smooth curves. The two highest curves were recorded at Wolf Trap, in the middle of Chesapeake Bay, where the water depth is 40 feet. The spectra were both recorded in early July, but in successive years. The frequency characteristics are almost identical, although the 1942 level is about 60 per cent higher than the 1943. The frequency at which the peak noise level occurs is here observed at about 350 c.p.s. This is more than an octave lower than the peak recorded at Cape Henry in 1942, about five weeks earlier, which is the third highest curve in Fig. 1. The difference is probably attributable to the time interval between the two measurements rather than to their geographical separation. Since the source of noise in all cases was almost certainly croakers or other members of the Sciaenidae family, the fish would have grown longer during the intervening period and thus would produce resonant vibrations having lower pitch.

Of the other two curves, one was recorded in June 1943 at Fort Macon, North Carolina. The peak occurs at about 600 c.p.s., and the source is very likely to be croakers. The other was recorded from a boat in the open Atlantic, approximately 20 miles off shore, south of Cape Lookout, North Carolina. The noise here, although of the same character as that of croakers, is of much higher pitch than any Sciaenidae noise recorded elsewhere, the peak occurring at about 2,400 c.p.s. A search of the literature revealed that the bastard trout (*Cynoscion nothus*) is common off shore in this area but has not been observed near shore. This species, attaining a minimum length of 3 inches, is smaller than other drumfish along the United States east coast, and would thus be expected to produce a noise of higher pitch.

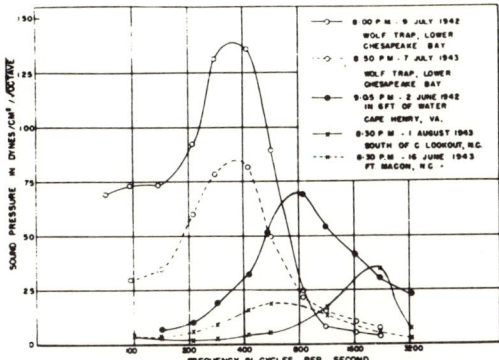

FIG. 1. Spectra of water noise caused by marine life at various points along the East Coast of the United States.

Seasonal variation. Distinct seasonal effects were observed in the water noise recorded in Chesapeake Bay. This variation was associated with seasonal movement of the croakers in

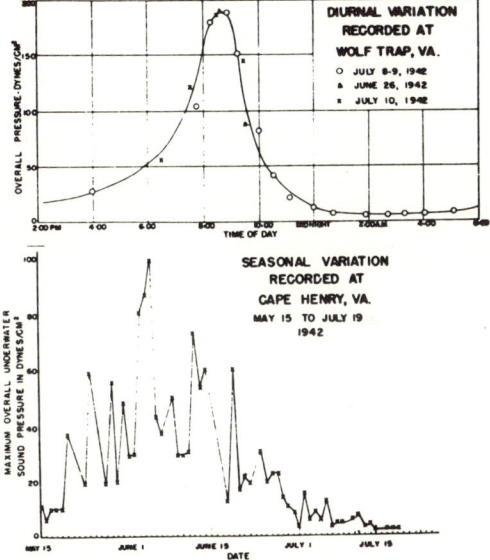

FIG. 2. Diurnal variation of water noise produced by marine life at Wolf Trap, Virginia, and seasonal variation of maximum daily water noise at Cape Henry, Virginia.

the Bay. The daily maximum nocturnal noise levels at Cape Henry are plotted in Fig. 2 over the period from the middle of

May to the middle of July. The peak of activity was during the early part of June, and by July 15 the activity was virtually over.

Diurnal effects. Diurnal variation in background noise level at Wolf Trap is also plotted in Fig. 2. The plot represents the composite of observations taken on three different days during the summer of 1942. Here the peak was reached at about 8:30 P.M., the curve being almost symmetrical around that time. At Beaufort and Fort Macon, North Carolina, the following year, the peak during June was reached at approximately 12:30 A.M. each night. The reason for this well-defined recurrent nocturnal appearance of the noise is not known definitely, although it is suspected that the peak coincides with feeding activity on the bottom.

Studies of Segregated Fish

Noises made by known fish species have been recorded and measured both at the Shedd Aquarium in Chicago and at the U. S. Fishery Biological Laboratory, Beaufort, North Carolina. The purpose of the tests in the aquarium was to determine which of the species in the collection was sonic, the conditions under which the sonic species could be induced to make noise, and the general nature of their noise. Although the noises were recorded on discs, it was not possible to measure absolute sound pressures or to make accurate frequency analyses. Most of the fish that produced noise in the aquarium belonged to families previously reported to be sonic in the biological literature. There were several species, however, for which no record of sound production could be found in the literature which turned out to be prolific noisemakers. Several members of the Pomacentridae family, such as the *Hypsypops rubicundus*, or garibaldi, of Southern California and the *Eupomacentrus fuscus*, or coral-reef fish, were among these, as were certain species of catfish.

At Beaufort, conditions of segregation were much more favorable for accurate quantitative measurement of fish noise. Species were put into separate enclosures set off by chicken-wire fence in a 75 x 85 foot experimental salt-water pond about 3 feet in depth, with a mud bottom. Hydrophones were planted in each enclosure. Fish were caught by commercial fishermen, who transported them to the experimental pond in a live car. The number of specimens investigated ranged from several hundred in the case of croakers to three in the case of sea robins. Results of the tests will be discussed briefly here. Fig. 3 shows the frequency characteristics of the various sonic species.

Croaker (Micropogon undulatus): The most common drum-fish in the estuarine waters of the U. S. East Coast is the croaker, and this is believed to be responsible for the greatest part of the noise observed in the open-water tests previously discussed. Its noise consists of rapid drum rolls resembling the sound of an electric drill being driven into asphalt. This sound is made by the action of special "drumming muscles" against the fish's air bladder, which is set into resonant vibration at a frequency that should be inversely proportional to its length. In captivity, inside a wire-net enclosure, croakers made noise spontaneously but with noticeably less vigor and intensity than when observed under entirely natural conditions. The noise came in bursts consisting usually of only two or three drum beats of lowered pitch instead of the rapid, vibrant trill heard in open water. Noises were audible under water as much as 25 feet from the source, as was evinced by moving the hydrophone away from the croaker enclosure until the characteristic noise could no longer be distinguished.

Toadfish (Opsanus tau): Most remarkable of all fish studied in the current survey was the toadfish. A sluggish, ill-tempered, nest-building bottom dweller, this genus produces a much more intense noise than any other form of marine life investigated. The sound is an intermittent, low-pitched musical blast of about $\frac{1}{2}$-second duration, somewhat similar to a boat whistle, and is concentrated at the low-frequency end of the spectrum, as shown in the typical curve of Fig. 3, which represents the noise emitted by a toadfish within a few inches of the hydrophone.

Unfortunately, the identification of this noise must be based on circumstantial rather than direct evidence. A direct identification was never possible, because no toadfish specimen would make noise in captivity. Hence, instead of following the usual procedure of capturing a specimen and inducing it to make noise, one first located the suspected noise in open water and tracked down its source. This was not difficult, because the characteristic musical sound was heard almost constantly, at least in the distance, wherever a hydrophone was lowered into the water. On moving the hydrophone toward the source of noise until maximum intensity was recorded, it was found that the source remained at one spot on the bottom for days at a time. This fact definitely suggested the toadfish as source, since it is the only sluggish species in this area.

To obtain direct proof, diving was tried but failed because of poor visibility. A baited crab trap was then lowered into the water at the point where the hydrophone indicated maximum noise. Shortly thereafter the blasts stopped for the first time in over a week; and when the trap was pulled up, it contained a toadfish.

Hogfish (Orthopristis chrysopterus): A close relative of the grunt, a common tropical offshore fish, the hogfish gets its name from the characteristic grunting noise it makes when taken from the water. This noise is produced by gnashing of the pharyngeal teeth and has a harsh, rasping quality. Under water the noise is made spontaneously in bursts of four or five rasps following each other in rapid succession.

The hogfish is common in the waters surrounding Beaufort, but its noise was heard only occasionally from the Fishery Laboratory Pier on Piver's Island or at Fort Macon. It was concluded on the basis both of tests at Beaufort and of previous aquarium studies that those fish making noise by gnashing of the pharyngeal teeth are not as important sources of underwater noise as those producing sound by action of the air bladder, viz., croakers and toadfish.

Spot (Leiostomus xanthurus): The spot, as a member of the Sciaenidae family, is closely related to the croaker, but its noise is of quite a different timbre. The sound might best be described as a series of raucous honks, having a volume level and frequency distribution typified in Fig. 3. Moving the hydrophone away from the enclosure containing the spot indicated that the noise was audible no more than 5 feet from its source, and hence is initially not as intense as that of the croaker.

Spot were heard occasionally around Piver's Island and in offshore waters around Cape Lookout, but the magnitude and character of the noise are such that it is not believed to be a significant contribution to the over-all background level in open water. Moreover, spot are reported to be solitary rather than gregarious, and hence sources of this noise would probably be dispersed.

Sea robin (Prionotus carolinus): The segregation tests on sea robins were not very satisfactory because specimens were seldom available, and the only recordings of their sounds were made in aquarium tanks rather than in the experimental enclosures. The noise of this species is so characteristic, however, that the aquarium tests made its identification in natural waters quite simple.

considerable difference between the noises, however, both in character and intensity, casts doubt upon the correctness of this observation.

Sea catfish (Felichthys felis): During the course of the segregation tests it was found rather unexpectedly that the common sea catfish is a significant noisemaker. It makes a rhythmic drumming noise like the beating of a tom-tom, differing from the drumming of the croaker in that it comes not in rolls but in rapid, evenly-spaced beats. This noise was heard under

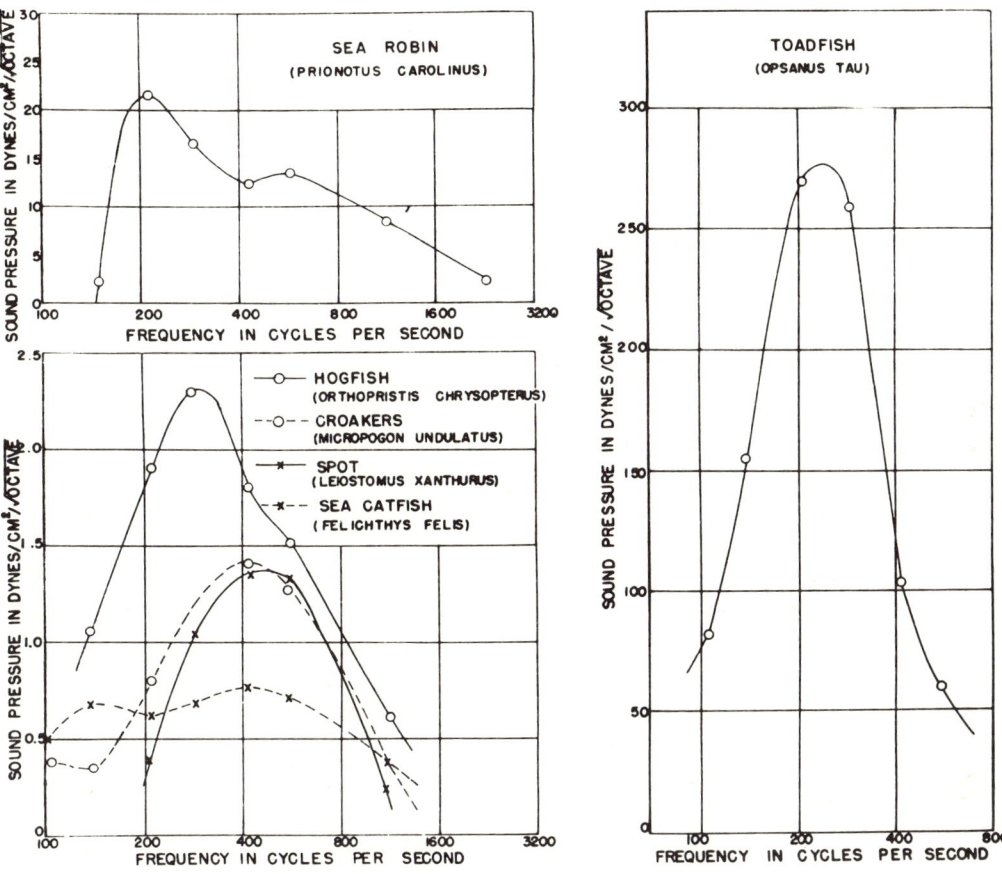

FIG. 3. Spectra of noise produced by segregated groups of sonic fish.

The sound of the sea robin might best be described as a modulated, rhythmic squawk, squeal, or cackle, resembling noises ordinarily associated with a barnyard. The curve in Fig. 3 gives the frequency characteristics and level for the case of a single specimen in a highly reflecting concrete pool.

Sea robin noises were frequently heard in the course of the open-water listening tests both at Piver's Island and Fort Macon, particularly at the latter location. They were also discernible during offshore listening tests past Cape Lookout. It was never possible in field measurements to determine sound levels due to sea robins alone, since they were always heard simultaneously with croakers and other sonic species.

The mechanism of sound production in the sea robin is, according to Tower (*1*), the same as that of the toadfish. The

natural conditions at Fort Macon and in Bogue Sound, near Morehead City. For this study measurements were made on four catfish in an experimental enclosure.

The motivation for noise production by marine life lies outside the scope of this paper. Further light on it, however, should make possible a more substantial knowledge of behavior patterns in sonic species.

The disc recordings made in connection with water listening described above as well as in the tests on segregated fish are available in the Naval Ordnance Laboratory's files. Dubbings can be made available to any biological laboratory which can put them to use.

Reference

1. Tower, R. W. *Ann. N. Y. Acad. Sci.*, 1908, **18**, 149–180.

9

Copyright © 1948 by The Sears Foundation for Marine Research, Yale University

Reprinted from *J. Mar. Res.*, 7(3), 410–429 (1948)

UNDERWATER AMBIENT NOISE[1]

By

VERN O. KNUDSEN
University of California, Los Angeles, California

AND

R. S. ALFORD AND J. W. EMLING
Bell Telephone Laboratories, New York

INTRODUCTION

Underwater sound continues to be a problem of high priority in naval warfare. Since atomic bombing may render obsolete most types of surface ships, the relative if not absolute importance of the submarine certainly has increased. Thus, the U. S. Navy logistics chief recently told a congressional committee, "By far the most important and difficult problem which confronts the navy in so far as ship characteristics and fleet operating technique are concerned is the problem of undersea warfare."

Quite apart from its military applications, the purely scientific investigation of sound in the sea presents a fascinating "front" between the known and the unknown which invites further exploration. The scientific advances along this front that were incidental to the researches in subsurface warfare of World War II are of considerable significance, and it is almost entirely from these researches that we have gained our present knowledge of underwater acoustics.

The present report deals with a small but significant part of underwater acoustics; it surveys and compiles the principal available data (to March 15, 1944) on underwater ambient noise. It is condensed from a much longer survey report[2] which contains 76 pages of text and 146 charts and figures, prepared by the authors as a comprehensive

---

[1] Based on a survey prepared by the authors under the direction of Section 6.1 of the National Defense Research Committee, Survey of Underwater Sound: Report No. 3—Ambient Noise, OSRD Report No. 4333, Sec. No. 6.1-NDRC-1848, September 26, 1944. The complete survey report has been declassified; photostatic or microfilm copies are procurable at the Publication Board, Scientific and Industrial Reports, Office of Technical Services, U. S. Department of Commerce, Washington, D. C. (The Pub. Board number of this report is 31021).

[2] See footnote 1.

reference work for naval and civilian groups concerned with underwater sound problems during World War II. The major part of the material of the original survey report was based on reports prepared by the following organizations: U. S. Naval Ordnance Laboratory; Massachusetts Institute of Technology—Bureau of Ships, U. S. Navy; Columbia University, Division of War Research; and University of California, Division of War Research. A complete list of reports consulted, together with a short summary of technical information applying to each, is given in the original survey.

TABLE I. GENERAL LOCATIONS AT WHICH AMBIENT NOISE WAS MEASURED

Boston Harbor and off Massachusetts
Long Island Sound
Off Block Island
New York Harbor and approaches
Lower Chesapeake Bay and off Cape Henry
Off Beaufort, North Carolina
Off Florida and the Bahamas
Puget Sound
San Francisco Harbor and approaches
Off San Diego
Off Oahu and Midway Islands
Off Portsmouth, England
Loch Goil, Scotland
Off Hebrides Islands

Table I lists the areas in which the measurements of underwater noise were made. The data reported are confined largely to the frequency range of 100 c.p.s. to 25 kc. They are expressed in terms of pressure levels in db relative to 0.0002 dynes/cm^2. Often the results are given in terms of the "pressure level spectra," *i. e.*, the pressure level in a band one cycle wide, as a function of frequency. These values are derived, since the measurements are usually made with filters having a band width of about one octave, although in some measurements the band width is either less or greater than one octave.

QUALITATIVE NATURE OF AMBIENT NOISE

The term *underwater noise* is used to describe unwanted underwater sounds which tend to impair the operation of acoustically operated devices. *Ambient noise*, sometimes referred to as background noise, is the sound normally prevailing in water, usually from a multiplicity of sources such as water motion, marine life, and unwanted ship sounds. Usually it is not possible to specify quantitatively the contribution of sound from each of the sources present, but frequently a particular type of source is known to be preponderant. In such cases it is convenient to refer to the ambient noise more specifically as water noise, noise from marine life, ship noise, etc.

There are three main sources of ambient noise:

(1) *Water Motion*—Usually the prevailing source of noise in open and deep water. The magnitude is largely determined by the motion of the sea surface, particularly the number and prominence of breaking waves and whitecaps. Therefore, noise from this source is related to weather conditions.

(2) *Marine Life*—Produced by a large number of vertebrates and crustacea. Usually found in shallow tropical or semitropical waters. Diurnal and seasonal variations are common.

(3) *Ship and Man-made Sources*—Found in busy harbors and connecting waters. The magnitude depends on local conditions and may vary greatly from time to time.

Water noise is caused by a large number of widely distributed sources at the water surface; for this type of noise the field is believed to be essentially isotropic. An isotropic distribution is also to be expected in the center of large areas containing marine life. At a distance from such areas, or near localized concentrations of marine life, noise levels will vary with orientation and with the distance from the localized source. A similar situation will exist where the noise is produced primarily by a few near-by ships.

Variability is an outstanding characteristic of ambient noise. The magnitude of the noise from an individual source usually varies from moment to moment, and many sources exhibit large diurnal and seasonal variations. Some sources, such as ships, change their geographical position with time. Ambient noise is usually produced by many sources, some near the point of measurement and others located at a distance. Consequently, variations in the transmission loss in the medium are also a cause of variability. For these reasons ambient noise cannot be specified as a constant quantity but must be described in statistical terms. That is, an estimate can be made of the most probable (or average) amount of noise that is to be expected under given circumstances. In addition, it is frequently possible to estimate the degree of variability in terms of a frequency distribution *vs.* time, or in terms of the standard deviation when the distribution is known to follow the normal law. A knowledge of the average noise and the magnitude of the variation from the average is essential for the correct application of quantitative data on ambient noise.

NOISE FROM WATER MOTION

The term *water noise* is used to designate the underwater noise (at a sufficient distance from the shore to avoid the sound of waves breaking near the shore) produced by the agitation of the sea surface. This agitation is usually produced by wind. Water noise is the principal type of noise encountered in open and deep sea water. In practically all parts of the ocean, water noise is an important component of the noise, at least in some portion of the spectrum.

Noise is also produced by breaking surf, by the impact of rain and hail on the water surface, and occasionally by a movement of the bottom material, such as rock, gravel, shells, etc. Underwater springs, volcanoes, and gas vents also have been suggested as possible sources of noise. Surface ice and icebergs are subjected to stresses which often give rise to audible sounds. No satisfactory measurements of such noises have been made.

It is generally assumed that water noise is produced principally by breaking wave crests at the sea surface. It is believed that the unbroken undulations of the surface water do not contribute appreciably to underwater noise. The magnitude of water noise probably is related in a complex manner to such variables as height of waves, steepness of waves, and the number and magnitude of whitecaps present. Ideally, the estimation of water noise levels should be based on the combined effects of all these variables. Practically, this is not possible

TABLE II. METEOROLOGICAL SCALES

State of Sea			Beaufort Wind Force			
Scale No.	Description	Height of Waves, Crest to Trough (Ft)	Scale No.	Description	Velocity m.p.h.	Knots
0	Calm	0	0	Calm	<1	<1
1	Smooth	<1	1	Light Air	1–3	1–3
2	Slight	1–3	2	Light Breeze	4–7	4–6
3	Moderate	3–5	3	Gentle Breeze	8–12	7–10
4	Rough	5–8	4	Moderate Breeze	13–18	11–16
5	Very Rough	8–12	5	Fresh Breeze	19–24	17–21
6	High	12–20	6	Strong Breeze	25–31	22–27
7	Very High	20–40	7	Strong Wind	32–38	28–33
8	Precipitous	>40	8	Fresh Gale	39–46	34–40
			9	Strong Gale	47–54	41–47
			10	Whole Gale	55–63	48–55
			11	Storm	64–75	56–65
			12	Hurricane	>75	>65

with the data now available. Fortunately, however, it has been found that a reasonably good estimate of noise level due to water motion can be based on either wave height or wind velocity. A summary chart for making these estimates (see Fig. 4) will be presented after considering some selected but typical results of the noise from water motion.

Table II presents some meteorological scales which are in current use, and which are used in this report, for describing the state of the sea (in terms of wave height) and the Beaufort wind force (in terms of wind velocity in knots or m.p.h.).[3]

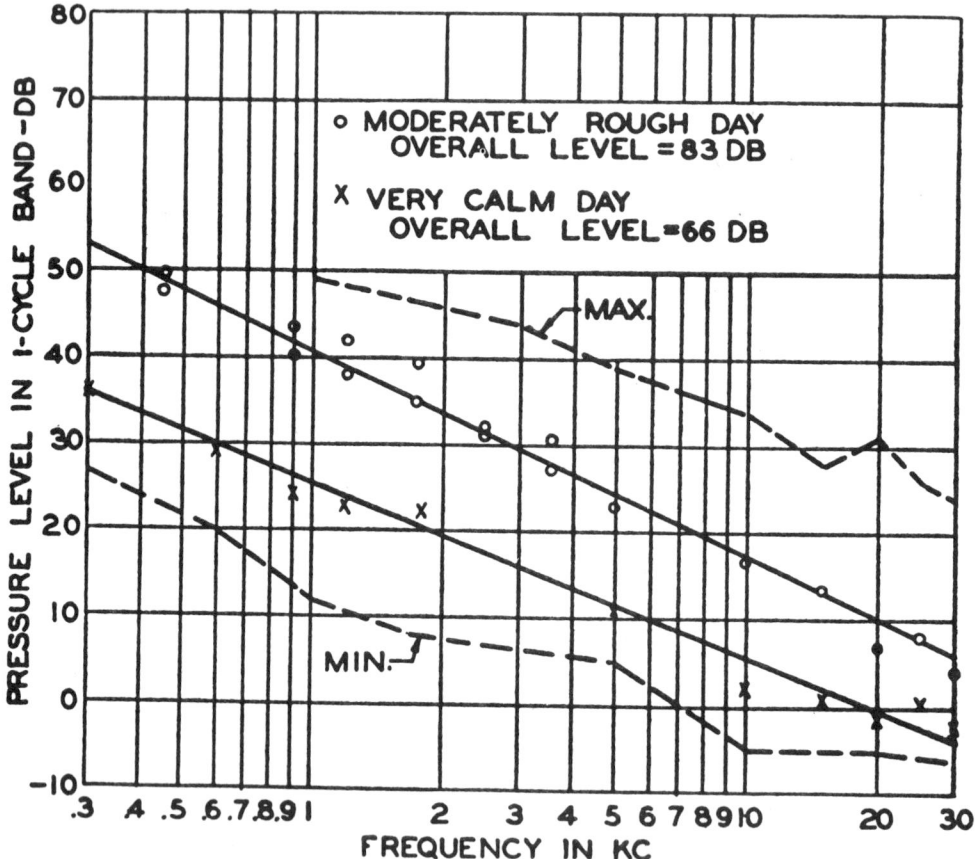

Figure 1. Pressure level spectra of ambient noise in Long Island Sound.

[3] In the original survey report, world charts prepared by the U. S. Weather Bureau are presented, one for each of the four seasons of the year, showing the distribution of average wind velocity, from which the average water noise can be estimated (with the use of Fig. 4). Thus, in one region of the North Atlantic, during December, January, and February, the average wind velocity is 24 knots, and, according to Fig. 4, the expected over-all (0.1-kc) noise level would be about 81 db.

Figs. 1–3 are illustrative of many results from measurements of noise, due mostly to water motion. Thus, Fig. 1 gives pressure level spectra measured in Long Island Sound on a calm day and on a moderately rough day. The over-all pressure levels (0.1–10 kc) corresponding to these spectra are 66 and 83 db, respectively, assuming a straight line extension of the curves to 100 c.p.s. On the same figure are also shown for comparison the maximum and minimum noise pressure levels measured from July to November 1942 in the waters adjacent to New York Harbor.

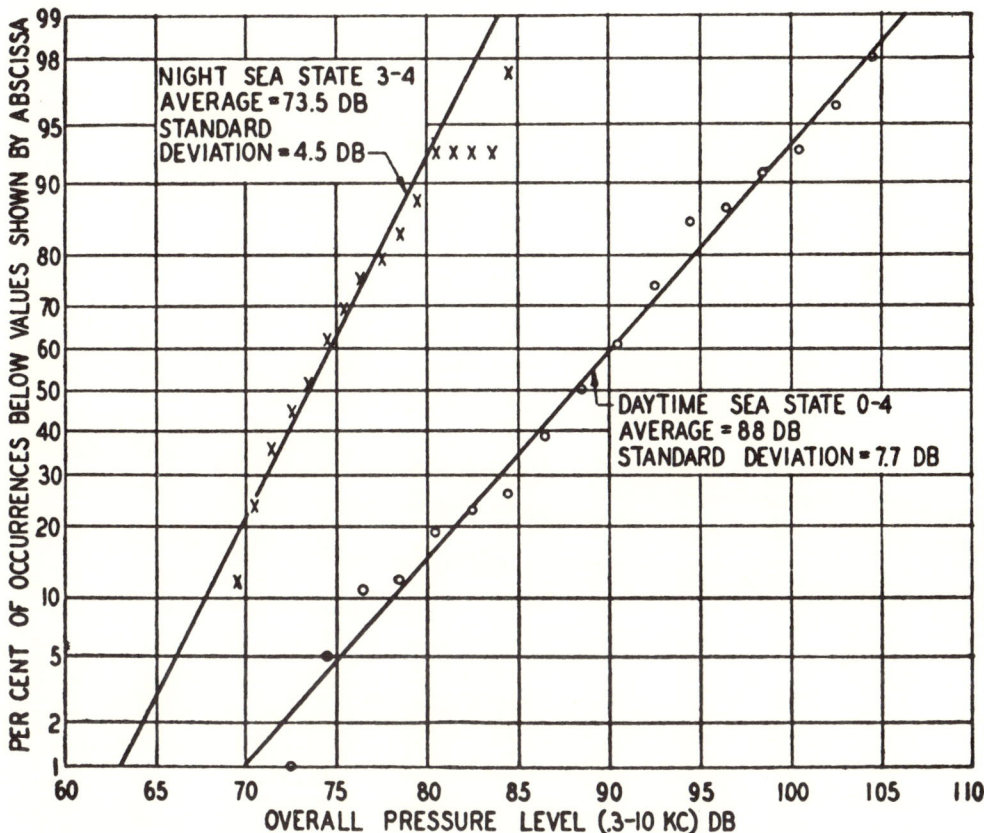

Figure 2. Ambient noise in approaches to New York Harbor; distributions of over-all pressure levels.

Fig. 2 is based on a noise survey of the area between Rockaway Beach and Sandy Hook conducted in March–April 1943. Pressure levels were measured throughout the frequency range of 150 c.p.s. to 25 kc, and over-all pressure levels were measured in a 0.3–10 kc band. Only the cumulative distribution of occurrence of over-all pressure levels for a series of daytime measurements (when ship sounds were predominant) and for a series of nighttime measurements (when ship sounds were almost completely absent) is shown in Fig. 2, plotted in the usual manner for determining the average value and the standard deviation. The curve for the nighttime measurements, when the sea state was estimated by the observers to be between 3 and 4, is believed to be free from man-made sounds, and probably therefore is representative of water noise. If the over-all pressure level measurements had been made with a band width of 0.1–10 kc instead of 0.3–10 kc, the pressure levels would have been about 2 db higher than those shown in Fig. 2. A comparison of the two curves of Fig. 2 reveals the marked difference between the daytime and nighttime underwater noise levels in the approaches to a busy harbor.

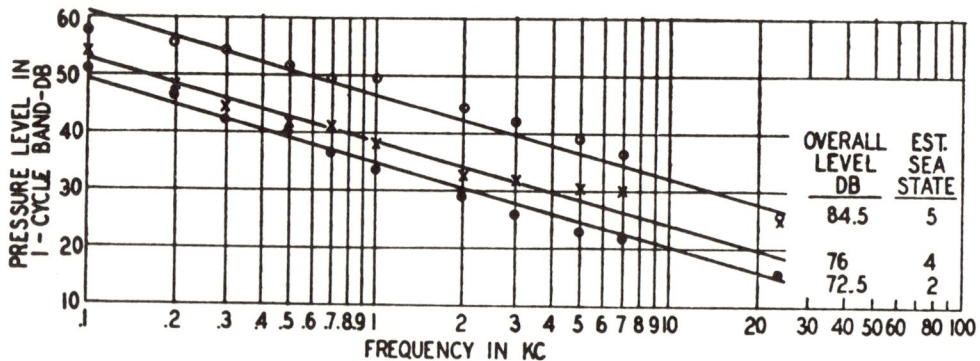

Figure 3. Pressure level spectra of ambient noise in open water about 700 feet deep.

Fig. 3 gives pressure level spectra of water noise in open and deep water five miles off Fort Lauderdale, Florida, for sea states 2, 4, and 5. The measurements were made by using a sensitive hydrophone having a low system noise. The hydrophone was carefully suspended from a cable by means of damped springs so that the movement of the hydrophone was negligible even in a very rough sea. The depth of submergence of the hydrophone was varied from 25 to 300 ft. No systematic variation of the over-all pressure level with hydrophone depth was observed; the small variations appeared to be random and probably resulted from the changing roughness of the sea and possibly from occasional sounds of distant ships or from marine life. Each point shown in Fig. 3 is the average of the data taken at all depths. There is a slight indication of a possible change in slope of the spectra for different states of sea, but the change is not systematic, and a reasonably good fit to the data is obtained with curves having the same slope as that of the spectrum determined by averaging all the data.

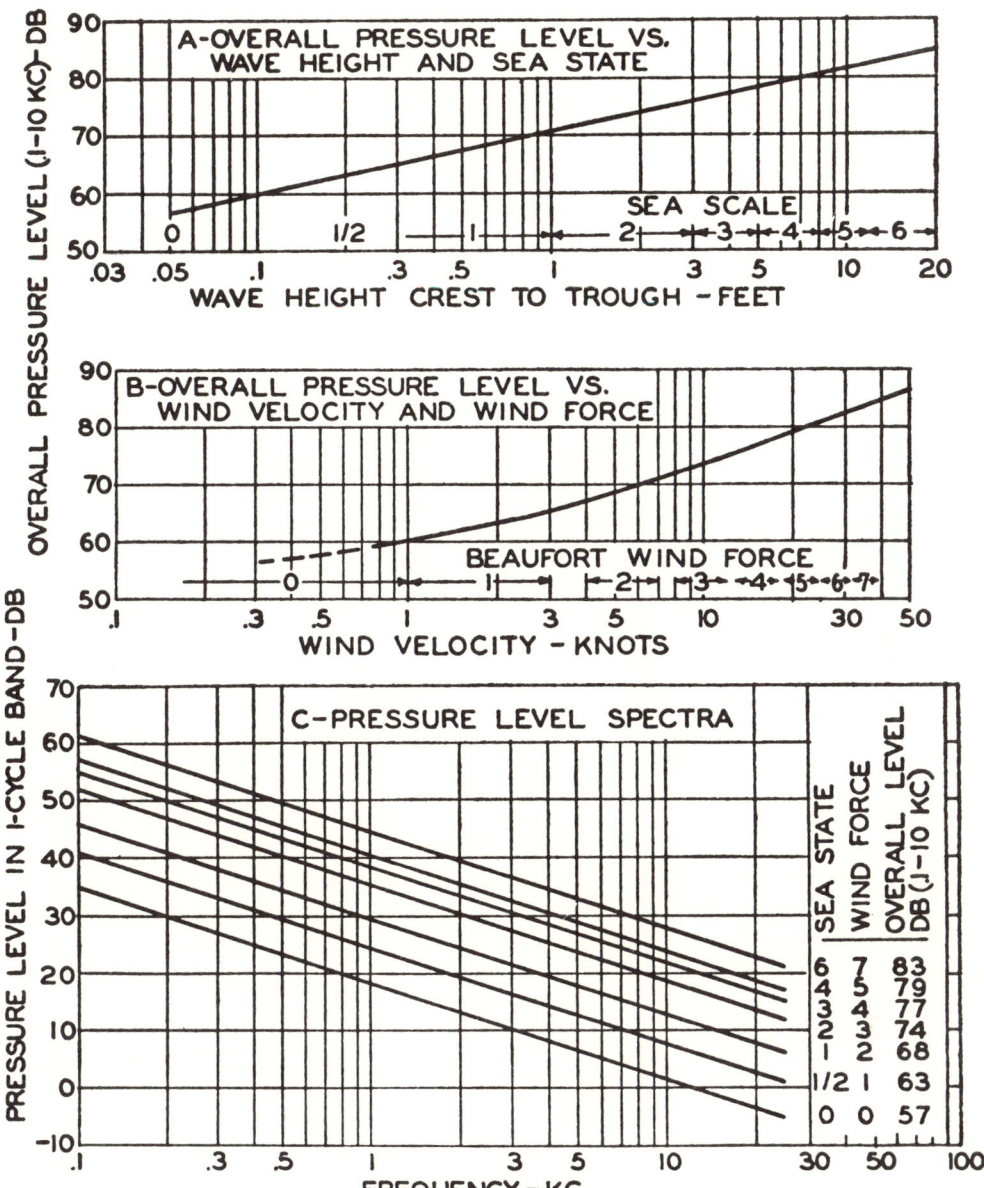

Figure 4. Ambient noise from water motion. Over-all pressure levels and pressure level spectra as a function of sea and wind conditions.

Fig. 4 presents composite summaries of ambient noise from water motion. The relation between over-all noise level and wave height (crest to trough) is shown in Part A. The "state of sea" scale corresponding to various wave heights is also shown along the abscissa. This sea scale provides a useful and an approximate method for describing the height of the waves.

The relation between over-all noise level and wind velocity is shown in Part B of Fig. 4. The Beaufort scale of wind force, which is frequently used for specifying approximate wind velocities, is also shown along the abscissa. The curves in Parts A and B are for average values. For a given wave height or wind velocity, considerable variation from the average noise level is to be expected. For example, noise levels higher than average are probable when whitecaps and spray are unusually prominent. On the other hand, noise levels lower than average are probable when the sea consists of long smooth crested waves. The standard deviation of observed noise levels with respect to the average values specified by the curves is of the order of 4 to 5 db. The variation is less at high wave heights and wind velocities than at low.

Frequently, when both wave height and wind velocity are known for a given location, the noise estimate based on wind and waves will differ. In such a case, an average of the two predicted noise levels usually will be more reliable than either one alone. An exception would occur near the lee shore of a large land mass. In this case, the estimate based on wave height will be the more accurate.

The average pressure level spectra to be expected for various wind and sea conditions are shown in Part C of Fig. 4. The slope of the spectrum appears to be independent of wind and sea and averages -5 db/octave. Experimental evidence indicates that random departures from this slope may occur, but usually the slope will not be more than -6 db or less than -4 db/octave. Neither noise level nor spectrum varies greatly with water depth so long as the water is sufficiently deep to prevent breaking of the waves. In deep water the average noise level and spectrum are essentially independent of the depth (20 to 300 ft.) at which the noise is measured. There is, however, a difference in the character of the noise. Near the surface, the noise from individual waves and whitecaps can be discerned, and the momentary variations in noise level are greater than at a greater depth.

Noise from Surf, Rain, Hail, Tide Rips, and Movement of Bottom Material

The roar of breaking surf, which often sounds very loud to a submerged swimmer, is clearly indicated or registered on underwater listening or sound recording equipment. However, there are almost no quantitative data on the noise produced by surf. Observers at Cape Henry reported that during "rough weather" the over-all pressure level (0.1–10 kc) measured near the bottom, 300 yards offshore, was 78 db (corresponds to sea state 3–4).

The impact of rain and hail is believed to produce a noticeable amount of noise in the absence of other prominent noise sources. There are no data available on the effects of hail and only a small amount of data on the effects of rain. One set of measurements indicates that with a sea of state 1, or greater, the impact of rain has a negligible effect at frequencies below 1 kc but, owing to the presence of noise from marine life, the effect above 1 kc could not be determined quantitatively. Measurements in the Thames River at New London, Connecticut, showed an over-all pressure level (0.1–20 kc) of 75 to 76 db in the presence of a "steady but not torrential" rain as compared to 57 db prior to the rain. This agrees qualitatively with observations made by swimmers in lakes and rivers while they were swimming under water during summer showers of rain and hail.

One set of measurements made near a tide rip (current 3.5 knots) in a calm sea showed a rise in the pressure level spectrum beginning at about 3 kc. Levels of about 38 db (in a one cycle band) were obtained at frequencies of 5–10 kc (corresponds to sea state 8–9).

Sounds attributed to the movement of gravel on the bottom have been reported, but there appear to be no quantitative data on the noise from this source.

NOISE FROM MARINE LIFE[4]

Many forms of marine life are capable of producing underwater noise. Table III lists some pertinent information regarding the noise-producing characteristics of vertebrate specimens investigated at Shedd Aquarium, and of other soniferous specimens of marine life. Thus, the croaker, by means of drumming muscles on its air bladder, gives bursts of sound in rapid succession, each burst having a short duration and a principal frequency of about 250 c.p.s.; the porpoise "barks" and follows this with a "gobble, gobble" sound, similar to that of a turkey cock; the toadfish emits intermittent "boops"; the spot produces a series of raucous "honks"; snapping shrimp, by snapping closed an enlarged pincerlike claw, emit a "crackling" sound, not unlike that of burning dry twigs, or, as heard from a distance, like the sizzle of frying fat. Among fresh water soniferous fish, the giant boom, or singing catfish, when played on an angler's taut line, is re-

---

[4] Several others have reported results of measurements on noise from marine life. See D. P. Loye and D. A. Proudfoot, Underwater noise due to marine life, J. Acous. Soc. Amer., *18*: 446–449 (1946); F. A. Everest, R. W. Young, and M. W. Johnson, Acoustical characteristics of noise produced by snapping shrimp, J. Acous. Soc. Amer., *20*: 137–142 (1948); and M. W. Johnson, F. A. Everest, and R. W. Young, The role of snapping shrimp (Crangon and Synalpheus) in the production of underwater noise in the sea, Biol. Bull. Woods Hole, *93* (2): 122–138 (1947).

TABLE III. NOISE-PRODUCING MARINE LIFE
Vertebrate Noise-Producing Specimens at Shedd Aquarium

Vertebrates

Name	Description of Sound	Sound Producing Mechanism	Principal Freq. CPS* 1	2	3	Length of Pulse Sec.*
Croaker	Bursts of drumming in rapid succession	Drumming muscles on air bladder	250	—	—	0.022
Spot-fin Croaker	Individual drum beats	Drumming muscles on air bladder				
Black Drum	Isolated groans	Drumming muscles on air bladder	50	80	150	0.270
Red Drum	Isolated groans	Drumming muscles on air bladder				
Garibaldi	Clicking, rasping	Pharyngeal teeth	7,400	1,000	150	0.011
Sea-anemone	Drumming, tapping	Pharyngeal teeth				
Single-striped Damozel	Drumming, tapping	Pharyngeal teeth	700	—	—	0.023
Coral-reef Damozel	Drumming, tapping	Pharyngeal teeth	700			
Common Triggerfish	Rasping, hissing, spitting	Pharyngeal teeth				
Surgeonfish	Sucking, rasping	Pharyngeal teeth				
Long-finned Pompano	Clicking, grinding	Hitting together dorsal fins				
Sheepshead	Crunching, grinding	Extrinsic food crushing	5,800	1,900	—	0.026
Razorfish	Low crunching	Extrinsic food crushing				
Hagfish	Loud crunching, grinding	Extrinsic food crushing				
Muttonfish	Crunching, sucking	Extrinsic food crushing				
Sting Ray	Crunching, grinding	Extrinsic food crushing				
Red Grouper	Clicking, grinding	Snapping teeth while eating				
Nassau Grouper	Clicking, grinding	Snapping teeth while eating				
Spadefish	Thump from rapid motion	Mechanical disturbance of water caused by quick rotation of body	220	150	—	0.150
French Angelfish	Thump from rapid motion					
Black Angelfish	Thump from rapid motion					
Diamond Flounder	Grating, scraping	Disturbance of gravel				
Sea Catfish	Popping, drumming	?				

Other Reported Noise Producers

Vertebrates

Name	Remarks
Porpoise	Bark and gobble (observed off Cape Henry)
Toadfish	Intermittent "Boops" (Beaufort, N. C.)
Foolfish	
Grunt	
Sand Perch	
Squeteague	Only males have drumming muscles
Sargo	
Midshipman	Probably like croaker
Gray Trout	Series of raucous honks (Beaufort, N. C.)
Spot	Squawk, squeal or cackle. Similar to croaker but more rasping (Beaufort, N. C.)
Sea Robin	

Crustacea

Name	Remarks
Snapping Shrimp	Crackling (principally *Crangon* [*Alpheus*] and *Synalpheus*)
Crabs	Especially *Cancer* and *Portunus*. Noise only while feeding
Barnacles	Occasional clicks of low intensity
Mantis Shrimp	Sharp click
Spiny Lobster	Grating

* From photomicro analysis of phonograph recording grooves.
Note.—The list given above may not be complete, not all noise producers being of equal importance; several other crustacea produce noise.

puted to sing "deep purring music that wanders up and down four full tones." Two of the forms of marine life, the snapping shrimps (*Crangon* and *Synalpheus*) and the croaker, are of great practical importance because they produce sustained noises of high level.

Snapping Shrimp. Snapping shrimp can be expected throughout the oceans at locations where environmental conditions are favorable. These conditions are:

Temperature—Limited by the 52° F winter surface isotherm. Some period of the year with about 60° F also required.

Depth—Generally less than 180 feet. The highest noise levels appear to occur in water between 30 and 150 feet deep.

Bottom—Rock, coral, shell, weed, or other material providing ready concealment. Relatively uncommon on mud or sand bottoms which are free from sheltering material.

The region in which shrimp are to be expected when depth and bottom conditions are favorable is roughly defined by a belt around the earth extending from latitudes 40° N to 40° S. Along the west coast of Europe extending north to Lands End, England, there is an additional area where the temperature is favorable. The inhabitable range along the west coast of South America south of Latitude 10° S and along the east coast south of Latitude 30° S is not known.

Since snapping shrimp are nonmigratory animals they can be considered as a constant characteristic of any region in which they have once been found. The noise produced by shrimp is continuous. There appears to be no pronounced seasonal variation and but slight diurnal variation. During the night hours, the noise level is usually a few db higher than during the daytime, and there is a peak in the noise level (3 to 4 db above daytime level) just before sunrise and another after sunset.

The sounds from snapping shrimp have been reported by others (ftn. 4, p. 419) and therefore they will not be considered further here, except for reference to Fig. 5 which gives summary curves typical of the ambient noise spectra to be expected in waters where snapping shrimp are prevalent. The data for this figure indicate that for frequencies up to about 1 or 2 kc the noise spectrum is largely determined by water noise. Above 2 kc the crackling of shrimp is the major source. Part A of Fig. 5 shows noise spectra with average shrimp noise and various sea conditions. Average shrimp noise is representative of the 24-hour average noise level found at water depths of 30 to 150 feet in the following regions: Off Florida, below Latitude 27° N, off Bahama and Cay Sal Banks, off San Diego, and off Oahu and Midway Islands.

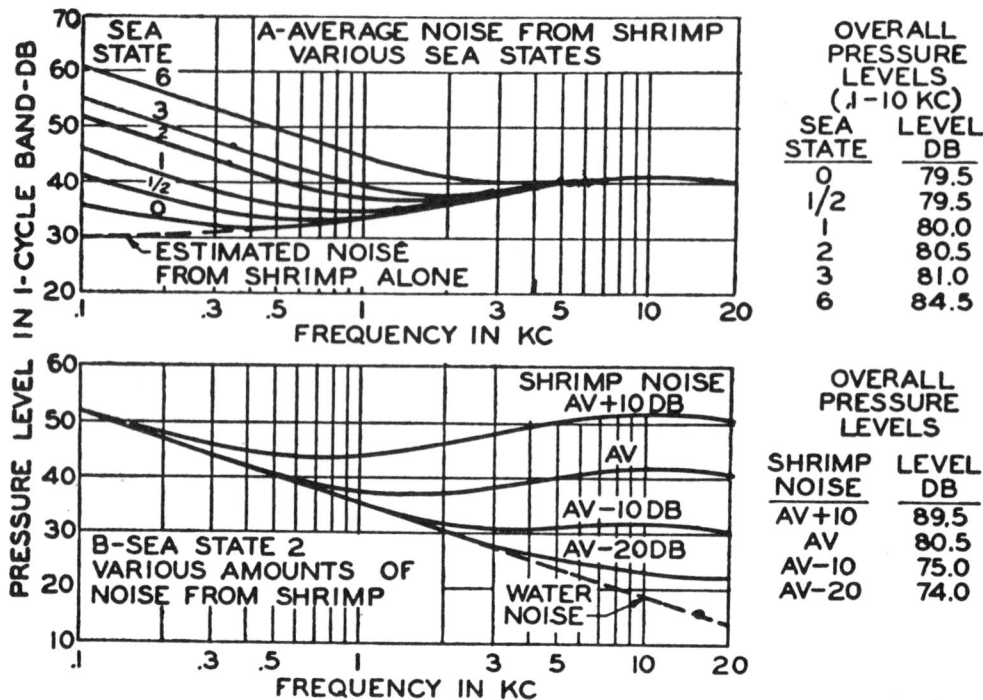

Figure 5. Pressure level spectra of ambient noise in the presence of snapping shrimp.

The noise levels at depths between 30 and 150 feet in these regions had a standard deviation of 5 db.

At depths greater than about 150 feet, shrimp crackle depends largely on the transmitted noise from near-by shallow water areas. At a distance of a nautical mile from such a shallow water area, the noise from shrimp is about 20 db lower than in the shallow water area. The colonies of shrimp appear to be smaller and more scattered at the colder margins of the areas which they inhabit. For example, along the eastern coast of the United States, shrimp have been found and their noise observed as far north as Latitude 35° N, but at offshore locations above Latitude 27° N the noise levels seem to be 15 to 20 db lower than the average level shown in Fig. 5. However, large colonies apparently occur in the harbors at Beaufort and Morehead City, N. C. Other isolated large colonies are probably to be found above Latitude 27° N.

The highest observed level of shrimp noise was measured near a pier at Kaneohe, Island of Oahu. The level at this location, at frequencies above 3 kc, was approximately 20 db above the average spectrum shown on Fig. 5. This is believed to be an exceptional condition caused by the presence of a large colony of shrimp living in the fouling material on the piling. Levels of this magnitude have not been observed in open water.

Spectra representative of a sea of state 2 and various amounts of noise from shrimp are shown in Part B of Fig. 5.

Croakers. Croakers (*Micropogon undulatus*), a variety of drumfish, are found during the late spring and early summer months in great numbers in Chesapeake Bay (estimated population 300 to 400 million). They also occur in considerable numbers at other east coast locations below Chesapeake Bay. The larger fish migrate to sea during the winter months. Different species occur on the west coast of the United States below Point Conception but apparently not in as great concentrations as along the eastern coast. Most of the available noise data have been obtained at east coast locations, and the discussion will be confined to the species found there.

Croakers produce noise by the contraction of drumming muscles attached to the air bladder. The sounds produced by an individual croaker consist of a series of "taps" that continue for about one and one-half seconds at a rate of about seven taps per second. The series is repeated at intervals of 3 to 7 seconds. The sounds resemble the tapping of a woodpecker on a dry pole. In an area with a high concentration of croakers, the sound during a period of great activity is a continuous roar, and the sounds of individuals may be heard only infrequently over the chorus of the whole population.

Croaker noise occurs principally during the feeding period, which starts in the evening as the bottom begins to darken. This, together with the migratory habits of the fish, accounts for the very pronounced diurnal and seasonal variation in croaker noise discussed in the following paragraph. Croakers are believed to feed in the shallow water on the slopes of banks.

Fig. 6 presents typical data on the seasonal and diurnal variations of over-all noise pressure levels from croakers at Cape Henry, Virginia. Note that the average over-all level in early June is 110 db, although levels as high as 119 db have been observed.

Typical spectra of ambient noise for various sea states in the presence of croakers (lower Chesapeake Bay) during hours of maximum activity are shown in Fig. 7 for late May to early June and for early July.

Other Marine Life. Other forms of marine life which may be of importance are toadfish, sea robins, some species of drumfish other than croakers, porpoises,[5] and a number of unidentified sources; one

[5] Some marine zoologists believe that the sounds here attributed to porpoises were produced by dolphins.

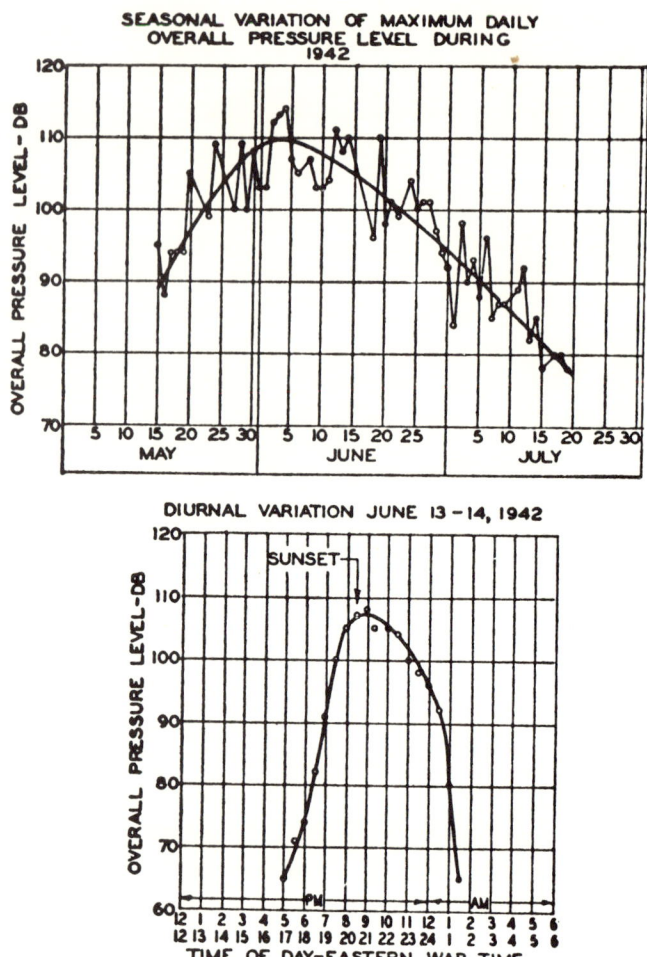

Figure 6. Seasonal and diurnal variation of over-all noise pressure levels from croakers.

such source is described as the sound of a mewing cat and another as an "awesome moaning."

Toadfish, individually, seem to produce higher noise levels than any other form of marine life thus far identified and reported, with the possible exception of the porpoise. The sound produced is an intermittent, low-pitched "boop" of about one-half second duration, similar to a boat whistle or sometimes like the cooing of a dove. The spectrum shown in Fig. 8 was measured very close to a single specimen. It will be appreciated that the levels decrease rapidly as the distance from the fish increases. These fish are shallow water bottom dwellers which nest under rocks, tin cans, and similar debris. Toadfish are not gregarious and apparently do not occur in sufficient concentration

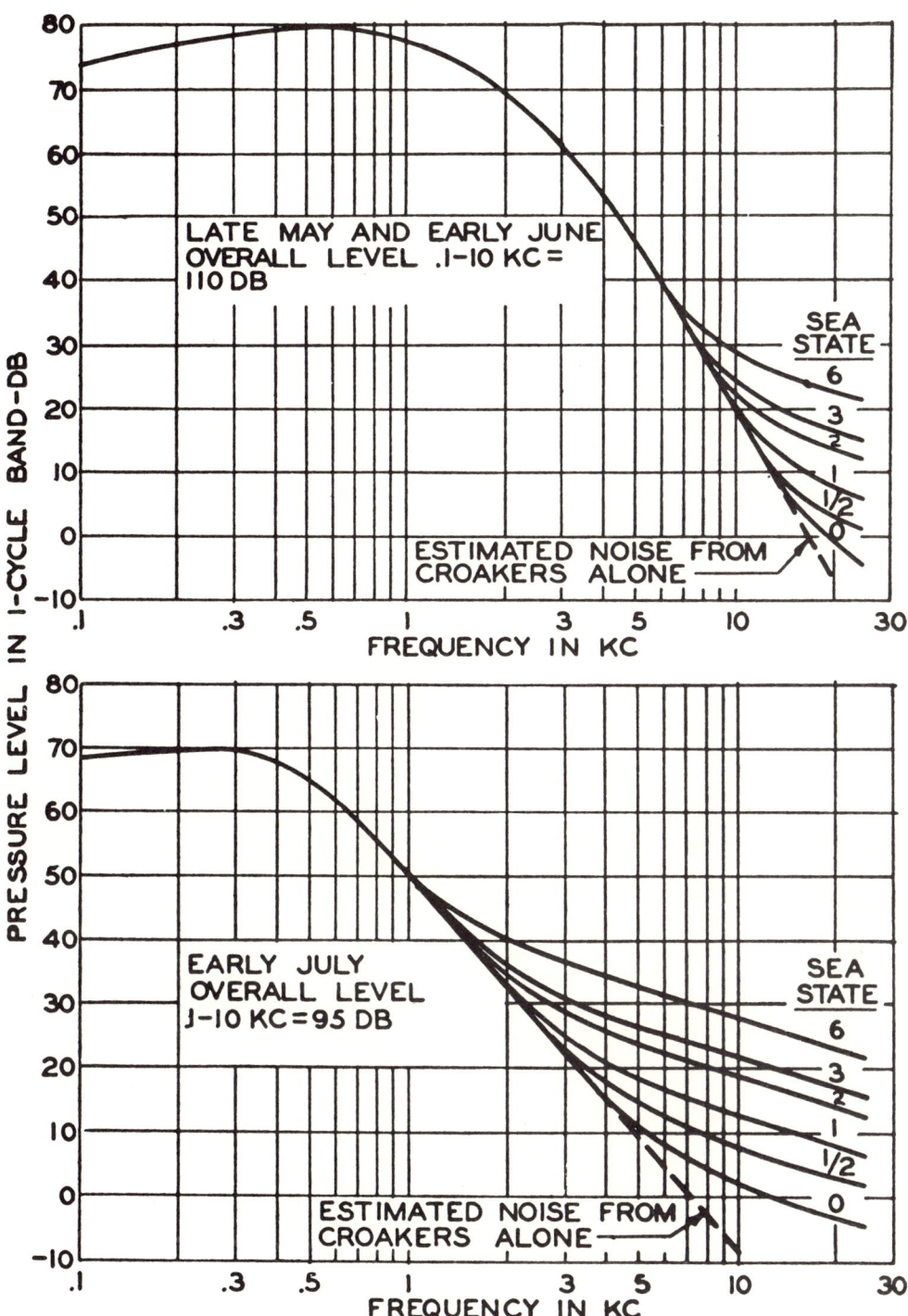

Figure 7. Typical pressure level spectra of ambient noise in the presence of croakers.

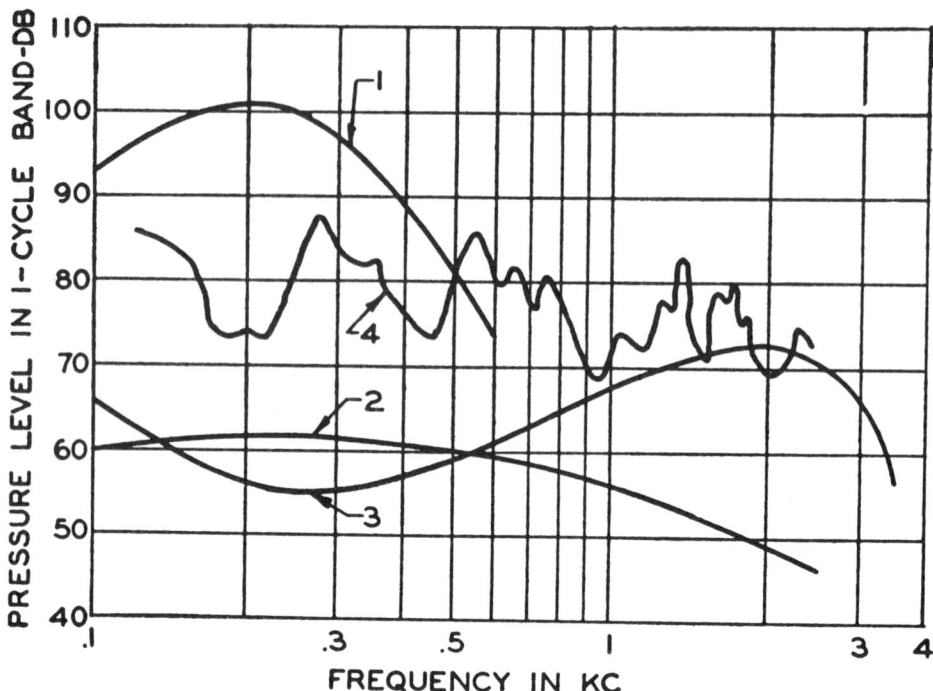

Figure 8. Illustrations of ambient noise spectra obtained in the presence of miscellaneous marine life. Curves: 1. Individual toadfish very close to hydrophone. 2. Principally from sea robins. 3. Unidentified "high pitched drumfish," possibly bastard trout. 4. Porpoises.

to produce the continuous roar generated by croakers during peak activity. Toadfish noise is subject to little diurnal variation. Other spectra shown in Fig. 8 are for (2) sea robins, (3) bastard trout, and (4) porpoises.

SHIP AND MAN-MADE SOURCES OF NOISE

In and near busy harbors and industrial centers, ships and manufacturing activities are the principal sources of ambient noise.

A wide variety of noise may be produced by industrial activities, and it is usually impossible to predict the noise from such sources in specific locations.

The noise produced by ships is more uniform in character than industrial noise, but it varies in magnitude over a wide range. It obviously depends on the number of ships present, their speed, their distance from the point of measurement, and the sound transmission characteristics of the water. Measurements in a number of east coast harbors showed an average over-all (0.1–10 kc) noise level of about 80 db. Location averages varied greatly, ranging from 65 db to 103 db. The standard deviation of the location averages was 8 db.

The average level of 80 db appears to be representative of locations with a moderate amount of ship traffic, such as the area near ship lanes at the upper end of Long Island Sound. An over-all pressure level (0.1–10 kc) of 90 db is more nearly representative of locations with a large amount of ship traffic, as, for example, in the approaches to New York Harbor during the daytime.

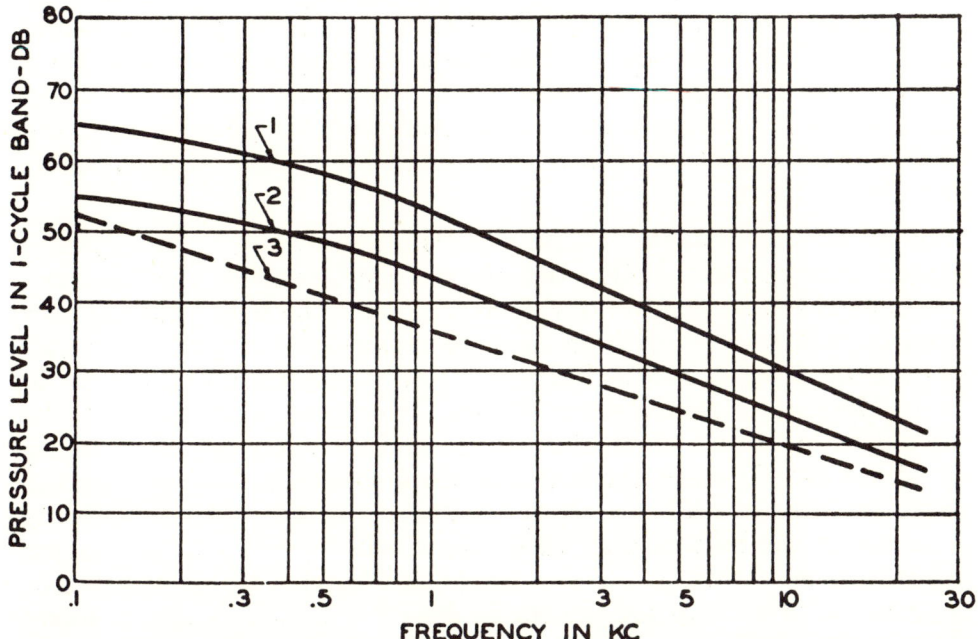

Figure 9. Pressure level spectra of ambient noise in the presence of ship sounds. Curves: 1. High noise locations; example, entrance to New York Harbor in daytime—over-all level 90 db. 2. Average locations; example, upper Long Island Sound near ship lanes—over-all level 80 db. 3. Water noise, sea state 2—over-all level 74 db.

Pressure level spectra representative of moderate and busy locations are shown in Fig. 9. A spectrum of water noise with a sea of state 2 is shown for comparison. In locations with light ship traffic, the average noise level is not likely to be lower than that produced by a sea of state 2.

PREDICTION OF NOISE CONDITIONS

The data compiled in the present report can be used for predicting the noise conditions to be expected in oceans, under various situations and in certain locations. Obviously, if survey data are available for a specific location they are preferable to estimates based on the general characteristics of the location. The principles which should be followed in predicting noise are outlined briefly as follows:

Open and Deep Water. Water noise is the major component. When sea state (wave height) and/or wind force are known, over-all pressure levels and spectra can be estimated from the data in Fig. 4. If, for a given wave height, the sea is unusually choppy with exceptionally prominent whitecaps and spray, the noise levels probably will be somewhat higher (4 to 8 db) than the average values shown in Fig. 4. If, on the other hand, the sea consists largely of long swells with exceptionally few whitecaps and little spray, the noise levels probably will be lower than average. When the prediction is to be made for a short time in the future, wind and sea conditions can be estimated from a series of weather maps. For prediction far in the future, the probable wind conditions can be estimated from U. S. Hydrographic Office Pilot Charts.

Shallow Coastal Waters. Noise from three sources is to be expected —water motion, marine life, and ships. Occasionally, surf and the movement of bottom material will contribute to the noise, but data for estimating the noise levels from these sources are not available. Water noise is estimated as outlined above. Noise from marine life is probable in tropical or semitropical waters. Snapping shrimp are to be expected at locations where the winter surface temperature is higher than about 52° F, where the water depth is less than 180 feet and the bottom consists of rock, coral, or similar material. Charts showing the nature of the bottom and water depth will be useful in estimating the prevalence of shrimp. Noise levels and spectra to be expected in the presence of shrimp are shown in Fig. 5. Croakers are a likely source of noise during the evening hours between late May and August in Atlantic coastal regions between Chesapeake Bay and Cape Canaveral. No doubt there are other regions where croaker noise is likely to be of local importance. The spectra shown in Fig. 7 are probably the best *a priori* estimates of noise conditions. Croaker noise is likely to be variable, and local surveys are advisable at locations where croakers are to be expected. The likelihood of ship sounds can be estimated from a knowledge of ship traffic. The data in Fig. 9 are probably representative of the underwater noise near active coastal ship lanes. During the absence of ships, water noise and marine life will, of course, determine the ambient noise level.

Bays, Rivers, and Harbors. The three main noise sources are possible contributors. The procedure for estimating noise conditions is the same as outlined above for coastal waters with these exceptions: In warm waters near piers with old piling, noise from shrimp may be 10 to 20 db above average. Recent dredging operations may reduce the shrimp population. In busy harbors ship noise probably will be the predominant source during the hours of shipping activity. Average noise conditions in harbors comparable with the approaches to New York City can be represented by the "High Noise" spectrum of Fig. 9. Local conditions such as industrial development at the shore or the presence of unusual man-made noise sources may necessitate a local noise survey.

The spectra of the ambient noises illustrated in Figs. 1 to 9 are typical of some of the noise levels that have been measured. They indicate the wide range and diversity of underwater ambient noise.

SUMMARY

The forgoing report presents the results of a survey of available data (to March 15, 1944) on underwater ambient noise at frequencies from about 100 c.p.s. to 25 kc. The data are presented in the form of pressure levels (for a wide band of frequencies or for a band one cycle wide) in db relative to 0.0002 dynes/cm^2. The principal sources of ambient noise are water motion, marine life, and ship sounds. In the absence of sounds from ships and marine life, the noise level in deep water is largely determined by the state of the sea; the over-all (0.1 to 10 kc) pressure level varies from about 57 db in a "calm sea" to about 83 db in a sea under a strong wind (about 35 m.p.h.). Similarly, the underwater noise from croakers may reach over-all levels as high as 119 db, and the sounds of ships in busy harbors produce underwater noises that often exceed 90 db. Data on the average pressure levels of the principal sources of ambient noise are summarized. The actual noise encountered may differ appreciably from the average values; standard deviations vary from about 4 to 8 db. Principles are deduced for predicting the noise levels that may be expected under usual conditions in specific locations.

STUDIES ON THE PRODUCTION OF UNDERWATER SOUND BY NORTH ATLANTIC COASTAL FISHES[1]

By

MARIE POLAND FISH
Narragansett Marine Laboratory

ALTON S. KELSEY, JR.
Brown University

WILLIAM H. MOWBRAY
Narragansett Marine Laboratory

ABSTRACT

Previous studies suggest that at least 42 families of Pacific Ocean fishes are potential sources of underwater sound. The present investigation has undertaken to ascertain experimentally which North Atlantic coastal species fit into that category and what stimuli are involved in sound production among fishes. Apparatus, test and analysis procedures are described. Recorded for 26 species are "biological" sound, the mechanism responsible for sound production, the frequency range, the principal frequency components, and a description of each sound. Relationship is noted between origin of sound and its frequency characteristics. The majority of species tested have principal frequencies between 75 and 300 cps; none exhibited sounds lower than 20 cps, and with three exceptions none produced sounds higher than 1600; stridulatory sounds of triggerfish, filefish, and burrfish reached the 2400-4800 cps octave. With the available filter system, nothing could be measured faithfully above that band. Noise of "mechanical" origin was recorded for 22 additional species. No sound was observed during routine testing of six species.

PREVIOUS INVESTIGATIONS

Widespread use of underwater acoustical systems during World War II made exigent a knowledge of ambient noise conditions. Early background measurements indicated that in the absence of extraneous noises (as from shipping, shore activity, and animal life) a magnitude of one dyne/cm^2 in an octave band would seldom be exceeded (8), whereas biological sources might be responsible for sustained noise levels of 200 to 300 dynes/cm^2 (6).

So little was understood of the noisemaking propensities of marine animals, however, that "a list of all known (worldwide) forms of marine life producing subsurface sound," prepared by the United

---

[1] Contribution No. 1 from the Narragansett Marine Laboratory of the University of Rhode Island. This paper is based on research conducted under contract with the Office of Naval Research.

States National Museum for the Navy in 1942, included only 14 families of fishes and 17 families of crustaceans. Collected data on their sound production were fragmentary, often mere hearsay, and they seldom included more than a few descriptive terms like "loud grunt," "hoarse croak," or "nasal whine." No information was available on the actual magnitude or frequency spectra of the various sounds.

Accordingly, investigations were initiated in such areas as Chesapeake Bay, where croaker interference at times masked the propeller sounds of ships, and San Diego, where observations as early as 1933 had spotlighted a crackling noise of great magnitude, identified (5) in late 1942 as snapping shrimps (*Crangon* and *Synalpheus*). Observations on noisemaking crustaceans continued throughout the war. Monitoring and disc recording were undertaken on the southeastern and southwestern coasts of the United States and in the vicinity of Hawaii, Ellice Islands, Noumea, Guadalcanal, and the Russell Islands. Also, during this period, six species (representing five families) of Virginia coastal fishes (9) and members of five families in captivity at the Shedd Aquarium (7) were isolated for sound recording and analysis. The only other experimental work of this type was done by Dr. Yoshio Hiyama in Japanese waters between 1944 and 1945 (1).

In 1946 the senior author undertook a Navy assignment to "assemble, analyze, and where possible correlate with environmental factors, available information from the central and western Pacific" on underwater sound of biological origin. The resulting reports (3, 4) added 31 families of fishes to the existing list of soundmakers in that area, bringing the total to 42. Since most of these families are represented by many species, it became evident that the potential sonic species might number thousands. For each known or suspected sonic family, available data were presented on geographical, seasonal and vertical distribution, size of individuals, abundance, and especially on the type of sound mechanism involved and a description of the sound produced.

PRESENT OBJECTIVE

Previous study (see REFERENCES) had thus indicated that noisemaking is common among many types of marine animals. It had shown which species are potential sources of underwater sound, what their characteristic sounds are, how produced, and when and where they may be expected. However, no attention had been given to the underlying causes of this ability. Thus the present investigation attempts to determine primarily the biological significance of soundmaking among fishes. From the physical viewpoint, the problems emphasized are: 1) the sound intensity possible from an individual fish; 2) the frequency spectrum of sound made by each species; 3) the duration and repetition rate of the sound; 4) the maximum distance at which it can be detected.

APPARATUS

Laboratory and vessel facilities were made available by the Narragansett Marine Laboratory, headquarters for the Project. Its location directly on Narragansett Bay, with no surf or industrial activity along shore and almost no ship traffic, provided an ideal low background level. A large wire enclosure and numerous live cars constructed in the Bay acted as storage space and test areas where specimens could be maintained under more normal conditions than those of indoor aquaria. Bay hydrophones were connected with recording instruments in the main laboratory, an arrangement which made possible protracted monitoring within a distance of 540 feet. Monitored also by hydrophone were many large glass aquaria inside the building as well as a concrete floor tank (19 x 3 x 2 feet), all with running seawater.

Marine sounds were recorded with a modified Naval Ordnance Laboratory Mk2 acoustic system. Type 1-A (Q4) quartz crystal hydrophones were used with the permanently installed equipment. These hydrophones have no significant temperature coefficient; they have a smooth, fairly flat frequency response from 10 to 5,000 cps and a rising characteristic from 5,000 to 10,000 cps. Sounds picked up by hydrophone were fed into an analyzer unit consisting of an audio amplifier with flat response from 40 to 20,000 cps, low and high pass filters, and power supply. The filter output was recorded by Esterline Angus Graphic Ammeters (EA) which indicated the root-mean-square pressure in the frequency bands below and above 500 cps. The response of the low pass EA is flat from 10 to 500 cps, while the high pass EA is flat from 500 to 30,000 cps.

During the earlier experiments, the putput of the audio amplifier was applied directly to a Presto 88A3S recording amplifier and Presto 6-N recording turntable, which made permanent disc recordings at 78 rpm. For playback, a Brush PL-20 pickup was used. Assembled thus, the response of the over-all recording system was from 20 to 5,000 cps, being flat within 4 db from 30 to 1,500 cps.

Later, a Webster Electric Model 111 tape recorder replaced the Presto equipment for direct recording. The tape recorder had not only the advantage of better frequency response, being ± 3 db from 35 to 7,500 cps, but it could be run continuously for long periods at low expense and the results could be edited later to yield a short recording of selected sounds.

In addition to this fixed apparatus, a portable outfit was assembled for use in the field. This consisted of a Navy JO Underwater Sound Receiving Equipment set and either a converted office type disc recorder (Gray Audograph Electronic Soundwriter) or the tape recorder (Webster Ekotape).

For permanent filing, all tapes made in the Laboratory, as well as field recordings, were transcribed to Presto discs.

ANALYSIS PROCEDURE

1. With HARMONIC WAVE ANALYZER (Hewlett Packard, Model 300A).

Frequency spectra of the sounds were obtained by applying the output of the disc recordings directly to this instrument. The meter element in the analyzer was replaced by an Esterline-Angus Graphic Ammeter to obtain permanent records. A half band width of 30 cycles was used for each measurement; that is, the selectivity of the analyzer was such that voltages 30 cps off the measured frequency were attenuated 40 db. Readings were taken at intervals of 25 cps throughout the entire audible frequency spectrum, each sound being replayed into the analyzer for measurement at each frequency. Deflections obtained on the EA are proportional to the rms sound pressure in a narrow band centered at the measured frequency. These points were plotted for typical and extreme individual cases, and for average sets of readings for selected fish of each species.

The assumption is made here that the fish sounds have a uniform energy distribution over the band width of the analyzer. In such a case, the ordinate of the curve at a given frequency would be proportional to the rms pressure in a band one cycle wide centered at that frequency. The area under each curve would then be proportional to the over-all rms sound pressure.

To obtain the proportionality factor, the area under each curve was measured by a planimeter. The factor K was then immediately given by the relation $P = \int_0^\infty p_f df = \int_0^\infty KD\, df$, where P = over-all rms pressure measured, p_f = pressure per cycle, f = frequency, and D = EA deflection.

Typical curves were plotted, giving pressure per cycle vs. frequency. Since the area under the corrected curve represents the total sound pressure, the pressure in any frequency interval could be calculated by obtaining the area under the curve in that interval.

2. With Octave Filter (Western Electric Type RA-363).

The sounds to be analyzed were played back through this filter set, and the a-c voltage output was measured by an Esterline-Angus recording d-c milliameter with full-wave bridge rectifier.

Recordings were first played with the octave filter switches in the over-all position, allowing all frequencies to pass equally; the a-c voltage measured was proportional to the over-all sound pressure. The recordings were then played once for each of the 14 filter sections, each of which passed frequencies within a one octave band and produced voltages proportional to the sound pressure in the chosen octave.

The actual over-all sound pressure was determined by direct measurement of the system gain from the hydrophone to the filter output with reference to the hydrophone calibration curves. The combination of these data furnished the value of sound pressure corresponding to the measured output voltage.

The sound pressure in each octave was obtained by applying a proportionality factor to the output voltage measured in that range. This factor was determined by plotting the voltages against frequency and by mechanically integrating the curve to find the area subtended. Since the area is directly related to the over-all pressure and can be expressed in terms of pressure per octave, frequency, and a constant, these quantities can be equated and the expression can be solved for the constant.

TEST PROCEDURE

Experimental material was obtained regularly from commercial traps in the vicinity of Pt. Judith and outer Narragansett Bay. Members of the staff who accompanied fishermen on their trap-hauling cruises, transported specimens as quickly as possible to the Laboratory, where they were tested immediately and again later after adjustment to captivity. Occasionally portable listening and recording equipment was taken out in the fishing boats in order to detect new noisemakers and to study the effect of crowding upon known sonic species. In some cases (e. g., *Opsanus tau*) recordings were made in the field in a natural environment, after which the captured noisemakers were delivered to the Laboratory for further observation.

Hydrophones permanently installed in the various test areas were valuable in picking up the general "chorus" of newly added specimens. Intensity levels were measured over 24-hour periods to investigate the diurnal cycle of sonic activity for mixed populations as well as for individual species.

There is always urgency in working with live specimens, and long periods of observation in captivity, which often decrease or even prevent normal sonic behavior, may be impossible. Thus, as laboratory study progressed, it became evident that some quick means of sorting the potential soundmakers from the silent ones must be found. Accordingly, a simple electric stimulator, consisting of a conventional glass and wooden aquarium (lined with insulating layers of rubberized hair to prevent knocking of the fish against bottom and sides), screen electrodes, and a variable autotransformer, was rigged. Currents up to three amperes were passed into the water. To date every species which has had the necessary soundmaking apparatus has responded to this electric shock by emitting sound. Whenever such capability was shown, the fish was earmarked for continued study.

Each species was subjected to a routine set of conditions to determine, if possible, what stimuli induce sound production. In brief, reactions to the following situations were studied:
1) When the fish was introduced into new surroundings.
2) After it became accustomed to the tank. (According to the species, this period varied from 10 minutes to several days. In each case, specimens were undisturbed until they seemed to be swimming and behaving normally.)
3) When food was offered.
4) When other fishes were added to the tank.
   a. One or more of the same species.
   b. One or more of a different species known to live peaceably in the same natural habitat.
   c. One or more of a known or suspected enemy species.
   d. Enough other specimens to create crowding.
5) When subjected to artificial stimulation, which varied from the lightest touch to extreme duress. (Sometimes sudden movement of the experimenter outside a glass tank was sufficient to provoke sound, whereas in other cases aggravation of a closely imprisoned specimen was necessary to bring about the same reaction.)
6) When electrically stimulated.

GENERAL RESULTS

Sound-recording apparatus was in operation between the following dates: 22 July to 1 September 1949; 22 to 25 March, and 19 June to 6 September 1950; 26 June to 11 September 1951. Sounds of biological origin were successfully recorded for 26 species, as indicated in Table I. The term "biological," used here to indicate sound, often purposeful, which originates in the body of the fish itself by action of either internal or external organs, is differentiated from "mechanical" noise, always accidental, which results from swimming, collision, feeding, or other activity of the fish.

TABLE I. FISHES WHICH PRODUCED SOUND OF APPARENT BIOLOGICAL ORIGIN.
IN THE SPECIES MARKED BY AN ASTERISK (*), THIS SOURCE IS QUESTIONABLE

SPECIES	NUMBER TESTED
Common eel (*Anguilla rostrata*)	10
Thread herring (*Opisthonema oglinum*)	5
Sea horse (*Hippocampus hudsonius*)	1
Hardtail (*Caranx crysos*)	29
Rudderfish (*Seriola zonata*)	13
Butterfish (*Poronotus triacanthus*)*	3
Striped bass (*Roccus lineatus*)	8
Black sea bass (*Centropristes striatus*)	22
Scup (*Stenotomus chrysops*)	75
Squeteague (*Cynoscion regalis*)	10
Kingfish (*Menticirrhus saxatilis*)	5
Cunner (*Tautogolabrus adspersus*)	7
Tautog (*Tautoga onitis*)	7
Spadefish (*Chaetodipterus faber*)	1
Common triggerfish (*Balistes carolinensis*)	1
Foolfish (*Monacanthus hispidus*)	6
Orange filefish (*Alutera schoepfii*)	12
Puffer (*Spheroides maculatus*)	45
Burrfish (*Chilomycterus schoepfii*)	21
Giant sunfish (*Mola mola*)	2
Longhorn sculpin (*Myoxocephalus octodecimspinosus*)	31
Common searobin (*Prionotus carolinus*)	55
Redwinged searobin (*P. evolans*)	69
Toadfish (*Opsanus tau*)	45
Silver hake (*Merluccius bilinearis*)	1
Sand flounder (*Lophopsetta aquosa*)*	3

The 22 species listed in Table II were responsible for noise of varying intensities attributable to accidental sources rather than to sound-making mechanisms within the body of the fish. In the case of large specimens, such as rays and scombroids, mechanical disturbance of the water caused by violent activity was audible, but it was easily distinguished from the air bladder knocking heard when striped bass, squeteague and others were agitated. Likewise, the frenzied rotation of a group of clupeoids could be detected by the hydrophone, but such swishing was quite unlike the pharyngeal rasping which hardtails produced under similar conditions. Sounds from collision against the tank or disturbance of gravelly bottom by swimming flatfish were commonly heard. Very small sounds made by sticklebacks might have been due to stridulation, but the evidence is not conclusive. Similarly with the scombroids and pilotfish, a suggestion of animal noise occurred, but, lacking verification, the record remains questionable. Specimens were sometimes injured or too exhausted by capture and transportation to react normally, and some died before they were sufficiently adapted to captivity for the full routine of experiments. In certain forms, therefore, the fact that biological sound has not yet been noted does not mean, necessarily, that the species is silent.

TABLE II. FISHES WHICH PRODUCED MECHANICAL NOISE AS A RESULT OF NORMAL OR INDUCED ACTIVITY WITHIN THE TEST AREA. IN THOSE CASES WHERE BIOLOGICAL SOUND MAY HAVE OCCURRED AS WELL, THOUGH RECORDS ARE NOT CONCLUSIVE, THE SPECIES IS MARKED BY AN ASTERISK (*). NONE WERE SUBJECTED TO ELECTRIC STIMULATION

SPECIES	NUMBER TESTED
Northern stingray (*Dasyatis centroura*)	2
Butterfly ray (*Gymnura micrura*)	1
Eagle ray (*Myliobatus freminvillii*)	1
Conger eel (*Conger conger*)	1
Bonefish (*Albula vulpes*)	1
Sea Herring (*Clupea harengus*)	15
Hickory Shad (*Pomolobus mediocris*)	2
Alewife (*P. pseudoharengus*)	12
Blueback (*P. aestivalis*)	1
Menhaden (*Brevoortia tyrannus*)	5
2-spined stickleback (*Gasterosteus aculeatus*)*	10
4-spined stickleback (*Apeltes quadracus*)*	3
Common mackerel (*Scomber scombrus*)*	7
Frigate mackerel (*Auxis thazard*)*	14
Spanish mackerel (*Scomberomorus maculatus*)*	4
Cero mackerel (*Scomberomorus regalis*)*	1
Pilotfish (*Naucrates ductor*)*	7
Barrelfish (*Palinurichthys perciformis*)	13
Shark sucker (*Echeneis naucrates*)	12
Sand dab (*Hippoglossoides platessoides*)	5
Northern fluke (*Paralichthys dentatus*)	23
Winter flounder (*Pseudopleuronectes americanus*)	6

TABLE III. FISHES WHICH PRODUCED NO MEASURED SOUND, EITHER OF BIOLOGICAL OR MECHANICAL ORIGIN, DURING ROUTINE TESTING. NONE WERE SUBJECTED TO ELECTRICAL STIMULATION

SPECIES	NUMBER TESTED
Pipefish (*Syngnathus fuscus*)	4
Silverside (*Menidia menidia notata*)	50
Waxen silverside (*M. beryllina*)	3
Mullet (*Mugil cephalus*)	2
Sand launce (*Ammodytes americanus*)	5
Mackerel scad (*Decapterus macarellus*)	1

TABLE IV. DATA ON SPECIES WHICH PRODUCED SUFFICIENT BIOLOGICAL SOUND FOR ANALYSIS AND STUDY

Species	Origin of Sounds	Description of Sounds	Frequency Range (cps)	Principal Frequency (cps)
Family Anguillidae				
COMMON EEL *Anguilla rostrata*	Air bladder. Escape of air bladder gas through pneumatic duct and branchial aperture.	Dull thud; thump. Clucking; bubbling "put-put."	<50–1200	(75–150) or (150–300)*
Family Clupeidae				
THREAD HERRING *Opisthonema oginum*	Air bladder.	Hollow knock of very low intensity.	<50–1200	(75–150) or (150–300)*
Family Hippocampidae				
SEA HORSE *Hippocampus hudsonius*	Stridulation of posterior margin of skull and coronet.	Loud click similar to snap of finger against thumb; single or 2–5 in series at c. 1 sec. intervals.	<50–1600	(400–800)
Family Carangidae				
HARDTAIL *Caranx crysos*	Pharyngeal teeth and air bladder.	Low thumps with electric shock.	20–850	(150–300) and (75–150)†
		Very loud rasps, as with a rough file, when netted.	325–1100	500
Family Seriolidae				
RUDDERFISH *Seriola zonata*	Pharyngeal teeth and air bladder.	Sharp knock.	<50–1200	(150–300)
Family Serranidae				
STRIPED BASS *Roccus lineatus*	Air bladder region, possibly "drummed" by operculum.	Low "unk" with tom-tom quality, single or in bursts of 3 or 4.	<50–1200	(75–150) or (150–300)*
BLACK SEA BASS *Centropristes striatus*	Air bladder.	Single small grunts.	<50–1200	225
Family Sparidae				
SCUP *Stenotomus chrysops*	Air bladder. Stridulation of upper and lower incisors.	Single guttural thumps. Scrape and rasp.	20–1400 350–1150	200–225 700
	Extrinsic feeding.	Loud crunching.	350–700	350

TABLE IV—(Continued)

Species	Origin of Sounds	Description of Sounds	Frequency Range (cps)	Principal Frequency (cps)
Family Sciaenidae				
SQUETEAGUE *Cynoscion regalis*	Air bladder and associated muscles.	Croaks; beats with deep, drum-like quality.	20–1200	(50–100) or 250*
KINGFISH *Menticirrhus saxatilis*	Pharyngeal teeth.	Clucks somewhat similar to above, but in bursts of higher pitch.	100–550	325
	Pharyngeal teeth.	Short rasp.	175–1475	250 and 350†
Family Labridae				
CUNNER *Tautogolabrus adspersus*	Air bladder.	Single low thump.	<50–800	(150–300)
TAUTOG *Tautoga onitis*	Air bladder.	Single deep thump similar to above but stronger.	<50–800	(75–150) or 200–225*
Family Ephippidae				
SPADEFISH *Chaetodipterus faber*	Air bladder and intrinsic muscles.	Low-pitched, drum-like beats, single or as a short burst.	<50–800	(75–150)
Family Balistidae				
COMMON TRIGGERFISH *Balistes carolinensis*	Pharyngeal teeth and air bladder. Stridulation in pectoral arch. Pectoral finrays drumming against taut membrane above air bladder. First dorsal fin.	Metallic scratching and spitting. Hissing and heavy humming. Humming Clicking.	<50–4800	(2400–4800)
Family Monacanthidae				
FOOLFISH *Monacanthus hispidus*	Specially adapted incisor teeth and air bladder. First dorsal spine. Extrinsic feeding.	Sharp, whining swish-swish. Low click. Considerable "chirp."	<50–800	(150–300)

TABLE IV—(Continued)

Species	Origin of Sounds	Description of Sounds	Frequency Range (cps)	Principal Frequency (cps)
Family Monacanthidae (cont.)				
ORANGE FILEFISH *Alutera schoepfi*	Specially adapted incisor teeth and air bladder. Feeding on soft food.	Toothy scratching; wheezing; "ha-chu." Very loud scratching similar to above.	<50–4800	700
Family Tetraodontidae				
PUFFER *Spheroides maculatus*	Grating of incisor teeth.	Long burst of "erk-erk"'s; nasal rasp, double like stroke and recovery of saw.	200–1600	300 and 800
Family Diodontidae				
BURRFISH *Chilomycterus schoepfi*	Grating of incisor teeth.	High-pitched whining scrape like puffer, but "erk" single, c. ½ sec. duration.	<50–4800	(150–300) or (2400–4800)*
Family Cottidae				
LONGHORN SCULPIN *Myoxocephalus octodecimspinosus*	Stridulation in pectoral arch.	Low drumming, like a generator hum.	20–650	50–70
Family Triglidae				
COMMON SEAROBIN *Prionotus carolinus*	Air bladder and intrinsic muscles.	Single squawk or series of rapid clucks, always with vibrant quality.	40–1400	300 150 450 600†
REDWINGED SEAROBIN *P. evolans*	Air bladder and intrinsic muscles.	Single grunt or burst of croaks, less staccato than above.	40–800	200 100 300†
Family Batrachoididae				
TOADFISH *Opsansus tau*	Air bladder and intrinsic muscles.	Growl or coarse grunt, single, of c. ½ sec. duration. Very loud, intermittent blast like a boat whistle; "boop"; c. ½ sec. duration.	80–650 220–1000	100 200 300 400† 330 625 950†
Family Gadidae				
SILVER HAKE *Merluccius bilinearis*	Air bladder.	Low knock or rap.	80–875	300

* Indicates variation among specimens.
† Indicates more than one principal frequency in an individual sound (listed in order of magnitude).

GENERAL CONCLUSIONS

The mechanism producing sound obviously determines its frequency range, and the principal frequencies involved may be a key to the origin of each sound. Thus species which use the air bladder for noisemaking exhibit a wide spread of frequency, extending very often from below 50 to 1,200 or 1,400 cps. Most of the sound energy, however, is concentrated toward the low end of the spectrum, with principal frequencies in the 75–150 cps octave band. Examples are the spadefish, squeteague, striped bass, tautog, common eel, thread herring, and toadfish (growl).

The size of the air bladder is roughly proportional to the over-all length of the fish. Thus, since the frequency increases as the size of the air bladder decreases, we may expect smaller specimens to exhibit higher frequency characteristics. Accordingly smaller fishes of the species listed above, as well as small sea bass, scup, cunner, common and redwinged searobins, have principal frequencies in the 150–300 cps octave band. Except for variations according to size, the frequency characteristics of most species are quite constant, and each individual maintains an almost identical sound pattern under repeated stimulation.

Often the curves are strikingly harmonic in nature. In the case of the common searobin, present records show a pronounced maximum at 300, with lesser signals at 150, 225, 450, and 600 cps. Harmonic content is apparent also in curves for toadfish grunts where the maximum is 100, with decreasingly strong components at 200, 300, 400, 500, and 600 cps. A second sound typical of toadfish during the spawning season, also produced by means of the air bladder and its intrinsic muscles, has a frequency range of 220 to 1,000 cps. Here there is a strong fundamental at about 330 in a very narrow band, with a second peak about one-ninth as high at 625, and a third peak at 950 cps one-twentieth as large as the fundamental.

Typically the air bladder sound has a hollow tom-tomish quality, which has been described as similar to the noise of a wet finger drawn across the surface of an inflated balloon. The words "thump," "knock," "thud," "grunt," "groan," "growl," "cluck," "bark," and "boop," used to define this type of sound, evidence its drum-like, vibrant, and often guttural property.

Where the sound is produced by stridulation, that is, by the scraping of some body part against another, the spread of frequency may be extremely wide, and the maximum energy is usually located higher in the spectrum than is the case for sounds originating in the air bladder. Typical are the seahorse clicks, where the principal frequency is in the

400–800 cps octave band, and the burrfish rasps, which reach the 2,400–4,800 cps band. In all cases of stridulatory origin, the sound is quite easily identified by its rasping, scratching, or whining characteristic.

In certain fishes more than one mechanism may function for sound production, but the frequency characteristics as well as the observed tonal qualities usually serve to differentiate between the various sounds. The hardtail, for instance, is capable of very noisy rasping when agitated, with its maximum at 500 cps. The loudness of the burst indicates that the stridulation of the upper and lower pharyngeal teeth may be amplified by air bladder reinforcement. When the same specimen is subjected to electrical stimulation, a typical air bladder thump concentrated in the 75–150 cps and 150–300 cps octave bands predominates. The knock produced by electrically stimulating the rudderfish also has the general frequency characteristics of an air bladder sound, but a sharp rather than a dull quality to this sound suggests pharyngeal teeth participation.

The scup is a notable example of a fish with the ability to produce more than one sound. Under certain conditions, single guttural thumps, spread between 20 and 1,400 cps with principal frequencies in the low 200's, indicate air bladder origin, whereas shrill rasping, limited to a narrower band between 350 and 1,150 cps with a maximum at 700, accompanied the observed scraping of upper and lower incisors.

Also the squeteague uses its air bladder for single, deep, drum-like croaks spread between 20 and 1,200 cps with a usual maximum in the 50–100 cps band; under different stimulation, by grating the pharyngeal teeth it emits bursts of higher-pitched clucking in a narrow frequency range of 100 to 550, the maximum peak being at 325 cps.

No fish of the North Atlantic coastal waters auditioned to date has exhibited a sound lower than 20 cps nor higher than the 2,400–4,800 cps octave band. The large majority of species has principal frequencies between 75 and 300 cps. The possibility of frequencies beyond the range indicated by present apparatus is to be explored.

A detailed analysis of sonic species, especially with reference to frequency pattern, amplitude and periodicity of the recorded sounds, mechanisms and stimuli necessary for sound production, and the general significance of each fish as a maker of underwater sound, will be presented in a separate paper (4).

ACKNOWLEDGMENTS

We are greatly indebted to the Naval Ordnance Laboratory for loan of much equipment, to Professors R. Bruce Lindsay and Robert T. Beyer for advice on the analysis of the 1949 physical data, and to

John S. Kelly, Jr., Marilyn P. Fish, Joseph B. Munro, Jr., and Walter A. Sturm, summer student assistants.

REFERENCES

(1) FISH, C. J.
    1946. Oceanography in Japan. Navy. F2M-1007-46 (1) Oceanography. NTJ Ser. No. 112 (Index No. X-40N) (Jap) 409-900 MS7-10 Mar. 1946. 114 pp.

(2) FISH, MARIE P.
    1948. Sonic fishes of the Pacific. Pacif. Ocean. Biol. Project, Tech. Rept., *2:* 1-144 (Woods Hole Ocean. Inst.)

(3)   1949. Marine mammals of the Pacific with particular reference to the production of underwater sound. Pacif. Ocean. Biol. Project, Tech. Rept., *8:* 1-69 (Woods Hole Ocean. Inst.)

(4)   Character and significance of sound production among the fishes of the North Atlantic. In preparation.

(5) JOHNSON, M. W., F. A. EVEREST AND R. W. YOUNG
    1947. The role of snapping shrimp (Crangon and Synalpheus) in the production of underwater noise in the sea. Biol. Bull. Woods Hole, *93* (2): 122-138.

(6) NAVAL ORDNANCE LABORATORY
    1942. Measurement of background noise in the water at the Wolf Trap Range. NOL Rept., *691:* 1-6 (U. S. Navy Yard, 9 Nov. 1942).

(7)   1943. Acoustic measurements at the John G. Shedd Aquarium, Chicago, Ill. NOL Memorandum, *3416:* 1-10 (U. S. Navy Yard, 30 Mar. 1943).

(8)   1943. Analysis of available data on underwater background noise, in the region from 50 to 10,000 cps. NOL Memorandum, *2923:* 1-12 (U. S. Navy Yard, 21 Jun. 1943).

(9)   1944. Investigation of biological underwater background noises in vicinity of Beaufort, N. C. (10B1). NOL Rept., *880:* 1-22 (U. S. Navy Yard, Mar. 1944).

Copyright © 1962 by The Zoological Society of London
Reprinted from *Symp. Zool. Soc. London*, No. 7, 45–60 (1962)

THE BIOLOGY OF SOUND-PRODUCING FISHES

BY

N. B. MARSHALL

British Museum (Nat. Hist.), London, S.W.7

CONTENTS

	Page
Introduction	45
Swimming sounds	46
Gas-bubble sounds	47
Sound-producing structures	48
Stridulation	48
Vibration of the swimbladder wall	49
Sounds produced by quick head movements	50
An ecological survey of sound-producing fishes	50
Freshwater fishes	50
Marine fishes	51
Neritic species	51
Deep sea fishes	55
Summary	58
References	59

INTRODUCTION

Water is a more stable medium than air for the propagation of sounds. Except for boisterous aquatic habitats, such as torrential streams and surf, water is less turbulent than the atmosphere, less disturbed by eddies and other discontinuities. Underwater sound paths thus tend to be less devious than those in the atmosphere. During the passage of sound waves, frictional losses are less in the denser fluid: hence, comparable sounds can be transmitted to greater distances in water than in air. (Richardson, 1957).

Knowing also the (evolutionary) capacity of living organism to explore and exploit their surroundings, it is hardly surprising that some of the more complex aquatic animals (crustaceans and vertebrates) are able to produce definite sounds. Many of these are fishes. Those with special sound-making organs appear to be confined to the Teleostii, which are by far the most diverse of all fishes. Certain sharks and rays may make sounds as they eat. Rays with a crushing dentition are particularly noisy when feeding on shellfishes (Fish, 1954). But it must be remembered that the elasmobranchs are without a swimbladder, which, as we shall see, can be an effective sound-producing organ. Moreover, except for the pectoral fins of rays, the fins of elasmobranchs are much less mobile than those of teleosts, less suited to produce stridulatory sounds. Lastly, is cartilage, even when calcified, hard enough for making sounds of this kind?

Most of the teleosts known to have sound-producing organs belong to the orders Ostariophysi (characins, carp-like fishes and catfishes), Anacanthini (cod-like fishes and macrouroids), Solenichthyes (sea horses and pipe-fishes, etc.), Berycomorphi (squirrel fishes etc.), Zeomorphi (John Dories, etc.), Percomorphi, (perch-like fishes), Scleroparei (mail-cheeked fishes), Plectognathi (trigger-fishes, file fishes, puffer fishes etc.), and Haplodoci (toadfishes etc.). Many more sound-making species are likely to be discovered, for underwater exploration by means of hydrophones and recording equipment is still in its infancy. Turning from systematic range to sonic range, teleosts produce sounds from less than 100 to about 11,000 cycles per second. There is no evidence that fishes produce supersonic sounds, nor does it seem likely that they respond to such sounds, not at least to those used in modern echo-sounders. The most stentorian teleosts are the toadfishes (*Opsanus*). At a distance of two feet from a hydrophone the sounds emitted by a single toadfish may reach an intensity of over 100 dB (reference level 0·0002 dyne/cm^2), a value that is "......comparable to the noise of a riveting machine or a subway train (Tavolga, 1960 a). At the other end of the scale, the sounds made by courting males of *Bathygobius soporator* could barely be picked up by a hydrophone very close to the producers. The sound pressure levels (0·001 to 0·002 dyne/cm^2) were well below those due to local populations of snapping shrimps (Tavolga, 1958). Lastly the sound patterns of fish voices usually range from single, isolated calls to series of short pulses of sound. Each call or pulse generally lasts from hundredths to tenths of a second.

SWIMMING SOUNDS

Before considering the special means of producing sounds, we must first remember that fishes may make sounds as they swim. Not only are these kinds of sound intrinsically interesting, but during any acoustic investigation of fishes, such sounds must obviously be distinguished from those due to special organs. If, for instance, a fish is stimulated electrically to elicit its proper voice, it is likely to make sudden, "noisy" movements that will be part of the sounds recorded.

Swimming sounds have recently been studied by Tokarev (1958), Shishkova (1958) and Moulton (1960). By means of a Kay Vibralyzer vibration frequency analyzer, Moulton investigated the sounds recorded from schools of anchovies (*Anchoviella choerostoma*) and jack (*Caranx latus* and *C. ruber*) in the Bermuda area. One large school of anchovies from which sound records were taken covered about half an acre in 2 to 4 fathoms of water. When the fishes were idling, no sounds were received, but as soon as the school began to stream or veer, considerable noise was generated. The greater the degree of streaming or veering, the louder the sounds. During streaming movements the frequency of greatest sound intensity was below 500 c/s, the maximum frequency being 1,600 c/s. The corresponding figures for veering movements were below 800 c/s and 2,000 c/s. Rather similar results were got by recording the veering sounds of small groups of jack (*Caranx latus*) in an aquarium.

Shishkova (1958) refers to swimming sounds as hydrodynamic noise, the noise made by the flow of water round moving fish forms. Moulton (1960), who considers that some of the noise might arise from inner activities, such as friction between parts of the skeleton and swimbladders resounding to the quickened play of axial muscles, points out that none of the sounds " are continuous, even during steady streaming movements, indicating that they stem from the bodies of individual fishes during muscular contraction, rather than from the hydrodynamic effects of continuous water flow round fish bodies." Frequency analysis of the swimming sounds of *Anchoviella choerostoma, Caranx latus, C. ruber* and *Trachinotus palometa* also suggests that the larger the fish the lower the frequency of greatest sound intensity. As the frequency of tail beats (for comparable levels of swimming activity) is lower in larger fishes than smaller ones (Bainbridge, 1958), there may well be some direct linkage between the frequency of muscular body waves and swimming sound frequency. At all events, it is clear that further observations and analyses are desirable. One factor to be considered is that fishes without a swimbladder can make thumping sounds as they veer. Indeed, any sudden movement by an underwater body (of sufficient size) will produce such sounds.

Swimming sounds appear to have a definite biological significance. On playing back the recorded swimming sounds from an anchovy school to young yellow jack (*Caranx latus*) in an aquarium, the latter quickened their movements in non-directional fashion. Now, Moulton (1960) found that young jacks may join *Anchiovella* schools and that they prey on members of the school. He suggests that the predators may be attracted to the school by its swimming sounds. Even more interesting are his observations on the behaviour of blinded *Anchoviella*. Blinded fish did not school (in an aquarium) with normal individuals when these were slowly milling around, but when they were startled (by a hand movement) and began to stream and veer, the blind ones immediately joined them in schooling fashion. Considering this and other evidence, Moulton (1960) reached this conclusion : " While movements of individual members of the fish school are important in maintaining the remarkable integrity of the school, the members of the school are probably sensitive to pressure waves created by the school and probably orient to these waves ; the movements of the school as a whole are probably important in maintaining its own integration."

GAS-BUBBLE SOUNDS

Sounds produced as a by-product of another vital activity are well known in teleosts with an open (physostomatous) swim-bladder, one connected by a duct to the oesophagus or stomach. The most diverse of the physostomes are the Ostariophysi, most of which live in freshwaters. Other freshwater groups include the osteoglossoids, mormyroids, certain salmonoids and Esocidae (pikes). The herring-like fishes (Clupeidae) and the eels (Apodes) comprise most of the marine physostomes.

When a physostome expels gas from its swimbladder, as it will on becoming over buoyant, bubbles emerge from the mouth or gill chambers, and the release of each bubble is associated with a mouse-like squeak. Fishermen spot ascending shoals of clupeids by the altered appearance of the sea surface, due to the bursting of myriads of released gas bubbles. In the sprat, at least, the bubbles emerge not from the pneumatic duct, but from the posterior opening of the swimbladder (Verheijen, 1956). A rising school of clupeids must thus be a bedlam of mouse-like squeaks.

Apart from maintaining neutral buoyancy, cyprinid fishes emit gas bubbles if they are excited or disturbed (Dijkgraaf, 1932, Plattner, 1941). But there is no evidence that the associated sounds have any biological significance. Dijkgraaf (1941) played Pan to minnows (*Phoxinus laevis*) by blowing bubbles of a suitable size through a piece of glass tubing, but there was no response to the sounds. However, many more observations are needed, particularly in natural surroundings, before we can be sure that gas bubble sounds are entirely without significance.

SOUND-PRODUCING STRUCTURES

Sounds are produced in teleosts by stridulation, by vibration of the swimbladder wall, and by sudden movements of the body.

Stridulation

Stridulatory sounds are produced by friction between neighbouring parts of the skeleton. Such sounds, as might be expected, have a rasping, scraping or scratching or whining character. " The spread of frequency may be extremely wide, often from 50 to 4,800 cycles, and as a general rule the maximum energy is located higher in the spectrum than is usual in air bladder sounds." (Fish, 1954). According to Tavolga (1960 b) stridulatory sounds are usually non-harmonic.

One of the commonest means of stridulation is by friction between the pharyngeal teeth. In a great many teleosts the upper pharyngeal bones (upper parts of the gill arches) bear teeth that bite against the lower pharyngeals, a pair of toothed bones derived from the fifth gill arches and situated in the floor of the throat just behind the gills. In various carangids e.g. (*Caranx hippo, C. crysos, C. latus, C. ruber, Chloroscombrus chrysurus, Vomer setapinnis, Alectis cilaris* and *Seriola zonata*), haemulids (e.g. *Orthopristis chrysopterus, Haemulon plumiere, H. sciurus*), sciaenids (e.g. *Cynoscion regalis, Menticirrhus saxatilis, Stellifer lanceolatus* and *Bairdiella chrysura*) and balistids (e.g. *Balistes carolineniss*), the sounds produced by the pharyngeal teeth cause the swimbladder to resonate (see Burkenroad, 1930) 1931 and Fish, 1954). When the swimbladder is deflated a faint grating sound is produced instead of the normal sound, which is a loud, deep, rasping kind of croak (Burkenroad, 1930, 1931).

Stridulation between the jaw teeth or tooth plates seems to be a particular noise-making means in various plectognath fishes (e.g. *Stephanolepis hispidus, Balistes, vetula, Melichthys piceus, Spheroides maculatus, Chilomycterus schoepfi* and *Diodon hystrix*) (see Fish, 1951 and Moulton 1958). Again, stridulation between the basal parts of anterior enlarged dorsal or pectoral rays and their supporting bones is well developed in one particular group, the Siluroidea (e.g. *Doras maculatus, Synodontis schal, Pseudaroides clarias* and *Pelteobagrus*) (Sørensen 1884, Uchida, 1934). In two marine catfishes, *Galeichthys felis* and *Bagre marinus*, stridulation appears to be due to the interplay of a dorsal spine of the cleithrum and a socket in the post temporal bone. These sounds are enhanced by swimbladder resonance (Tavolga, 1960 b).

In sea horses (*Hippocampus*), stridulatory sounds are generated from friction between a star-shaped ossification on the nape (the coronet) and the posterior margin of the skull. Lastly, the bones of the pelvic girdle are involved in the stridulation of the long horn sculpin (*Myoxocephalus octodecimspinosus*) (Fish, 1954).

Vibration of the swimbladder wall

Apart from resounding to stridulatory sounds, the swimbladder wall may be thrown into vibrations by special drumming muscles, by the axial musculature and by the pectoral fins being used as brushes. Swimbladder sounds are usually low pitched, gutteral, vibrant and drum-like. The frequency range extends from below 50 c/s to 1,200 or 1,400 c/s, but most of the energy is concentrated towards the low end of the spectrum (Fish, 1954). Swimbladder sounds have several harmonics above a fundamental in the 100 to 300 c/s range (Tavolga, 1960 b).

In various catfishes of the families Doradidae, Pangasiidae, Auchenipteridae, Synodontidae, Malapteruridae and Ariidae, the drumming muscles are associated with an " elastic spring apparatus ". (Sørensen, 1884, Bridge & Haddon, 1893). In *Auchenipterus nodosus*, for instance, " ......the modified transverse processes of the fourth vertebrae are bent downwards and backwards to terminate in oval, bony plates attached to the anterior wall of the swimbladder. Inserted on the anterior face of each transverse is a powerful muscle which runs to the occipital region of the skull. The activity of the muscles, combined with the spring-like nature of the processes, throws the swimbladder into vibrations, which produce sounds (Sørensen, 1884, 1895)."

Continuing this quotation from a review of drumming mechanisms by Jones and Marshall (1953) : " In the second type one attachment of the external muscles is directly on the swimbladder wall, and the other on a neighbouring structure. The latter attachment is to the skull in species of the catfish family Pimelodidae (Bridge and Haddon, 1893), *Therapon* spp (Hardenburg, 1934), *Ophidium* spp and *Triacanthus brevirostris* (Sorensen, 1884), while in the gadoids, *Raniceps raninus* and *Gadus callarias* (Hagman, 1921), it is to the ribs adjacent to the anterior part of the bladder. In other gadoids, *Molva molva* and *Lota lota* (Hagman, 1921) and in the sciaenids *Cynoscion regalis, C. nebulosum, Micropogon undulatus* and *Bairdiella chrysura* (Tower, 1908), the outer part of each drumming muscle is attached to the lateral wall of the body cavity.

" In the third type of sound-producing swimbladder, the attachments of the external muscles are entirely on the bladder wall, to which the muscles are closely applied. Examples are the gadoids, *Brosmius brosme, Gadus pollachius, G. aeglefinus, G. virens* and *Phycis mediterraneus*, the macrurid fish, *Macrurus fabricii* (Hagman, 1921), the John Dory, *Zeus faber* (Dufossé, 1874), gurnards belonging to the genera *Trigla* (Dufossé, 1874 ; Rauther, 1945) and *Prionotus* (Tower, 1908 ; Burkenroad, 1931), the singing midshipman, *Porichthys notatus* (Green, 1924) and the toad-fish *Opsanus tau* (Tower, 1908)."

Contraction of axial muscles, parts of which are closely associated with the swimbladder wall, appear to be responsible for sound production in : *Merluccius bilinearis, Holocentrus ascensionis, Roccus saxiatilis, Centropristes striatus, Tautoga onitis, Epinephelus striatus, Pomacanthus arcuatus* and *Angelichthys ciliaris*. (Fish, 1954, Moulton, 1958).

Lastly, in certain trigger fishes (*Balistes carolinensis, B. vetula* and *Melichthys piceus*) there is a thin drumming membrane, covered with large, plate-like scales, above the base of each pectoral fin and this lies close to the lateral wall of the swimbladder. When a triggerfish flutters the pectoral fins over this membrane, throbbing swimbladder sounds are produced. (Fish, 1954, Moulton, 1958).

Sounds produced by quick head movements

In *Bathygobius soporator* (see also page 55) and various blennies, (*Chasmodes bosquianus, Hypleurochilus geminatus* and *Hypsoblennius hentz*), thumping sounds are made by sudden downward or sideways movements of the head. These sounds are definitely not incidental, for they play an important part in prespawning behaviour (Tavolga, 1960 b).

AN ECOLOGICAL SURVEY OF SOUND-PRODUCING FISHES

Freshwater fishes

Three quarters of the kinds of primary freshwater fishes belong to the Ostariophysi, fishes in which the swimbladder is linked to the inner ears (pars inferior) by Weberian ossicles. In another rather diverse group, the Mormyroidea, each inner ear has a closely fitting air sac, derived during development from a forking of the forward end of the swimbladder. The suprabranchial, air-breathing chambers of the anabantoid fishes can also act as hearing aids.

Concerning their hearing powers, Dijkgraaf (1960) calls such fishes " specialists ", as opposed to " normal " fishes without such extra means of sound reception. Compared to " normal " species the " specialists " have keener hearing, an extended frequency range and better powers of pitch discrimination.

Perhaps the enhanced hearing capacities of the Ostariophysi may have contributed to their remarkable success in freshwater environments (Winn & Stout, 1960). Certainly, the Weberian apparatus is always perfectly developed, even though it may be modified in some ostariophysans that have the anterior chamber of the swimbladder partly enclosed in a bony capsule (the posterior chamber being reduced or absent). As we wrote elsewhere concerning these species : " In certain of the catfishes, homalopterids, cobitids and gobiodontine cyprinids, there is a close connection between the outer walls of the anterior chamber and the skin, which looks remarkably like a sound receiving device. Everything possible seems to have been done to increase the density in connection with a bottom-living habit and yet keep in being the accessory auditory channel " (Jones & Marshall, 1953). But virtually nothing is known of the part played by sound in the lives of ostariophysan and other kinds of freshwater fishes.

The prevalence of sound production among catfishes (Siluroidea) is thought to be correlated with means of assembling the sexes in turbid waters, but underwater records have yet to be taken. Winn & Stout (1960) have recorded the sounds made the satin fin shiner (*Notropis analostanus*) and related North American cyprinids. Male and female satin fins produce sharp knocking sounds lasting from 11 to 60 milliseconds and having a main frequency range of 85 to 2,000 cycles. There are occasional extensions to frequencies as high as 11,000 c/s. These knocking sounds are most frequently uttered during the breeding season, and they are particularly evident during fights between male fish. When a male is courting a female the knocks are not so loud and they are produced so frequently that the result is a purring sound. Production of sound by the males is highest at (breeding) temperatures between 25 to 30° C and after treatment with testosterone (at a temperature between 25 and 27°C). It is interesting that in nature these sounds may be produced in the fairly noisy conditions obtaining in small rapid streams, but the authors remark that the calls are uttered when the fish are close together.

Delco (1960) has obtained interesting results on sound discrimination by males of two other species of *Notropis*, the blacktail shiner (*Notropis venustus*) and the red shiner (*N. lutrensis*). Only the ripe females appear to be vocal. Sound spectrograph analysis of the calls of female *Notropis lutrensis* revealed that they are trilled and each lasts for 0·84 sec., the dominant frequency range being 200 to 275 c/s. The calls of female *N. venustus* are not trilled and their duration is much shorter (·047–·07 sec), the dominant frequency band (125 to 250 c/s) also being lower.

After playing back these sounds in an observation tank and noting the relative times spent by male fish in the part of the tank containing the loudspeaker, Delco concluded that the males of these two kinds of shiner showed " ......a positive approach reaction to female calls of their respective species, and that the effective discrimination by the males, even when simultaneously offered calls of the other species, operates as an important sexual isolating mechanism ". An intriguing feature of the behaviour of the males was that on swimming round the playback speaker they would sometimes nip it with their jaws. They are evidently well aware of a sound source.

These two investigations of sound production and behaviour in cyprinid fishes are welcome. Study of underwater sounds, both those made by moving water and those due to fishes should be particularly rewarding in freshwaters, which are populated by so many auditory " specialists ". To adapt a classic of English misusage : Here is almost a virgin field, pregnant with possibilities.

MARINE FISHES
Neritic species

Precise study of the sounds made by marine fishes began—and it began of necessity—during the last world war—" ......when listening for enemy ships became a specialized employment for Navy men and their gear. Listening personnel on warships mistook the calls of fishes for the sound of enemy ships, harbour stations monitored these calls and defence forces were marshalled on false alarm ; sonic mines dropped into the Sea of Japan were discharged by animal sounds. As a result of these experiences, biologists were sent to the Pacific area, where the sounds were particularly troublesome, to study the distribution and sources of these sounds. In this study, of forty-two families of fishes studied, seventeen were found to contain noise making members. More recently, Dr Marie Fish has described fifty-four additional sound-producing species of our own (American) coastal waters." (Moulton, 1956). Some of the more important neritic teleost families containing sound-making members are : the Gadidae (cod-like fishes), Syngnathidae, (sea horses), Zeidae, (John Dories), Serranidae (groupers, etc) Carangidae (jacks etc.,) Haemulidae (grunts), Chaetodontidae (butterfly fishes and angel fishes), Sciaenidae (drum fishes), Gobiidae (gobies), Scorpaenidae (scorpion fishes), Cottidae (bullheads), Triglidae (gurnards), Balistidae (trigger fishes), Monacanthidae (file fishes), Diodontidae (porcupine fishes), Tetraodontidae (puffer fishes) and the Batrachoididae (toad fishes and singing midshipman).

Wartime listening at hydrophone stations along both coasts of the U.S.A. revealed that there were daily and seasonal rhythms in fish sounds, the bulk of which were traced to various kinds of drum fishes (Sciaenidae). The seasonal changes were evidently linked to the reproductive migratory habits of three fishes, while sharp diurnal rises in sound level coincided with feeding periods. For instance, croakers (*Micropogon undulatus*) begin to feed at dusk and it was from this time until midnight that their choruses were loudest. The sound pattern consists of bursts of drumming in rapid succession, which are not unlike the noise of a pneumatic drill (pulse length 0·02 sec, principal frequency 250 c/s) (Knudsen, Alford & Emling, 1948).

Why croakers and other sciaenids should drum as they feed is not at all clear, but perhaps some light has been thrown on this behaviour in other fishes by Moulton's (1960) work in the Bermuda area. When recorded pharyngeal tooth rasps of jack (*Caranx latus*) were played back to members of this species in an aquarium, the response, and it was a decided one, appeared to be a feeding reaction. " The fish became exceedingly active, swimming about furiously, and facing the transducer to nibble at its rubber surface." Clearly, this is but a preliminary observation, inviting further study, but it is not impossible that noisy feeding might have a biological function. One can imagine a group of foraging fish gathering round an individual that had chanced to find a good supply of food, the discovery of which would be plainly audible. In this way feeding becomes communication and food resources would tend to be shared. But there is another aspect of audible feeding activities. When playing back these sounds (from *Caranx latus*) into the sea ...... " an adult barracuda (*Sphyraena barracuda*, Sphyraenidae) came abruptly to a stop about eight feet from the suspended transducer and lay quietly facing it for about three minutes " (Moulton, 1960). Even at best underwater vision is very restricted, but if a predator can hear his prey and find the sound sources, which are preoccupied with filling their own stomachs, what a fine opportunity he has. On the other hand, a prey species may be alerted by the feeding sounds of a predator : When the tooth rasps of jack (*Caranx latus*), were played to a school of anchovies (Anchoviella), these were quick to clear the area around the transducer (Moulton, 1960).

Swimbladder structure and sound patterns vary from one kind of teleost to another, which suggests that members of a species could use their " voices " as a means of recognition. The swimbladders of four drum fishes, and they are common off the eastern shores of the U.S.A. are shown in Fig. 1. Such individual expression of swimbladder form, ranging from the simple, spindle-shaped sac of the spot (*Leiostomus xanthurus*) to the diverticular elaboration of the bearded drum (*Pogonias cromis*) must surely be related to individual acoustic expression. When the drumming muscles are in action, the acoustical effects of swimbladder appendices must be added to those of the main cavity. But a characteristic resonant response, the modal frequency of which depends on the length of the swimbladder (larger fishes having deeper voices than smaller ones), and extra qualities of timbre, are not the only means of individual sonic expression. Sequences of sound pulses may also have an individual, code-like character. This is well seen in two species of toad fishes.

Toadfishes (*Opsanus*) have a heart-shaped swimbladder, and in *Opsanus tau* which ranges from the Gulf of Maine to the West Indies, the cleft of the heart reaches almost to the middle of the sac. In *Opsanus beta*, a species found in the Gulf of Mexico, the swimbladder is not so deeply cleft. (Tavolga, 1960 b).

These fishes make two kinds of noises. They growl as when disturbed by man or fish (a male guarding eggs being particularly vocal), and they make loud hooting sounds rather like blasts of a ship's siren. The growls are virtually identical in both species (fundamental frequency 100 c/s with harmonics up to 1,000 c/s), except that those of *Opsanus tau* are a little more abrupt. There is a much greater contrast in the hooting calls. In *Opsanus tau* the fundamental frequency is 140 c/s with harmonics up to nearly 2,000 c/s : the corresponding figures for *O. beta* are 350 c/s with a harmonic range to 3,850 c/s. These differences are presumably related to divergences of swimbladder structure and size. *Opsanus tau* is a larger fish than *O. beta* (average sized *tau* are about 240 mm. in length, whereas those of *beta* are about 130 mm.), and we have seen that the swimbladder of tau is more deeply cleft. However, there is a further difference in sound pattern. Whereas *tau* emits single hoots, *beta* produces double hoots, each hoot being prefaced by a short grunt (Tavolga, 1958 a, 1960 a & b).

The part played by these sounds in the life of a toadfish has still to be explored. Growling sounds are clearly associated with aggressive behaviour. It seems unlikely that the hooting noises are mating calls, for they are by no means confined to the spawning season. Tavolga (1960 b) is inclined to think that these sounds may be linked with territorial habits, and perhaps to species discrimination. More observations are clearly required.

Besides specific divergence in swimbladder structure there is sexual dimorphism. In drum fishes such as *Bairdiella chrysura Leiostomus xanthurus*, *Pogonias chromis*, *Stellifer lanceolatus* and *Cynoscion* (*nothus, nebulosus* and *arenarius*) only the males possess drumming muscles on the swimbladder. The females are limited to croaking sounds, made by stridulation of the pharyngeal teeth, (Burkenroad, 1931). However, both male and female croakers (*Micropogon undulatus*) have drumming muscles.

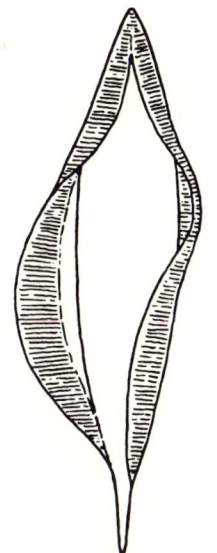
Leiostomus xanthurus ♂

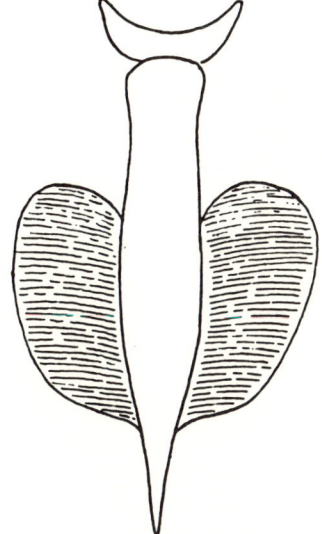
Bairdiella chrysura ♂

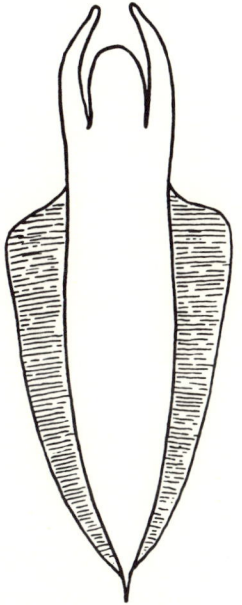
Cynoscion nebulosus ♂

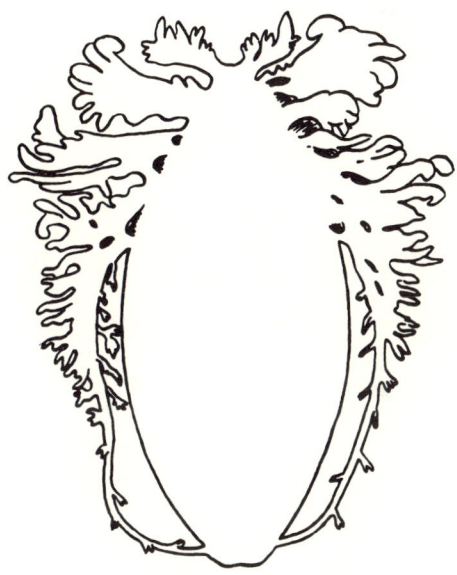
Pogonias chromis ♂

Fig. 1—Variation of swimbladder form in four common Eastern American drumfishes. In *Leiostomus xanthurus*, *Bairdiella chrysura* and *Cynoscion nebulosus*, which have been redrawn from Tower (1908), the drumming muscles are shown.

This is cogent anatomical evidence that sounds must be linked with some phase of sexual congress in drum fishes, but close study of any one species has yet to be made. Dufossé (1874) observed that the sounds made by sea horses

(*Hippocampus brevirostris*) were particularly evident during the breeding season. One kind of sound is made by stridulation between the back of the skull and the front of the coronet, a star-shaped ossification on the nape. When a sea horse tosses its head, friction between skull and coronet produces snapping sounds, and recent observations by Fish (1954) have charmingly shown the association between these sounds and mating activities in *Hippocampus hudsonius*. These are her words : " Later observations on the mating of a pair of captive sea horses revealed that preliminary activity consisted of slow swimming, either together or apart, accompanied by noisy snapping of the head. Clicks were often produced alternately by the two fishes and during their actual embrace, these sounds were loud and almost continuous."

Sound making in *Bathygobius soporator* is confined to the male, who only appears to call when he is courting a female. (Tavolga, 1958 b, 1960 b). Outside his shelter the male produces low pitched grunts which coincide with sudden, downward thrusts of the head and the emission of strong jets of water from the upper parts of the gill openings. The sounds are non-harmonic ; they last from 150 to 350 m secs and have a main frequency range of 100 to 500 c/s. Most of the sound energy is contained in a band from 110 to 150 c/s. On approaching a gravid female, the male utters these sounds three or four times in a second, and they are also frequent when a female is following a male to his shelter.

When these calls were played back, male fish would come out of their shelter and approach the hydrophone. Similar sounds recorded from a blenny (*Chasmodes bosquianus*), artificially generated sounds (from pure sine waves) and even human imitations of their grunts were sufficient to elicit decided approach responses. Females would only approach the hydrophone if a captive goby was placed nearby. Sound reception, as Tavolga concluded in an earlier (1956) paper, is but one of the sensory means concerned with prespawning behaviour in *Bathygobius*. Vision and olfaction are also involved, and they combine with hearing to facilitate successful pairing and spawning in this interesting species of intertidal goby.

As already implied, the blenny, *Chasmodes bosquianus* makes very similar sounds (they are almost identical in frequency) and they are synchronous with a quick, sideways shake of the head. At six to eight inches from the hydrophone the sound pressures ranged from 0·005 to 0·008 dyne/cm^2, which may be compared with 0·001 to 0·002 dyne/cm^2 that were recorded from *Bathygobius* under the same conditions. Tavolga (1960 b) estimates that in terms of decibels above human auditory threshold, these figures would be equivalent to about 30 dB and 20 dB respectively. Again, these sounds are confined to the males and they are associated wih courtship, which is also true of two other blennies, *Hypleurochilus geminatus* and *Hypsoblennius hentz* (Tavolga, 1960 b).

Deep sea fishes

There is plenty of recorded evidence that the deep ocean may be full of sounds, but the producers, other than cetaceans, have yet to be identified. Concerning fishes, one approach to this problem is anatomical, to see if the swimbladder has special drumming muscles. Extensive survey of swim-bladder structure in bathypelagic teleosts did not reveal such muscles (Marshall 1960), but they are found in the rat-tailed fishes (Macrouridae), which are benthic in habit. (Marshall, 1954). (This does not, of course, prove that the bathypelagic species are silent, for contractions of the axial muscles and nearby stridulations can elicit resonant sounds from the swimbladder, see p. 49).

Since this earlier account, I have now found drumming muscles in eight other species of rat-tails, making eleven in all. (Many other species will need to be examined if this survey is to be representative). Some of these dissections are drawn in Fig. 2. which shows that in some species (e.g. *Nezumia bairdi*, *N. hildebrandi* and *Coelorhynchus caribbaeus*) the muscles are completely attached to the swimbladder wall : in others, one attachment is to the body wall (e.g. *Malacocephalus laevis*, *M. occidentalis* and *Hymenocephalus cavernosus*). These two kinds of attachment are very like those found in drumfishes (see Fig. 1) and other groups (see p. 49), and it is reasonable to assume that the muscles, which like all drumming muscles are red, are concerned with sound production. Direct evidence of sound making in rat-tailed fishes is limited to a Mediterranean *Coelorhynchus*, which when caught utters a croaking sound.

So far, I have found that drumming muscles are confined to male rat-tailed fishes, which is not to say that the females are mute. These fishes are very diverse in the warmer regions of the ocean, but some are found in temperate and cold waters. Most species live over the upper reaches of the continental slope, the populations being largely contained within a depth range of 200 to 1,000 metres. Between such levels there is an underwater twilight, and down to 4,000 metres, at least, there is a background of luminescent flashes (Clarke and Hubbard 1959). Moreover, many macrourids have their own means of luminescence, yet water is a poor transmitter of light, and at best underwater vision is very limited compared to that on land. But in the vastness of deep waters, and remembering that water is the acoustically superior medium, the extra means of sound signalling possessed by male rat-tails can be imagined to be vital for assembling the sexes during the breeding season. It may also be significant that all but the abyssal species have an extremely large otolith in the sacculus, which in most fishes is the part of the inner ear concerned with hearing.

Certain deep water sound sequences that were studied by Griffin (1950) are of interest here. By reconstructing the sound paths of single loud calls and their echoes from the deep sea floor, Griffin was led to wonder whether the caller might have been engaged in echo-sounding. If a fish, as it well may have been, it could easily have heard the echoes, but would it then be able to gain some awareness of depth ?

It is interesting to speculate whether the fish might have been a macrourid. The main frequency of the calls (500 c/s) is what might be expected from the swimbladder of a rather large rat-tailed fish. Now these fishes certainly live along the deep-sea floor, as cameras and observers in bathyscaphes have shown, but they can also be caught at mid-water levels. Perhaps they move off the bottom during the spawning season and then lay their eggs at some favourable mid-water level. If so, echo-sounding would be useful.

THE BIOLOGY OF SOUND-PRODUCING FISHES 57

To return to harder concerns, what is needed is a means of identifying the sound makers. One possibility, which I hope to explore, is a deep-sea camera which can be triggered of by a receiver tuned to relevant sound frequencies. To take clear pictures, the receiver will probably have to be rather insensitive.

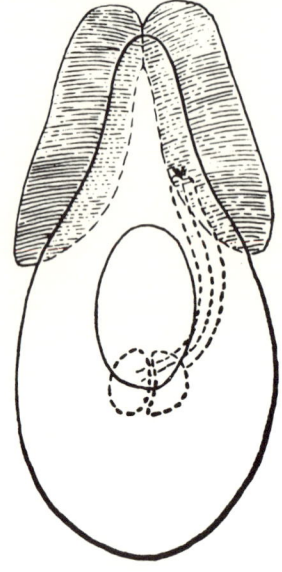

Hymenocephalus cavernosus ♂

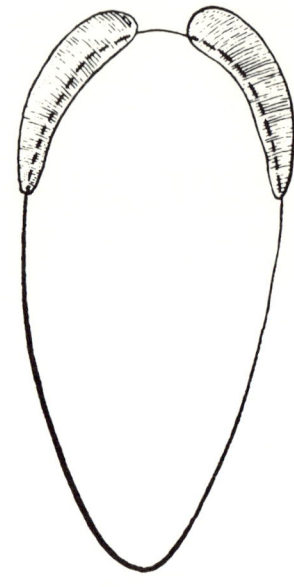

Malacocephalus occidentalis ♂

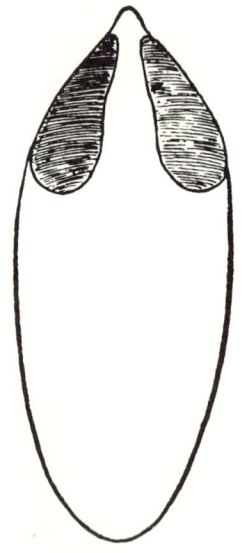

Nezumia hildebrandi ♂

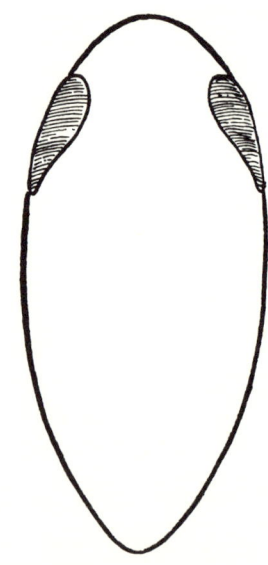

Cariburus zaniophorus ♂

Fig. 2—Four rat-tailed fishes with special drumming muscles on the swimbladder. In *Hymenocephalus cavernosus* and *Malacocephalus occidentalis* one attachment of each drumming muscle is to the body wall, the other being on the swimbladder. In species such as *Nezumia hildebrandi* and *Cariburus zaniophorus* these muscles are intrinsic to the swimbladder.

Lastly, macrourids are unlikely to be the only sound producing fishes in the deep ocean. The Brotulidae, another diverse and common group of benthic fishes, and the deep-sea cods (Moridae) may well have direct or indirect means of making swimbladder sounds. Moreover, the morid fishes appear to be unique among deep-sea species in having connections between the swimbladder and the inner ears. What is the significance of this extra auditory channel?

SUMMARY

Fishes may make (incidental) sounds as they swim or by emitting gas bubbles from the swimbladder. Special sounds making structures appear to be confined to the Teleostii, which produce sounds by stridulation, by vibration of the swimbladder wall and by sudden movements of the head.

Swimming sounds may lead to responses by predators (or prey) and they seem to be involved in schooling behaviour. Gas bubble sounds appear to be without biological significance.

Stridulatory sounds are generated by friction between neighbouring parts of the skeleton, those made by the pharyngeal teeth often inducing resonant response in the swimbladder. Direct means of eliciting swim-bladder sounds are by way of special drumming muscles, which in certain catfishes are associated with an elastic spring mechanism. Contractions of nearby parts of the axial musculature and pectoral fin movements (certain trigger fishes) may also lead to sound production by the swimbladder. Swimbladder sounds tend to be lower in pitch than those due to stridulation. Quick head movements (in *Bathygobius* and certain blennies) give rise to sharp thumping sounds.

A great many fresh fishes, most of which belong to the order Ostariophysi, are auditory "specialists": they have some form of linkage between the swim bladder and the inner ears. This accessory auditory channel enhances their hearing capacities, but little is known of the part played by sounds in their lives. Numerous catfishes produce sounds (by stridulation or by vibration of the swimbladder wall), which could be an important means of assembling the sexes in turbid waters. Cyprinid fishes (*Notropis* spp. and related kinds), are particularly vocal during the breeding season. Sounds may be uttered during fighting (among males) and during courtship, and it seems likely that they are used as species recognition signs and as a means of isolating the sexes.

There is anatomical and acoustical evidence that many of the shelf-dwelling marine fishes are sound producers. Studies are now beginning to reveal the biological roles of sound. Those made by the pharyngeal teeth may lead to feeding activities (and attract predators or alert prey). Sounds are certainly associated with aggressive display and mating activities, and, most likely, with species recognition and territorial behaviour.

Sounds are also prevalent in the deep ocean. Anatomical studies, which are still in progress, indicate that the rat-tailed fishes must be conspicuous sound producers. Swimbladder drumming muscles appear to be confined to the males. Means of communication (and perhaps of echo-sounding) would seem to be particularly important in the deep sea environment.

REFERENCES

Bainbridge, R. (1958). The speed of swimming of fish as related to size and to the frequency and amplitude of the tail beat. *J. exp. Biol.* **35** : 109.
Bridge, T. W. & Haddon, A. C. (1893). Contributions to the anatomy of fishes. II. The air bladder and Weberian ossicles in the siluroid fishes. *Philos. Trans.* (B) **184** : 65.
Burkenroad, M. D. (1930). Sound production in the Haemulidae. *Copeia* **1930** : 17.
Burkenroad, M. D. (1931). Notes on the sound producing marine fishes of Louisiana. *Copeia* **1931** : 20.
Clarke, G. L. & Hubbard, G. J. (1959). Quantitative records of the luminescent flashing of oceanic animals at great depth. *Limnol. Oceanogr.* **4** : 163.
Delco, E. A. (1960). Sound discrimination by males of two cyprinid fishes. *Texas J. Sci.* **12** : 48.
Dijkgraaf, S. (1932). Über Lautäusserungen der Elritze. *Z. vergl. Physiol.* **28** : 389.
Dijkgraaf, S. (1941). Haben die Lautäusserungen der Elritze ein biologische Bedeutung ? *Zool. Anz.* **136** : 103.
Dijkgraaf, S. (1960). Hearing in bony fishes. *Proc. roy. Soc.* (B) **132** : 51.
Dufossé, M. (1874). Recherches sur les bruits et les sons expressifs que font entendre des poissons d'Europe. *Ann. Sci. nat.* (Zool.) ser. 5, **19** : art. 5 ; **20** : art 3.
Fish, M. P. (1954). The character and significance of sound production among fishes of the Western North Atlantic. *Bull. Bingham Oceanogr. Coll.* **14** : art 3.
Greene, C. W. (1924). Physiological reactions and structure of the vocal apparatus of the California singing fish, *Porichthys notatus*. *Amer. J. Physiol.* **70** : 496.
Griffin, D. R. (1950). Underwater sounds and the orientation of marine animals. Project N.R.162–429, Contract N.6 onr. 264 t.o.g. between the Office of Naval Research and Cornell University, no. 3 pp. 1–26.
Hagman, N. (1921). *Studien über die Schwimmblase einiger Gadiden und Macruriden.* Akad. Abhand. Lund.
Hardenburg, J. D. F. (1934). Ein Tone erzeugender Fisch. *Zool. Anz.* **108** : 224.
Jones, F. R. H. & Marshall, N. B. (1953). The structure and functions of the teleostean swimbladder. *Biol. Rev.* **28** : 16.
Knudsen, V. O., Alford, R. S. & Emling, J. W. (1948). Underwater ambient noise. *J. mar. Res.* **7** : 410.
Marshall, N. B. (1954). *Aspects of deep sea biology.* London : Hutchinson.
Marshall, N. B. (1960). Swimbladder structure of deep-sea fishes in relation to their systematics and biology. *Discovery Rep.* **31** : 1.
Moulton, J. M. (1956). *Fishes and sound in the sea.* Bowdoin Aluminus (February).
Moulton, J. M. (1958). The acoustical behaviour of some fishes in the Bimini area. *Biol. Bull. Woods Hole* **114** : 357.
Moulton, J. M. (1960). Swimming sounds and the schooling of fishes. *Biol. Bull. Woods Hole* **119** : 210.
Plattner, W. (1941). Études sur la fonction hydrostatique de la vessie natatoire des poissons. *Rev. suisse Zool.* **48** : 201.
Rauther, M. (1945). Über die Schwimmblase und die zu ihr in Beziehung tretenden somatischen Muskeln bei den Triglidae und anderen Scleroparei *Zool. Jb.* (Anat.) **69** : 159.
Richardson, E. G. (1957). Propagation of sound in the atmosphere and the sea. in *Technical aspects of sound* **2** : 1. Amsterdam.
Shishkova, E. V. (1958). Recording and study of sounds made by fish. *Trud. VNIRO* **36** : 280.
Sørensen, W. (1884). *Om lydorganer hos fiske. En physiologisk og comparativ—anatomisk undersøgelse,* Copenhagen.
Sørensen, W. (1895). Are the extrinsic muscles in the air bladder in some Siluridae and the " elastic spring " apparatus of others subordinate to the voluntary production of sounds ? What is, according to our present knowledge, the function of the Weberian ossicles ? *J. Anat., Lond.* **29** : 109 ; 205 ; 399 ; 518.
Tavolga, W. N. (1956). Visual, chemical and sound stimuli as cues in the sex discriminatory behaviour of the gobiid fish, *Bathygobius soporator*. *Zoologica, N.Y.* **41** : 49.
Tavolga, W. N. (1958 a). Underwater sounds produced by two species of toadfish, *Opsanus tau* and *Opsanus beta*. *Bull. mar. Sci. Gulf. Carib.* **8** : 278.

TAVOLGA, W. N. (1958 b). The significance of underwater sounds produced by males of the gobiid fish, *Bathygobius soporator*. *Physiol. Zool.* **31** : 259.
TAVOLGA, W. N. (1960 a). Foghorn sounds beneath the sea. Nat. *Hist. N.Y.* **69** : 44.
TAVOLGA, W. N. (1960 b). Sound production and underwater communication in fishes. *In Animal sounds and communication*. Pp. 93–136. *Publ. Amer. Inst. biol. sci. Washington.* No. 7.
TOKAREV, A. K. (1958). On biological and hydrodynamic sounds produced by fish. *Trud. VNIRO* **36** : 272.
TOWER, R. W. (1908). The production of sound in the drum fishes, the sea robin and the toadfish. *Ann. N.Y. Acad. Sci.* **18** : 149.
UCHIDA, K. (1934). On the Japanese species of sound-producing fishes. *Nipon Gakujitsu Kyokai Hokotu* **9** : (2) May, 1934, (in Japanese).
VERHEIJEN, F. J. (1956). On a method for collecting and keeping clupeids for experimental purposes together with some remarks on fishery with light sources and a short description of free cupulae of the lateral line organ on the trunk of the sardine, *Clupea pilchardus*, Walb. *Pubbl. Staz. Zool. Napoli* **28** : 225.
WINN, H. E. & STOUT, J. F. (1960). Sound production by the satin fin shiner, *Notropis analostanus*, and related fishes. *Science, N.Y.* **132** : 222.

Part IV

MECHANISMS OF SWIM BLADDER SOUND PRODUCTION

Editor's Comments
on Papers 12 Through 15

12   SKOGLUND
Abstract from *Neuromuscular Mechanisms of Sound Production in* Opsanus tau

13   PACKARD
Electrophysiological Observations on a Sound-producing Fish

14   TAVOLGA
Mechanisms of Sound Production in the Ariid Catfishes Galeichthys *and* Bagre

15   HARRIS
Considerations on the Physics of Sound Production by Fishes

In species where sound appeared to be important in communication, the sonic mechanisms often involved the swim bladder and associated specialized muscles. Studies in functional anatomy of such sonic mechanisms became a major area of research in the field, and the articles at the beginning of this volume review the large literature on the subject. A few articles that can be considered major contributions to the field are reprinted here.

The first neurophysiological evidence of the rapid responsiveness of sonic muscles was presented in a short abstract by C. R. Skoglund, and a confirmation of these properties in an entirely different species was reported by A. Packard. Sonic muscles in many divergent species, e.g., toadfish, ariid catfish, squirrelfish, etc., may well be homologous as evidenced by a common innervation. The muscles appear to possess a remarkable resistance to tetanization, coupled with their fast-acting properties.

The next two papers are essentially studies in functional anatomy, in which the skeletal, muscular, and neural morphology of the sonic mechanisms in certain marine catfish were described. The sound output was shown to be a direct translation of the vibratory rate of the sonic muscles, with swim bladder resonance contributing little to the system. Based, in part, on the paper by Tavolga, G. G. Harris showed how the physics of acoustical energy in water is related to swim bladder sonic mechanisms, and, in addition, he presented a summary of the principal equations dealing with underwater sound generation and propagation. Both these papers are now out of print and thus difficult to obtain.

NEUROMUSCULAR MECHANISMS OF SOUND PRODUCTION IN *OPSANUS TAU*

C. R. Skoglund

The functional activity of the intrinsic striated muscles of the swim bladder, which are engaged in sound production of the toadfish (Tower, 1908; Fish, 1954), has been studied by simultaneous recordings of action potentials, mechanical effects, and sound. Most experiments have been performed on the excised bladder kept in air at room temperature (21°C).

The motor nerve was found to conduct at above 25 m/sec. The shape of the compound action potential indicated fairly uniform fiber sizes, which was confirmed by preliminary histological analysis (fiber diameter 10 μ).

The muscle consists of short fibers running transversely to the long axis of the muscle. A single maximal nerve volley caused a well synchronized contraction as verified by simultaneous recording from different points by steel microelectrodes.

When the muscle contraction was recorded by an RCA transducer with stylus placed against the muscle surface, the earliest detectable movement occurred during the falling phase of the muscle action potential. Contraction peak was reached within 5 –8 msec, and relaxation was complete in an additional 5 –7 msec. The changes in bladder pressure during contraction, measured by a capacitance manometer, showed a similar time course.

The sound recording with an electrodynamic microphone (60 –10,000 cycles) was limited for technical reasons to the initial high-amplitude change. This was recorded as a di- or triphasic wave lasting about 2.5 msec, and audible as a "pop" sound. The first deflection was synchronous with the falling phase of the muscle action potential, and thus, the second occurred during the early phase of recorded muscle movement.

The data are used to interpret the natural sound production of the toadfish, of which some comparative studies have also been made; the results may be relevant to the general problem of muscle sound generation.

REFERENCES

Fish, M. P. (1954). The character and significance of sound production among fishes of the western North Atlantic. *Bull. Bingham Oceanographic Coll.*, **14**, 1–109.

Tower, R. W. (1908). The production of sound in the drumfishes, the sea robin and the toadfish. *Ann. N. Y. Acad. Sci.*, **18**, 149–180.

ELECTROPHYSIOLOGICAL OBSERVATIONS ON A SOUND-PRODUCING FISH

A. Packard

Zoological Station, Naples, Italy

SIMULTANEOUS sound and action potential records have been made from a New Zealand marine teleost, *Congiopodus leucopaecilus* (Perciformes), the pigfish, which produces noises in the water by resonance of the swimbladder. The muscles responsible for the sound are a pair of delta-shaped intracostal striated muscles running obliquely between the vertebral intercentra and back border of the pectoral girdle of either side of the body, and lying against the side walls of the swimbladder. In a fish 24 cm. long the swimbladder is 4·5 cm. and the muscles 3 cm.

The records were made on the two beams of a double-beam oscilloscope, the top beam recording the action potentials from a platinum electrode touching the surface of the exposed muscle, and the bottom beam recording the wave form of the sounds in air from a crystal microphone (of the type used in hearing aids) placed against the opposite side of the body. The fish was mounted on a block of wood and perfused through the mouth with aerated seawater.

The records show the expected correlation between the moment of muscle contraction and the onset of the sound produced. They were obtained without artificial electrical stimulation of any kind, and from behaviour observations and tape recordings of the fish in aquarium tanks can be taken as representing the normal pattern of sound production in this fish. Each sound may be thought of as a single drum beat, and they occur singly or at intervals of more than a second, or as a train of beats at rates of 4–10 per sec. when their function probably includes that of threat. These repetitive sounds are accompanied by raising of the dorsal fin armed with stout spines, and dissection reveals that the occipitospinal nerve supplying the drumming muscle also has fibres running dorsally to the spine-raising trunk musculature.

The muscle spike is of extremely short duration and consists of a number of summed muscle action potentials. Fig. 1 shows the recruitment of fibres during the first 6 beats in a train; the synchrony of firing of the different motor units is such that the total duration of the summed spike is increased by less than 25 per cent of the initial one.

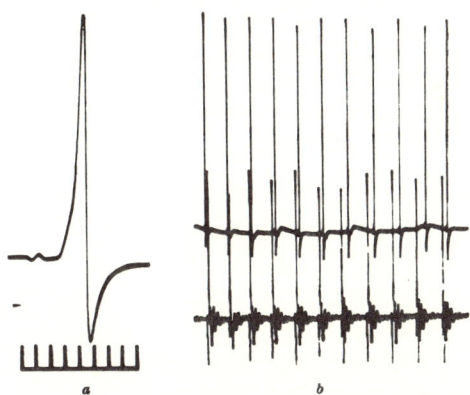

Fig. 2. *a*, Typical summed muscle action potential preceded by the incoming nerve volley (scale in msec.): *b*, record similar to Fig. 1 (but with altered gains) made during a prolonged burst of drumming

The electrical events associated with each contraction (Fig. 2a) last 9 msec. The initial deflexion in Fig. 2a is the incoming motor nerve volley recorded in volume. The muscle spike, arising after a neuromuscular delay of 1·4 msec., has the following characteristics: rising phase, 2·1 msec.; rapid downward phase, 0·5 msec., and relatively slow (6 msec.) return to resting level from the positive side of the base-line. The end-plate potential appears as a shallow slope (1·1 msec.) at the beginning of the rising phase. Recorded summed spike values are of the order of 6 mV.

There is synchrony of contraction between the drumming muscles of either side. Occasionally during recording one of the two muscles would fail to contract; when this occurred with the exposed muscle, the action potential of the other could still be seen on the recording side and the sound was still maintained (in Fig. 1, the third, fourth, tenth, fourteenth and sixteenth drum-beats are of this kind); in these instances the sound vibrations showed a less-complicated and more-consistent wave-form than when the muscles of both sides contracted. Analysis of the wave-form of the sounds indicates that a 25-cm. fish produces a frequency of about 200 cycles per sec.

It has thus been possible to investigate an overt

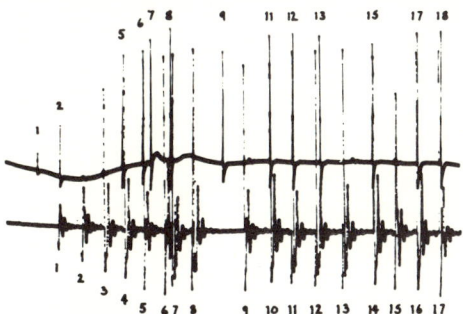

Fig. 1. Muscle action potentials (top trace) and sound vibrations (lower trace) recorded by moving-film camera during the start of a train of sounds at an average of 8/sec. The two beams were staggered in position: the first sound corresponds to the first muscle spike, etc.

piece of fish behaviour at a basic physiological level, and it is to be hoped that further features of the behaviour on one hand and its nervous basis on the other will be studied in this preparation. The intracostal muscles of *Congiopodus* and the pattern of impulses they receive from the central nervous system are highly specialized for sound production. The rapidity and constancy of the twitch contraction ensure the production of a specific sound of fairly regular wave-form, though the constancy and loudness of the signal must be limited by the restricted properties of the swimbladder as a resonator. Within these limitations it would seem, both from the steady maximum recruitment-level recorded (Fig. 2b) and the fact that the observed rate of drumming is many times slower than the refractory period of the muscle, that the system achieves a near optimal and consistent signal relatively free from interference by a preceding drum-beat. In each of the drum-beats illustrated, except the sixth in Fig. 1, the vibrations of the swimbladder recorded in air have almost completely disappeared from the lower trace before the onset of the next beat. Interference is clearly seen in the seventh drum-beat.

Finally, the cerebellum is notably much larger proportionately in *Congiopodus* than in *Trigla*—a related fish the sounds of which are restricted to the breeding season—and in the absence of information about the role of fish sounds in orientation and communication it is worth speculating that the cerebellum is being used in the integration of the drumming response possibly in relation to reception of its own sounds or those of other members of the school.

This work was done while at the Portobello Marine Biological Station, supported by the Nuffield Foundation. I am indebted to Dr. G. H. Satchell for assistance with the recording, which was carried out in the Department of Physiology, University of Otago Medical School, New Zealand.

MECHANISMS OF SOUND PRODUCTION IN THE ARIID CATFISHES *GALEICHTHYS* AND *BAGRE*

WILLIAM N. TAVOLGA
Research Associate, Department of Animal Behavior
The American Museum of Natural History

INTRODUCTION

DESPITE THE FACT that many catfishes are well known to emit sounds, reports in the literature are very sparse and sporadic. Agassiz (1850) briefly discussed the fact that catfishes and other fishes grunt by forcing air out of the swim bladder through the pneumatic duct. Dufossé (1874) mentioned the sounds produced by *Silurus glanis* as being such eructations or *"bruits de souffle."*

The squeaking or grating sound of the base of the pectoral fin spine as it rotates within its socket was described by Sörensen (1894) and by Burkenroad (1931) for a number of catfishes. In *Galeichthys* and *Bagre*, at least, such sounds have not been recorded from freely swimming animals under water, do not occur under normal circumstances, and are very unlikely to serve any communicatory function (Tavolga, 1960).

The grunting or humming sounds ascribable to the "elastic spring" mechanism were described by Sörensen (1894) for *Doras*. In this form, the sound was reported loud enough to be audible in air up to a distance of 100 feet. In addition, Sörensen demonstrated that, in the pimelodid catfish *Platystoma*, the extrinsic "compressor" muscles were responsible for sound production. Aside from some general comments on fish sounds by Aristotle, probably the earliest identified sonic catfish was *Synodontis*, reported by Geoffroy St.-Hilaire in 1829. In Egypt, this fish was commonly called "porcus, *parce que, disent les auteurs, il* grogne comme le cochon." One of the sonic catfishes, not specifically identified, was reported by Smith (1927) to be a common noisemaker in Siam. The "singing fish" of Ceylon, heard mainly at night and sounding like a distant automobile horn, was identified by Lange (1953) by its common name in Tamil, but from the account it was probably a catfish. The grunting sounds of *Galeichthys* were described by Burkenroad (1931), and the mechanism of the "elastic spring" was thought to be responsible.

Dobrin (1947) was one of the earliest investigators to record and measure under-water catfish sounds and, indeed, probably any fish sounds. The species named in his paper was *Felichthys felis*, which is now a synonym of *Bagre marinus*, but it is quite certain that he meant the common sea catfish *Galeichthys felis*. He reported a rhythmic drumming noise, with a sound pressure of about 0.8 microbar and a fundamental frequency of around 150 cycles per second. Knudsen, Alford, and Emling (1948) reported the sea catfish as producing a "popping" or "drumming" noise. On a commercially available record, Kellogg (1955) presented a sample of sounds produced by a large chorus of thousands of individuals of *Galeichthys*. He very aptly described it as sounding like the "bubbling of a giant percolator."

In a previous report (Tavolga, 1960), the under-water sounds of both *Galeichthys* and *Bagre* were described in detail, with the aid of sound spectrograms and accompanying recordings. The descriptions are summarized below in the present paper.

This report seeks to establish the morphological basis of sound production in the ariid catfishes *Galeichthys* and *Bagre*. Both the sea catfish, *Galeichthys felis* (Linnaeus), and the gaff-topsail catfish, *Bagre marinus* (Mitchill), are common estuarine and shore-line forms in Florida, where all this work was done. The skeletal structures involved are described first, then the muscles and their innervations. In the process of determining the precise muscles and nerves responsible for sound production, I obtained data on the physiology of these structures that enabled me to form some interpretation of the mechanics of the apparatus. It can be assumed that the structure of the swim bladder and associated sonic organs determines the quality (or timbre), pitch, and other properties of the sounds that are emitted (Fish, 1954; Tavolga, 1960), but the exact acoustics of the sound-producing mechanisms have not yet been satisfactorily explained, nor has the reason for the high efficiency of these low-frequency, under-water loud speakers.

ACKNOWLEDGMENTS

The research for this paper was supported by a contract between the Office of Naval Research, Department of the Navy, and the American Museum of Natural History, Contract No. Nonr 552 (06) NR 301–322.

The author is greatly indebted to Mr. F. G.

Wood, Jr., Director of Exhibits at Marine Studios, St. Augustine, Florida, for making the facilities of the Marineland Research Laboratory available for this project. Mr. Clifford Townsend and the entire Collecting Department of Marine Studios were extremely helpful in obtaining the fish used in this work.

The assistance of Mr. G. Scott Johnson, presently at Indiana University, is gratefully acknowledged. Mr. Johnson was a participant in the Undergraduate Research Training Program of the National Science Foundation, sponsored by the American Museum of Natural History.

The author is also grateful for the comments and criticisms of Dr. Lester R. Aronson and Dr. Bobb Schaeffer, of the American Museum of Natural History.

KEY TO SYMBOLS AND ABBREVIATIONS ON THE PLATES

aort. can., aortal canal formed by the overgrowth of a superficial layer of ossification covering the ventral and lateral surfaces of the first several vertebral centra
aort. grv., aortal groove (as above)
ART, articular
BOC, basioccipital
CENTR-1, first vertebral centrum
CENTR-6, sixth vertebral centrum
Cer. hemi., cerebral hemisphere (frontal lobe)
CERHY, ceratohyal
CLEITH, cleithrum
COR, coracoid
DENT, dentary
DETH, dermethmoid
DSOC, dermosupraoccipital
EPHY, epihyal
EPOT, epiotic
EPOT-LAM, epiotic lamina
EXOC, exoccipital
EXOC-Col, short column of bone connecting the exoccipital with the base of the horizontal support of the Müllerian ramus
Fac. lobe, facial lobe of medulla
FR, frontal
horiz. supp., horizontal support along the anterior edge of the Müllerian ramus shown in plates 11 and 19
HYOM, hyomandibular
HYPHY, hypohyal
inc. oss., areas of incomplete ossification in the expanded fourth transverse process
IOP, interopercular
lat. supp., lateral supporting lamina of the fourth neural spine in *Bagre* (pl. 18)
MPTER, metapterygoid
MüR, Müllerian ramus; the distal end of the anterior ramus of the fourth transverse process
N. VII, facial nerve (VII)
N. IX, glossopharyngeal nerve (IX)
N. X, vagus nerve (X)
NAS, nasal
NS4, neural spine of fourth vertebra
nuch. sh., nuchal shield
Occip. n. gang., dorsal root ganglion of occipital nerve
Occip. n. dors. branch, dorsal branch of occipital nerve (to protractor muscle)
Olf. lobe, olfactory lobe
Olf. tract, olfactory tract
OPERC, opercular
Opt. lobe, optic lobe
Opt. n., optic nerve (II)
ORSP, orbitosphenoid
PAL, palatine
PASP, parasphenoid
pect. sp., enlarged first pectoral fin spine
PFR, prefrontal
PMAX, premaxillary
POP, preopercular
PROT, pro-otic
Protractor mus., protractor muscle of *Springfederapparat*
PT, posttemporal (supracleithrum of some authors)
PT-Inf, inferior limb of posttemporal
PT-Sup, superior limb of posttemporal
PTOT, pterotic
PTSP, pterosphenoid (alisphenoid of some authors)
PVOM, prevomer
QUAD, quadrate
Ramus recur. VII, ramus recurrens branch of facial nerve (VII)
SCB, scalebone (posttemporal of some authors)
SOC, supraoccipital
Spin. n. 1, 2, 3, spinal nerves 1, 2, and 3
SPOT, sphenotic
subv. proc., subvertebral process formed at the point of juncture of the basioccipital and the anteriormost vertebrae
sup. oss., superficial layer of ossification covering the ventral and lateral portions of the first several vertebrae
TP4, transverse process of the fourth vertebra; in the plates this label points to the terminus of the posterior ramus
TP5, TP6, TP7, transverse processes of the fifth, sixth, and seventh vertebrae
TRIP, tripus; the first and largest of the series of Weberian ossicles
URHY, urohyal
Vag. lobe, vagal lobe of medulla

IN SKELETAL STRUCTURE, and other features as well, the Suborder Siluroidea (of Regan, 1911; or Nematognathi of Jordan, 1923) is a very distinct group of fishes. The Order Ostariophysi (Cypriniformes of Berg, 1947), which includes the cyprinoids, characinoids, and gymnotoids, as well as the siluroids, is also a clear and natural group characterized by the presence of the Weberian apparatus. This series of four ossicles, which connect the cavity of the swim bladder with that of the inner ear, was first described by Weber in 1820. He named three of the ossicles "malleus," "incus," and "stapes." The fourth and innermost was called the "claustrum." To avoid erroneous implications of homology, Bridge and Haddon (1889) first proposed the terms "tripus" (instead of malleus) for the largest, crescent-shaped element; "scaphium" (instead of stapes) for the usually spoon-shaped inner ossicle that overlies the lateral surface of the sinus impar; and "intercalarium" (instead of incus) for the small ossification in the ligament between the tripus and scaphium. "Claustrum" remained as the term for the bone on the median side of the sinus impar, between it and the neural canal. The bilobed sinus impar is a posterior extension of the perilymphatic cavity. The ossicles themselves are derived from portions of the neural arches and transverse processes of the first three vertebrae, and their precise embryonic origins are still a matter of some dispute (De Beer, 1937; Krumholtz, 1943).

Weber originally postulated that the ossicles served a function of transmitting sound to the inner ear in a manner analogous to that of the middle ear ossicles of mammals. Dijkgraaf (1949, 1952, 1960) has clearly shown the broader frequency response and lower auditory threshold of ostariophysine fishes, as opposed to those that lack a Weberian apparatus. Similar data were reported by von Frisch (1923), Stetter (1929), and Evans (1925, 1935). By means of extirpation methods, von Frisch and Stetter (1932), von Frisch (1936), and Poggendorf (1952) were able to prove the auditory function of the ossicles, which does not imply, however, that the Weberian apparatus is exclusively auditory in function (see reviews by Jones and Marshall, 1953, and Jones, 1957).

The skeletal characteristics of the catfishes (Siluroidea) include a non-protractile mouth with a reduced maxillary, a heavy broad cranial roof, and the ankylosis of the first several vertebrae to the occiput. The centra of the first four vertebrae are usually fused into a heavy complex, rigidly attached to the basioccipital. The fifth, sixth, and seventh vertebrae are more typical but are usually immovably joined to the first four. The dermosupraoccipital is extended caudad and participates in the support of the nuchal shields and enlarged first dorsal fin spine. The Weberian apparatus varies from that of other Ostariophysi in the elimination of the claustrum from the functional chain of ossicles and the reduction in size of the intercalarium (Chranilov, 1929; Krumholtz, 1943).

The modifications of the fourth vertebra are of particular interest here, because the transverse processes are enlarged and invariably form the support for much of the swim bladder, and, in some forms, are associated with sound production. Bridge and Haddon (1893) described the skeletal and swim-bladder structure for most of the siluroid genera, although their arrangement of families and genera does not fit the later systematic schemata of Regan (1911) and Berg (1947). The primitive condition with respect to skeletal support of the swim bladder seems to be exemplified best by the Siluridae and the Bagridae. In these families, the large swim bladder is supported dorsomedially by the enlarged and fused centra of the first five vertebrae, dorsolaterally by a flattened shelf of bone formed by the transverse processes (parapophyses) of the fourth and fifth vertebrae, and anteriorly by a decurved extension of the anterior ramus of the parapophysis of the fourth vertebra. The latter firmly abuts the inferior limb of the posttemporal bone. Essentially, the same structure is also present in the siluroid families Plotosidae, Ameiuridae, and Chacidae, and in some members of the Schilbeidae (Bridge and Haddon, 1893; Wright, 1884; Kindred, 1919).

Johannes Müller (1842, 1843) described a

special modification of the anterior ramus of the fourth vertebra. In the genera *Auchenipterus*, *Doras*, *Euanemus* (Doradidae), *Synodontis* (Synodontidae), and *Malapterurus* (Malapteruridae), he found that this ramus was free of attachment to the posttemporal, distally enlarged into a plate attached to the forward wall of the swim bladder, and supported by a thin, spring-like parapophysis. He also found that the presence of this *Springfederapparat* was associated with a pair of protractor muscles which originate on the occipital region and insert on the anterior surfaces of the enlarged rami. He theorized that the function of this complex was to control air pressure within the swim bladder, in keeping with the Cartesian diver theory of swim-bladder function as proposed by Borelli in 1680 and originating with Robert Boyle in 1675. Bridge and Haddon (1893) described the "elastic spring" apparatus in a number of additional genera and species and added the genus *Pangasius* (family Pangasiidae) to the list. They agreed fundamentally with Müller's hypothesis of the function of this structure. Sörensen (1894) presented strong evidence that, in the Doradidae, at least, the elastic spring acted as a sound-producing mechanism, an hypothesis with which Bridge and Haddon (1894) later concurred.

The anterior ramus of the fourth vertebra shows considerable significant variations which are important in family distinctions. Its modification into a *Springfederapparat* is also quite variable, as is its "elasticity." Considering the systematic and functional importance of this structure, I propose that, regardless of its shape and function, the anterior ramus of the transverse process of the fourth vertebra in siluroids be named the "Müllerian ramus." This term also has the values of brevity and historical interest.

Many of the siluroid families possess variously specialized swim bladders of transverse tubular, bilobed, or bipartite shapes. In most of these the Müllerian rami, and in some cases other bony elements, form a pair of investing capsules of cylindrical or globular shape around the greatly reduced swim bladder. This type of modification is present in the genus *Ageniosus* of the family Doradidae (family Ageniosidae, according to Berg, 1947), in some genera and species of the Schilbeidae and Pimelodidae, and in all members of the Amblycepidae, Sisoridae, Clariidae, Hypophthalmidae, Trichomycteridae, Bunocephalidae, Callichthyidae, and Loricariidae. In no case in which the bladder is so reduced is the Müllerian ramus modified into an elastic spring or has sound production by means of the swim bladder ever been reported.

An interesting modification is present in many species of the Pimelodidae. Members of this family that possess a large swim bladder have a Müllerian ramus of the silurid or bagrid type, in which it is firmly joined to the posttemporal, and there is no "elastic spring." Bridge and Haddon (1893) described a pair of compressor muscles originating from the occiput and inserting on the anteroventral surface of the swim bladder. They also described a pair of small muscles that always accompany the compressors. Each of these originates medially on the occiput and inserts on the tripus. They hypothecated that the tensor tripodes muscles function as dampers on the tripus against too violent air movements within the bladder. Such movements would be produced by the compressors. Sörensen (1894) clearly demonstrated, in the genus *Platystoma*, that the compressors do not compress the bladder but function in sound production. Bridge and Haddon (1894) subsequently concurred with Sörensen.

The family Ariidae is usually placed in a primitive phylogenetic status (Regan, 1911; Berg, 1947) because of the supposed fossil antiquity of the genus *Arius*. The hind part of the skull and the vertebral complex were described in detail by Bridge and Haddon (1893) for *Arius*, and other genera were briefly mentioned. The skull of *Arius* was described by Koschkaroff (1905) and figured by Gregory (1933). Karandiker and Masurekar (1954) reported on the skull of *Arius platystomus* in some detail, but did not mention the vertebral complex. Merriman (1940) figured and described some selected aspects of the osteology of *Galeichthys felis* and *Bagre marinus*. The Müllerian ramus was not described in any case, nor was its significance as a *Springfederapparat* recognized.

At this point, a nomenclatorial digression seems to be in order. It may be confusing to the reader, as it was initially to me, to have a catfish family called Bagridae and the genus *Bagre* in the family Ariidae. The name *Bagre* Cuvier was listed as preoccupying the names *Felichthys* Swainson and *Ailurichthys* (emended to *Aelurichthys*) Baird by Jordan in 1917. *Bagre* was adopted by Hubbs (1936) and most subsequent authors. The genus *Bagrus* Valenciennes, however, is the type of the family Bagridae. It was therefore suggested by Jordan and by Hubbs that the names *Bagre* and *Bagrus*, because of their similarity, be considered homonyms and that the next available name for *Bagrus* be substituted. This would be *Porcus* Geoffroy St.-Hilaire, which would make the family name Porcidae. Jayaram (1956) interpreted the International Rules as permitting both *Bagre* and *Bagrus* to remain, the former in the family Ariidae and the latter in the Bagridae. To add complication, some authors, including Jayaram, refer to the Ariidae as the Tachysuridae.

Prior to a description of the ariid skulls, it is appropriate to give a detailed account of the condition as found in the Siluridae. The following descriptions are limited in scope to the occipital region of the skull and the anterior vertebral complex.

The nomenclature of the bones described here and labeled in the plates is based mainly on that used by Harrington (1955) and, to some extent, that of Gregory (1933).

Wallago sp.
Plates 1–3

This account is based on a specimen in the collection of the American Museum of Natural History. The skull lacks a few elements, such as a Weberian apparatus on the left side, and has a few cracked bones, but it is otherwise in good condition. *Wallago* is a member of the Siluridae, and its general cranial osteology closely resembles that of most silurids and bagrids (Bridge and Haddon, 1893; Joseph, 1960).

Supraoccipital
Plates 1, 3

Dorsally the dermal part of this bone is broad and slightly humped in the middle, and its surface is prominently ridged with parallel and interlacing rugosities, as is the entire dorsal cranial surface. Anteriorly it is sutured to the frontals; laterally, to the sphenotics, pterotics, and scalebones (parietals are absent, as in all siluroids). Posteriorly the dermosupraoccipital extends as a stout process that is involved in the support of the nuchal shields characteristic of the order. Under the posterior process a median vertical supporting ridge expands at the occiput, where it sutures with the epiotics laterally and exoccipitals ventrally (pl. 3). Under the base of the posterior process is a pair of large foramina. The nerves that pass through these apertures are the rami lateralis accessorius of the facial (VII) (Herrick, 1901), also called the rami recurrens (Berkelbach van der Sprenkel, 1915, and most subsequent reports). This pair of large cutaneous sensory branches is common to siluroids. The enlarged neural spine of the third vertebra projects into a median groove on the ventral posterior surface of the supraoccipital.

Exoccipital
Plates 2, 3

A posterior, vertical ridge divides this bone into two wings. The anterolateral wing is sutured to the basioccipital, pro-otic, sphenotic, and epiotic, and it possesses a large, multiple foramen for the glossopharyngeal, vagus, and occipital nerves. The posterior, median wing is sutured to the basioccipital, epiotic, supraoccipital, and contralateral exoccipital. It forms the arch of the foramen magnum and possesses several nerve and nutrient foramina at its base. The vertical ridge extends dorsad onto the epiotic and ventrad to the base of the inferior limb of the posttemporal.

Basioccipital
Plate 2

This bone shows few special modifications from that of the normal teleost. It is sutured to the parasphenoid, pro-otics, exoccipitals, and inferior limbs of the posttemporals. It forms the floor of the foramen magnum and possesses a small, median, nutrient foramen on its posterior ventral surface.

Epiotic
Plates 1, 3

The vertical ridge of the exoccipital continues dorsad onto the epiotic to form an acute angular ridge projecting caudad. Dorsally the epiotic is joined to the superior limb of the posttemporal; laterally it is sutured to the pterotic, medially to the supraoccipital, and ventrally to the exoccipital. A large dorsolateral foramen is present.

POSTTEMPORAL
Plates 1–3

This is the supracleithrum of Regan (1911) and others. The superior limb of this V-shaped bone is immovably joined to the scalebone (posttemporal of Regan, 1911) dorsally and the epiotic ventrally. This limb is also supported by a ventral process of the pterotic, and it fits into a fossa formed by the three supporting bones. The inferior limb of the posttemporal is columnar and firmly sutured to the basioccipital and ventral edge of the exoccipital. The distal, apical portion of the posttemporal is broadened and possesses a deep notch into which the dorsal point of the cleithrum fits. This joint is a loose one and permits some anterior and posterior swinging of the pectoral girdle, as well as some vertical sliding movement. The posterior surface of this apex has a facet at which it is firmly laced with connective tissue to the Müllerian ramus (see below).

WEBERIAN APPARATUS

A detailed account of these ossicles is not given here, for they vary little among the siluroids and are not an important element involved in sound production. Bridge and Haddon (1893) described the ossicles for most of the siluroids, and mentioned them in *Wallago* briefly. A number of more recent and detailed reports concerning these ossicles include those by Krumholz (1943) and Chranilov (1929). These structures appear to be relatively conservative, and among the various siluroid families there is little difference from the form described originally in *Silurus* by Weber (1820).

FIRST VERTEBRA
Plate 2

Only a thin centrum is present and distinguishable. Laterally and ventrally it is fused with the following vertebral complex. Midventrally an aortal groove continues caudad.

SECOND AND THIRD VERTEBRAE

The centra of these vertebrae are indistinguishably fused with the centrum of the fourth vertebra. A pair of small, lateral, winglike projections extend from what is probably the second centrum. These serve to support, in part, the anterior wall of the swim bladder. Dorsally the arch and spine of the third vertebra project forward. The broadly compressed spine fits immovably into a groove in the supraoccipital. Posteriorly this spine joins that of the fourth vertebra by means of a thin, median lamina.

FOURTH VERTEBRA
Plates 1–3

This structure is highly modified, characteristically so in the siluroids, and it forms an important support for the anterior chamber of the swim bladder, the Weberian apparatus, and the spine of the dorsal fin. The centrum is elongate and not distinguishable from that of the fifth. It is invested by a layer of superficial bone (pl. 2, sup. oss.) which extends ventrally to form a deep aortal groove (pl. 2) along the ventral midline. The neural spine is large and inclined caudad (pls. 1, 3). It is grooved posteriorly and receives the bony supports for the spine of the dorsal fin. Anteriorly the neural spine is connected to that of the third vertebra by a median lamina. Laterally a pair of strengthening ridges extend out onto the transverse processes.

The transverse process (parapophysis) is greatly expanded and flattened to form the roof of the anterior chamber of the swim bladder (pls. 1–3). The anterior, Müllerian ramus is stout, sharply decurved, and expanded distally into a thick, rugose wing which is firmly laced to the distal end of the inferior limb of the posttemporal. The decurved portion of the Müllerian ramus supports the anterior face of the swim bladder. The posterior ramus is broad, slightly arched, inflexible, and continuous with the Müllerian ramus. Distally it fans out and is immovably joined to the parapophysis of the fifth vertebra.

FIFTH VERTEBRA

Because of the layer of superficial ossification, the fifth centrum is not distinguishable from the fourth. The neural spine is a low, median ridge. Anterior and posterior rami of the parapophysis are visible as thickenings at the base of the transverse process. Distally the two rami fuse to a point (pls. 1–3). The sixth and following vertebrae show no special modifications.

Galeichthys felis
Plates 4, 5; 6, figure 1; 7–12

The skull and vertebral complex show few modifications from the typical ariid form as described by Bridge and Haddon (1893) for *Arius pidada*, Koschkaroff (1905) for *Arius thalassinus*, Bhimacher (1933) for *Arius* and related forms, Gregory (1933) for *Arius* sp., and Karandikar and Masurekar (1954) for *Arius platystomus*. The following description is based on several specimens of various sizes collected in the vicinity of Marineland, Florida.

SUPRAOCCIPITAL
Plates 4; 6, figure 1

Dorsally the dermosupraoccipital is a broad, rough-surfaced shield, sutured to the frontals, sphenotics, pterotics, and scalebones. Posteriorly a broad, flat process supports the nuchal plates. The posterior face is sutured laterally to the epiotics, ventrolaterally to the exoccipitals, and medially to the neural spine of the third vertebra. On this posterior face, immediately below the projecting posterior process, is a pair of ramus recurrens (VII) foramina.

EXOCCIPITAL
Plates 5, 11

This bone is smoothly convex, unlike that of *Wallago*. It is sutured to the basioccipital, pro-otic, sphenotic, epiotic, and contralateral exoccipital, where it forms the arch of the foramen magnum. Along its ventral edge are three foramina. The middle and largest of these is for the passage of the glossopharyngeal and vagus nerves. The ventral branch of the occipital nerve passes through the posterior of these foramina. An extension from the posterior ventromedial angle of the exoccipital forms a bony column which is immovably fused to the base of the Müllerian ramus (pl. 11). A small foramen just dorsal to this column is for the passage of the dorsal branch of the occipital nerve.

BASIOCCIPITAL
Plates 5; 6, figure 1

As in *Wallago*, the basioccipital is sutured to the parasphenoid, pro-otics, exoccipitals, and inferior limbs of the posttemporals and forms the floor of the foramen magnum. Posteriorly it is indistinguishably fused with the anterior vertebral complex of centra and possesses a large, midventral, aortal foramen. Posterior to this foramen is a prominent ventral projection, bifid at the tip. This projection, called the "subvertebral process" by Bridge and Haddon (1893), is composed of the basioccipital and at least the first two vertebral centra. The prominent subvertebral process is considered to be characteristic of the Ariidae.

EPIOTIC
Plates 4; 6, figure 1; 10-12

Except for a projecting lamina (see below), this bone is smoothly convex, sutured to the supraoccipital, exoccipital, and pterotic. Dorsally it forms a groove together with the scalebone for the reception of the superior limb of the posttemporal. From its dorsal edge a prominent lamina of stout, flat bone projects ventrocaudad. The epiotic lamina (erroneously considered part of the supraoccipital by Bridge and Haddon, 1893) extends to and is sutured to the dorsal ridge of the posterior ramus of the fourth vertebra (pls. 4, 12). Laterally the lamina has a caudally directed, pointed process, and medially it fuses with the base of the third neural spine (pls. 10, 11). The presence of this lamina is characteristic of the ariids (Bridge and Haddon, 1893), although Regan (1911) stated (in error) that in *Galeichthys* this structure does not reach the parapophysis. The epiotic lamina forms a roof over the Müllerian ramus and serves as the surface of origin for the "protractor" muscle (pl. 12; pl. 21, fig. 1). The Doradidae, which include many sound-producing species with a highly developed *Springfederapparat*, also possess posterior extensions of the epiotics, but they serve as supports for the nuchal plates (Regan, 1911) and do not appear to be involved in the sonic mechanisms.

Posttemporal

Plates 4, 5; 6, figure 1; 10–12

The superior limb is short, stout, and immovably joined to the pterotic, scalebone, and epiotic. In older specimens, there is often a small foramen between the portion joined to the pterotic and that which fits into the groove between the scalebone and the epiotic. The inferior limb is long and cylindrical, joined to the basioccipital. The distal apex has a deep notch, into which the dorsal spine of the cleithrum fits loosely (pls. 7, 9). There is no direct connection with the distal end of the Müllerian ramus (see below).

First, Second, and Third Vertebrae

The centra are indistinguishably fused to the basioccipital and participate in the subvertebral process. The neural spine of the third vertebra is shaped like an I beam, and it is inclined forward to join the base of the supraoccipital.

Fourth Vertebra

The fourth, fifth, sixth, and seventh centra are fused together and covered ventrally and laterally by an investing layer of bone (pls. 5; 6, fig. 1; 11, 12). This superficial ossification forms a large, mid-ventral, aortal canal, which extends from the basioccipital to the base of the seventh vertebra (pl. 5). The presence of such a canal, rather than a groove, was considered an ariid character by Bridge and Haddon (1893).

The Müllerian ramus (pls. 4, 5; 6, fig. 1; 10–12) is decurved, pointed, and stiffened along its anterior edge by a thin horizontal ridge (pl. 11, horiz. supp.) extending from the base of the third arch. At its base it is attached by a short column to the exoccipital (pl. 11). The distal tip of the Müllerian ramus is freely movable within the limits of elasticity of the transverse process as a whole. In *Arius*, Bridge and Haddon (1893) stated that the distal tip is "applied to" the posttemporal. Gregory (1933), in his figure, showed it to be free, while Regan (1911) said that it was "rigidly attached" in the Ariidae. In all specimens that I have seen, the distal tip is only very loosely attached to the posttemporal by a small portion of areolar connective tissue. In dried and partially cleaned skeletons, however, this tissue sometimes remains and hardens and resembles a ligament, which may explain the discrepancies in the accounts. Conceivably, considerable specific and generic variation may also exist. The matter is important, because the freedom of movement of the Müllerian process is an essential feature in the sound-producing mechanism. In freshly dissected specimens, the distal tip of the Müllerian ramus is movable only in a dorsoventral arc. The horizontal ridge described above prevents movement in any but the dorsoventral direction.

The *Springfederapparat* itself is a thin, fragile shelf forming an arched fan, ventrally concave, connecting the anterior and posterior rami of the fourth vertebra (pl. 12). Even in young specimens, less than 4 inches in total length, this region is thoroughly ossified, while other parts of the cranium are still partially cartilaginous. Sounds can be elicited from such individuals.

The posterior ramus is rigidly supported by the epiotic lamina (pls. 10–12). Thus a deep, roughly tetrahedral cavity is formed, bounded on three sides by the *Springfederapparat*, epiotic lamina, and lateral occipital region (epiotic and exoccipital; pl. 12). Within this cavity is the "protractor" muscle (pl. 21, fig. 1).

The fourth neural spine is formed as in *Wallago*, inclined caudad and supporting the underpinnings of the spine of the dorsal fin (pls. 6, fig. 1; 10, 12).

The divisions between the fifth and sixth and sixth and seventh vertebrae are visible from above. Each has a pair of parapophyses (pls. 4, 5). The fifth pair is the longest and broadest and is firmly joined to the posterior edge of the fourth.

Bagre marinus

Plates 6, figure 2; 13–20

The major features of the skull and other aspects that distinguish this species from *Galeichthys* were described by Merriman (1940). In the following description, the points of difference between *Galeichthys* and *Bagre* are emphasized. The account is based on several specimens collected in the vicinity of Marineland, Florida.

Supraoccipital
Plates 6, figure 2; 13, 18, 20

The bone is basically the same as that of *Galeichthys* and that of *Arius*, except for a mid-ventral ridge under the posterior projecting process. This ridge is fused with the neural spine of the third vertebra. The foramina of the rami recurrens (VII) are lower in position.

Exoccipital
Plate 14

A broad, thin, medial process projects back over the neural canal and meets, but does not fuse with, the third neural spine. Along the ventral edge there are a small anterior foramen and a large middle foramen (glossopharyngeal and vagus), followed by a smaller foramen for the ventral branch of the occipital nerve. Behind the last-mentioned, a short extension supports the anterior portion of the tripus. Dorsal to this extension is the foramen for the dorsal branch of the occipital nerve. The exoccipital process in *Galeichthys* which projects dorsal to the tripus and joins the base of the Müllerian ramus is represented in *Bagre* by a small point on the median edge of the exoccipital.

Basioccipital
Plates 6, figure 2; 14

This bone is almost identical in form to that of *Galeichthys* and contributes to a prominent subvertebral process.

Epiotic
Plates 6, figure 2; 13, 18, 20

In basic form and in the shape of the projecting lamina, the epiotic is like that in *Galeichthys*. The lamina, however, is narrower and is not joined medially to the neural arch. The effect therefore is to reduce the surface area available for the origin of the protractor muscle.

Posttemporal
Plates 6, figure 2; 13, 14, 18–20

The superior limb is short and proximally biramous. Its anterior ramus is sutured to the pterotic, and the posterior ramus fits into a deep groove formed by the scalebone and the epiotic. The notch at the distal apex of the posttemporal is very deep and wide. In dissections, the cleithral spine is found to fit very loosely into this notch, thus permitting vertical sliding as well as swinging movements of the pectoral girdle (pls. 15, 17). There is no connection with the Müllerian ramus.

First, Second, and Third Vertebrae

The neural spine of the third vertebra resembles that of *Wallago* in being compressed into a thin median ridge ankylosed to the supraoccipital. This thin ridge is continuous caudad with the fourth neural spine. The first and second vertebrae are not distinguishable.

Fourth Vertebra

The Müllerian ramus is stoutly supported along its anterior edge by a horizontal shelf leading from what appears to be the base of the third arch (pls. 14, 19, 20). This structure is also present in *Galeichthys*, but in *Bagre* it is slightly less flexible. The thin, curved lamina that joins the Müllerian ramus with the posterior ramus is even more delicate and fragile than in *Galeichthys*, particularly in the region lateral to the juncture of the epiotic lamina, where the bone has lacy areas of incomplete ossification, even in mature specimens (pls. 13, 14, 19). A similar area is usually present on the dorsal, posterior surface of the *Springfederapparat*, just medial to its juncture with the epiotic lamina (pl. 19). Because of the stout horizontal support, the dorso-ventral flexibility of the *Springfederapparat* is less in *Bagre* than in *Galeichthys*. The distal tip of the Müllerian ramus is free of the posttemporal, as is probably true of most ariids.

The neural arch of the fourth vertebra is vertical (pls. 6, fig. 2; 20) and is supported laterally by a pair of flat, triangular lamina, the bases of which are fused to the posterior rami (pl. 18). A large vertical fossa is thus formed on each side of the supraoccipital process (pls. 13, 20), bounded anteriorly by the occiput, laterally by the epiotic lamina, posteriorly by the flange of the fourth spine, and ventrally by the base of the Müllerian ramus. A portion of the epaxial musculature fits into this fossa. The insertion of this muscle is the skin of the dorsum and the base of the dorsal fin.

THE MUSCULAR BASIS OF SOUND PRODUCTION

AMONG THOSE SILUROIDS that possess a *Springfederapparat* as a modification of the Müllerian ramus, there is invariably a muscle that originates on the occipital region of the skull and inserts on the anterior face of the expanded Müllerian ramus. Müller (1842, 1843) first described the presence of such a muscle in a number of genera of the families Doradidae, Synodontidae, and Malapteruridae. Bridge and Haddon (1893) called this a "protractor" muscle and described its presence in some additional forms, including the family Pangasiidae. The exact origin of the muscle was reported as being the posterior face of the epiotic and exoccipital.

Sörensen, in his doctoral dissertation in 1884 (not seen by me), first postulated the function of the protractor muscle and the *Springfederapparat* in sound production and also presented some experimental evidence.

Ever since Sörensen's (1894) and Bridge and Haddon's (1893, 1894) reports, the sound-producing potential of the *Springfederapparat* has been recognized in all the above families. In the Ariidae, Burkenroad (1931) described the grunt-like sounds of *Galeichthys milberti* (= *G. felis*) as being produced by a mechanism similar to the "elastic spring." He reported the presence of dorsoventrally oriented muscle fibers inserting on the thin shelf of bone over the dorsal face of the anterior swim-bladder chamber. Although his description was brief, it is now clear that he was discussing the "protractor" muscle. Despite the fact that sound production has probably been known from the first moment that a man caught a sea catfish, Burkenroad's is the earliest published account that I can locate of sound production in this family (Ariidae).

The following descriptions are based on dissections of fresh and preserved specimens, and on serial sections of juvenile individuals. The latter were fixed in 10 per cent formalin in sea water, decalcified in formic acid, sectioned at 10 microns, and stained with Delafield's hematoxylin and eosin. Small portions of the protractor muscle, with other muscles from the same individuals, were fixed in Gilson's fluid, sectioned at 2 microns, and stained with hematoxylin and eosin. As controls, portions were taken from the levator pectoralis (trapezius), the pectoral fin adductor, and the epaxial muscle from the midbody region.

Galeichthys felis
Plates 21, figure 1; 22

In dissection, the protractor muscle can best be approached from the side. A soft, triangular area can be discerned by palpation just behind the cleithrum. The anterior, vertical leg of the triangle is formed by the posttemporal; the dorsal leg, by the outer edge of the epiotic lamina; and the ventral leg, by the outer edge of the fourth parapophysis. The last of these is not palpable from the surface, because it is deeper and thinner than the others. After the skin and superficial muscle in this region are peeled away, the triangular area occupied by the protractor is easily evident, especially in a fresh specimen. The muscle is quite visible because of its deep red color. It is obviously much more highly vascularized than any of the neighboring tissues, and indeed more so than any other muscle tissue in this fish. The protractor muscle is soft and spongy in texture, and further dissection is best continued after the muscle is hardened in fixative.

Most of the volume of the tetrahedron formed by the occiput, epiotic lamina, and Müllerian ramus is occupied by the protractor muscle (pls. 12; 21, fig. 1). Its surface of origin is an oval that extends over the entire ventral surface of the epiotic lamina, including the medial portion that unites with the third neural arch. A few bundles of fibers also originate from the portion of the epiotic proper just beneath the lamina. The almost circular area of insertion is the thin layer of bone from the ridge along the anterior edge of the Müllerian ramus to the site of fusion of the epiotic lamina with the posterior ramus, i.e., the insertion covers a large portion of the dorsal surface of the *Springfederapparat*.

The general shape of the protractor muscle is that of a greatly truncated cone. The fiber bundles from the surface of origin converge slightly as they approach the insertion surface (pls. 21, fig. 1; 22). The anterior fibers are longest and converge most, whereas the posterior fibers are short and are almost parallel to one another. In the posterior portion, the fiber bundles extend unbroken from origin to insertion, while in the anterior portion, a few columns of fibrous tissue run from points of convergence of muscle bundles to the surface of insertion.

Sections perpendicular to the long axis of the fibers show the presence of numerous capillaries between the fibers and a central core of loose fibrous tissue and larger blood vessels. The diameter of the fiber bundles ranges from 600 to 700 microns, and the bundles are all roughly circular in cross section.

In tissues taken from mature specimens (more than 10 inches in length), fixed in Gilson's fluid, the diameter of the muscle fibers ranges from 25 to 45 microns (average about 30 microns) and the cross striations are distinct, with the sarcomere size about 1 micron. Nuclei are peripheral in position. Myofibrils are very fine, closely packed, and with relatively little sarcoplasm around them. In tissues from juveniles (less than 4 inches in length), the fibers are thinner (6 to 15 microns, average 12 microns) and the striations very sharp (p. 22, fig. 2). The sarcomeres are almost 3 microns in length, and the Q bands appear finely granular under ×1000 bright-field magnification. These tissues were fixed in 10 per cent formalin in sea water, and the differences in fiber size and striations may be in part the result of a different fixative. The Gilson's fixed material is probably more reliable and in general shows fewer artifacts.

In comparison, similarly treated (Gilson's fixation) muscle tissue from other parts of the same individuals shows clearly larger fibers, with more variability in size. Diameters range from 75 to 150 microns (average about 100 microns). The striations appear less distinct, but have the same spacing, and the myofibrils are coarser, with the intervening sarcoplasm more visible.

It is well known (Prosser, 1950) that fast-acting muscles tend to have finer, more closely packed myofibrils with less sarcoplasm than slower-acting types. The so-called "dark" muscles are generally slow acting, and their color is the result of the accumulation of myoglobin in the tissue. The deep red coloration of the protractor muscle is probably caused not by myoglobin but by the high degree of vascularization. The tissue, in a freshly dissected animal, bleeds profusely when damaged even slightly, and the coloration is quickly washed out during fixation, which is not true of "dark" muscles.

Bagre marinus

Plate 21, figure 2

In dissections, the protractor muscle is more difficult to locate than is that of *Galeichthys*. The posttemporal extends laterad, and the dorsal spine of the cleithrum protrudes dorsad so as partially to cover the triangular area within which the protractor is placed (pl. 20). A considerable amount of superficial muscle and connective tissue must be removed before the protractor can be exposed. Once revealed, it can be easily seen and recognized because of its triangular shape and deep red color.

In *Bagre*, the protractor muscle is smaller than that of *Galeichthys*, and it is conical in shape (pl. 21, fig. 2). The surface of origin is an oval on the distal two-thirds of the ventral side of the epiotic lamina. All the fibers converge to a small area of insertion just median to the tip of the Müllerian ramus dorsal to its horizontal supporting ridge. The surface along which the muscle lies is always well ossified. The sites of incomplete ossification are never those involved with the attachment of the protractor.

Except for the greater convergence of fibers and additional fibrous tissue connecting these, the protractor muscle of *Bagre* is identical in microscopic anatomy to that of *Galeichthys* (see above).

NERVE SUPPLY OF THE SOUND-PRODUCING MECHANISM

The innervation of the protractor muscle is virtually the same in *Bagre* and *Galeichthys*, so that one description serves for both. Slight differences, where they occur, are mentioned. The description is based on dissections of both fresh and preserved specimens, and on serial sections of juvenile individuals. The protractor muscle is supplied entirely by a branch of the occipital nerve (nomenclature according to Addens, 1933, but see discussion below). The innervation was established not only by anatomical observations (pls. 23, 24) but by stimulation experiments.

On exposure of the cranial cavity, the occipital nerve roots can be seen just posterior to the roots of the vagus (pl. 23). The occipital nerve possesses a dorsal root, with a ganglion, and a ventral root. Posteriorly the next nerve is clearly a true spinal nerve, with its roots just posterior to the foramen magnum. The occipital nerve penetrates the lateral floor of the exoccipital and emerges through two foramina posterior to the vagus-glossopharyngeal foramen. The upper of the two foramina serves the dorsal branch of the occipital nerve. The dorsal branch runs caudad along the outside of the exoccipital portion of the auditory capsule almost up to the ventral surface of the epiotic lamina. Here it turns laterad and ramifies into the protractor muscle along its surface of origin. In *Bagre*, the nerve passes through, but does not innervate, a large mass of epaxial muscle before reaching the protractor.

The course of the ventral branch of the occipital nerve is also of interest, and some ancillary problems are touched on as it is described. It is a larger nerve than the dorsal branch and presumably contains sensory as well as motor fibers. The ventral branch runs laterad and follows the anterior surface of the inferior limb of the posttemporal for about one-third of its length, then turns abruptly ventrad. At this turn, a small twig is given off laterally, along the posttemporal, to the anterior surface of the cleithrum. Here this twig ramifies into a large, oval muscle, the origin of which is the ventral surface of the pterotic and insertion on the anterior surface of the dorsal limb of the cleithrum (see pls. 7–9, 15–17 for skeletal parts). Stimulation of the muscle shows that it functions as a levator pectoralis. It is not clear, however, whether the twig of the occipital nerve is a motor nerve or not. Innervating this muscle is also a twig from the last branchial nerve (vagus). This was found in the study of serial cross sections. What is most probably the same muscle has been described in *Ameiurus* as the trapezius by McMurrich (1884). Wright (1884) stated that it was innervated by a branch of the first spinal nerve (=occipital nerve). Herrick, in his work on *Menidia* (1899) and *Ameiurus* (1901), claimed that Wright was in error and that the trapezius muscle is innervated by the posteriormost branch of the vagus, i.e., a precursor of the eleventh nerve. In the codfish, *Gadus*, Herrick (1900) located a functionally comparable muscle innervated by spinal nerves and concluded that in this form a true trapezius was absent. According to Addens (1933), the trapezius of teleosts is probably homologous to that described in selachians and ganoids. The fact of differences in innervation does not preclude common origin, according to Black (1917), and, based on embryological studies, Edgeworth (1911) concluded that the trapezius of teleosts is derived from the upper edge of the fourth levator arcuum branchialum, regardless of innervation.

Below the posttemporal, the ventral branch of the occipital nerve runs within a sheet of connective tissue that forms the septum between the pericardial and perivisceral cavities. Here the nerve splits into a lateral and a medial branch. The lateral is the larger of the two, and it joins branches from the first and second spinal nerves. Together, these nerves form a portion of the brachial plexus or ramus cervicalis. Stimulation of the occipital nerve at this point shows that it is a motor supply to the adductors and abductors of the spine of the pectoral fin.

The smaller medial branch runs ventrad and forward along the dorsomedial edge of the cleithrum. As it turns forward, it sends a few short twigs to some small slips of muscle. These short muscles run from the edge of the cleithrum where it unites with the coracoid

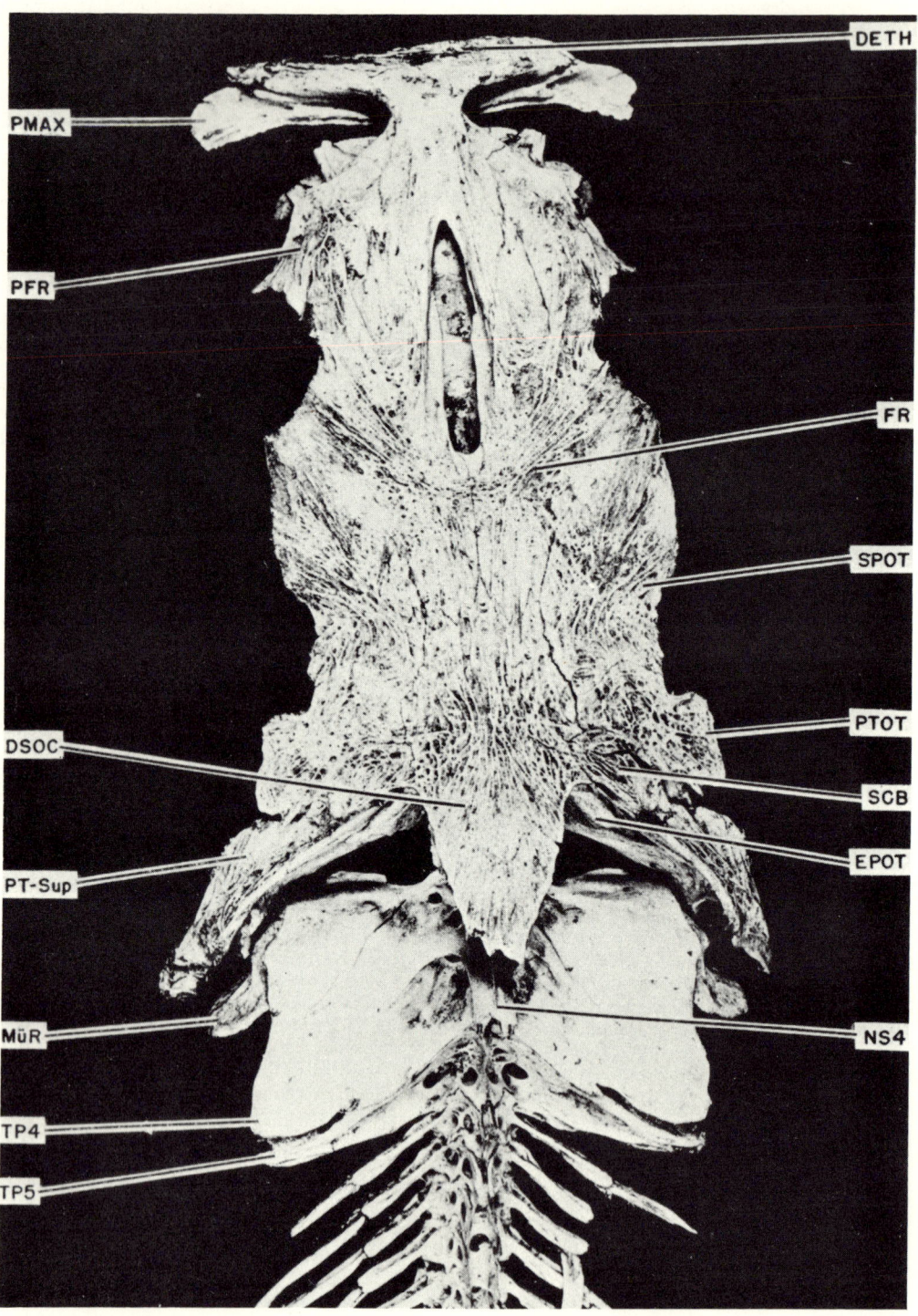

Dorsal view of neurocranium and anterior vertebral complex of *Wallago* sp. × ½

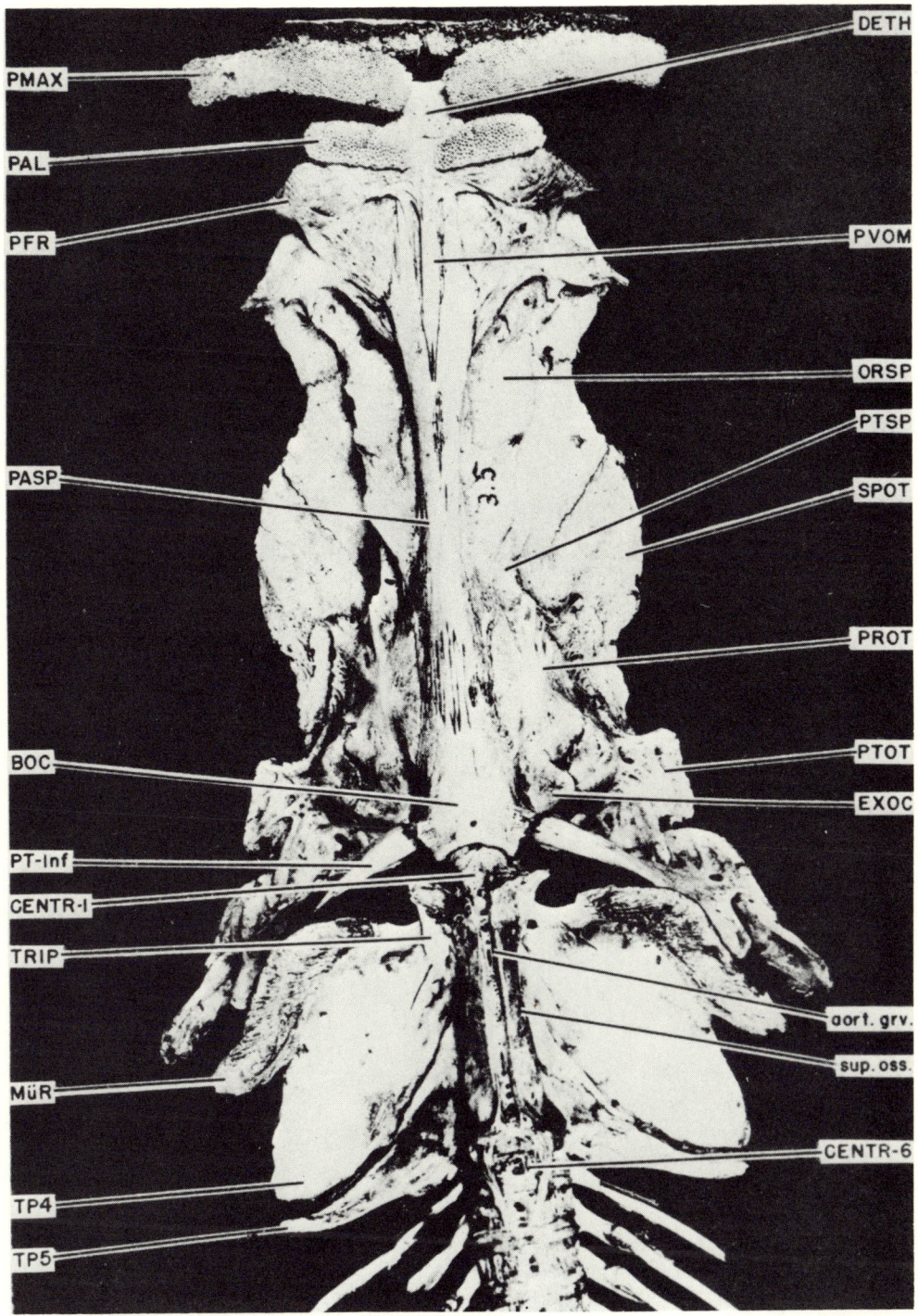

Ventral view of neurocranium and anterior vertebral complex of *Wallago* sp. × ½

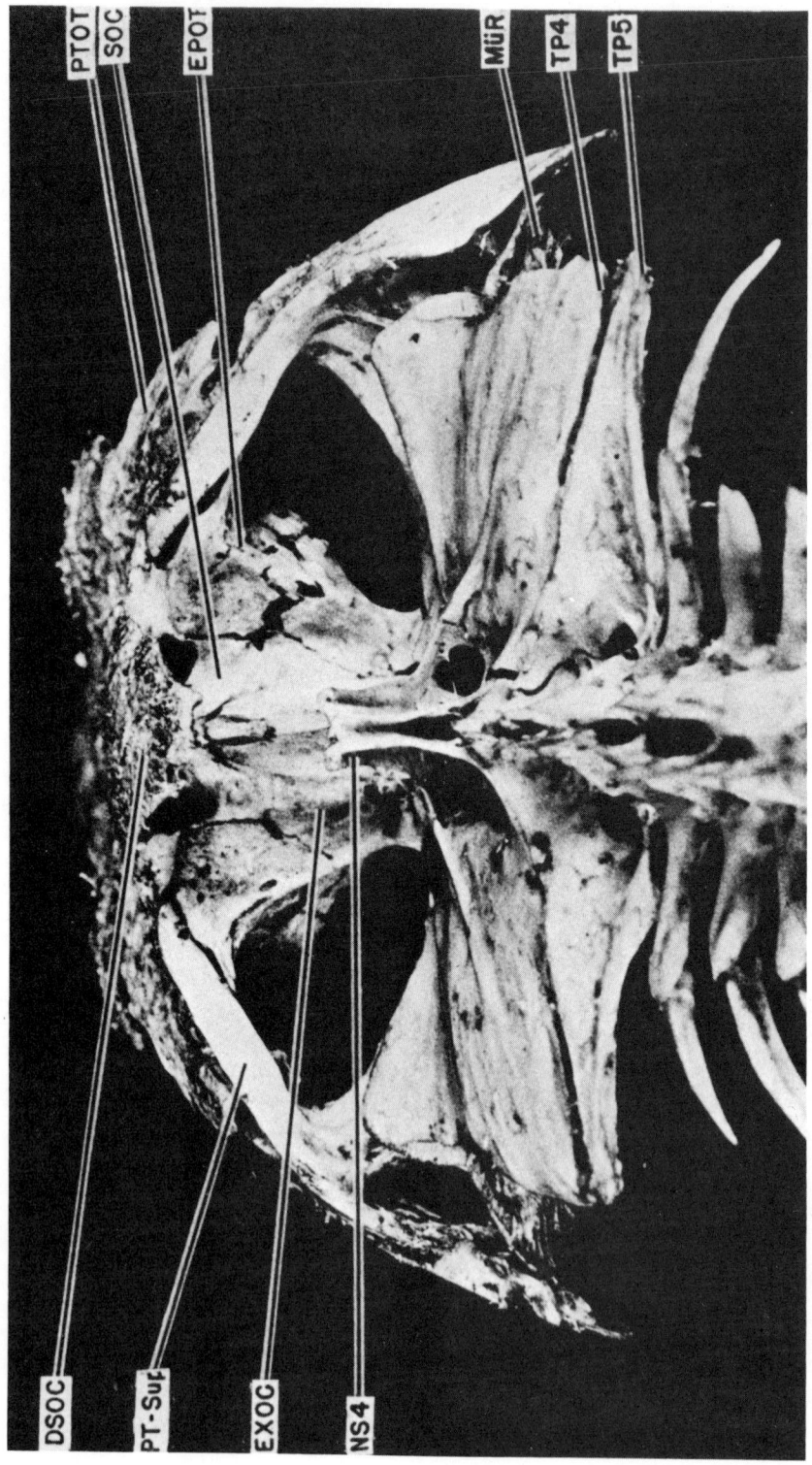

Occipital view of neurocranium and anterior vertebral complex of *Wallago* sp. × 1

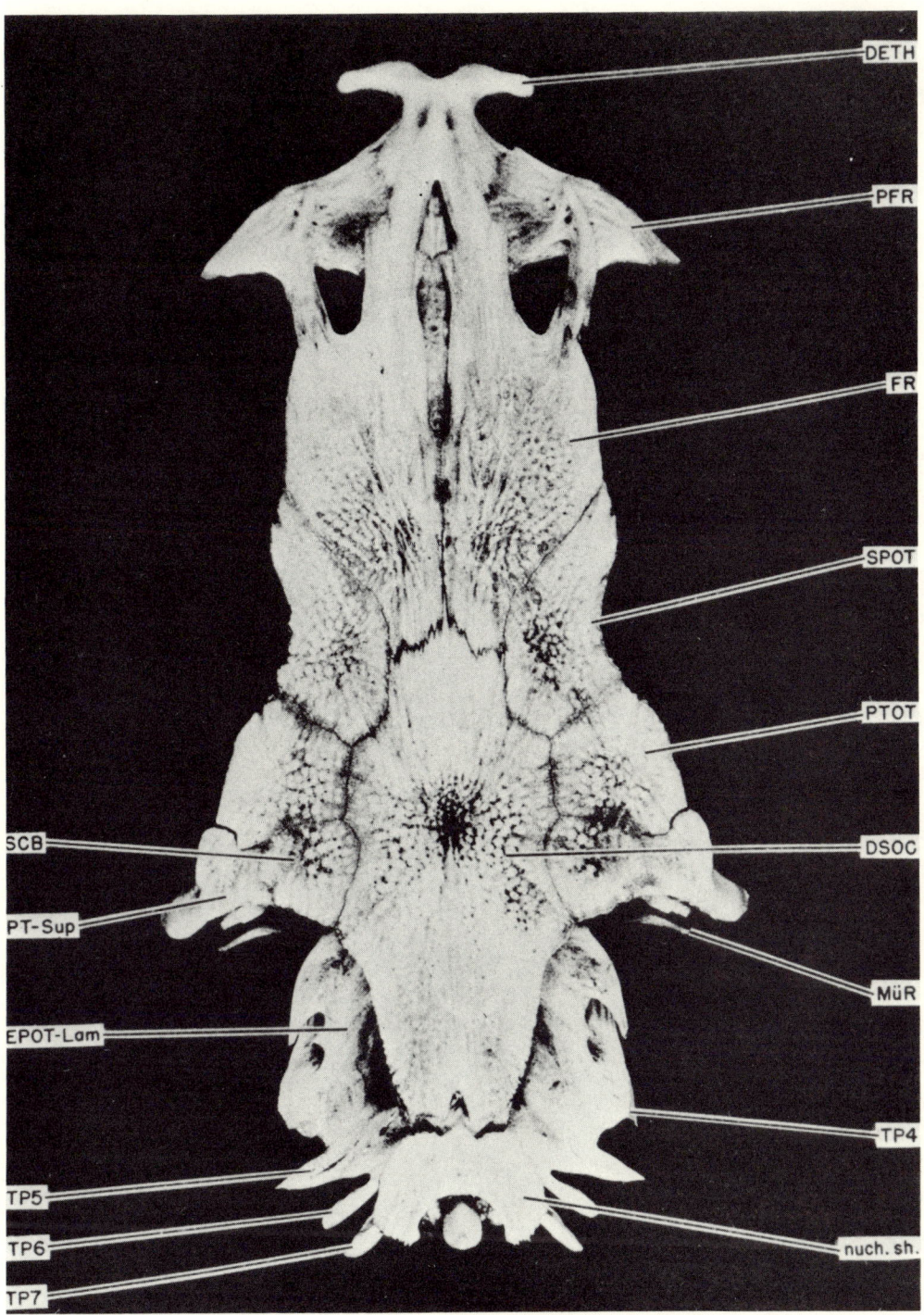

Dorsal view of neurocranium and anterior vertebral complex of *Galeichthys felis*. × 2

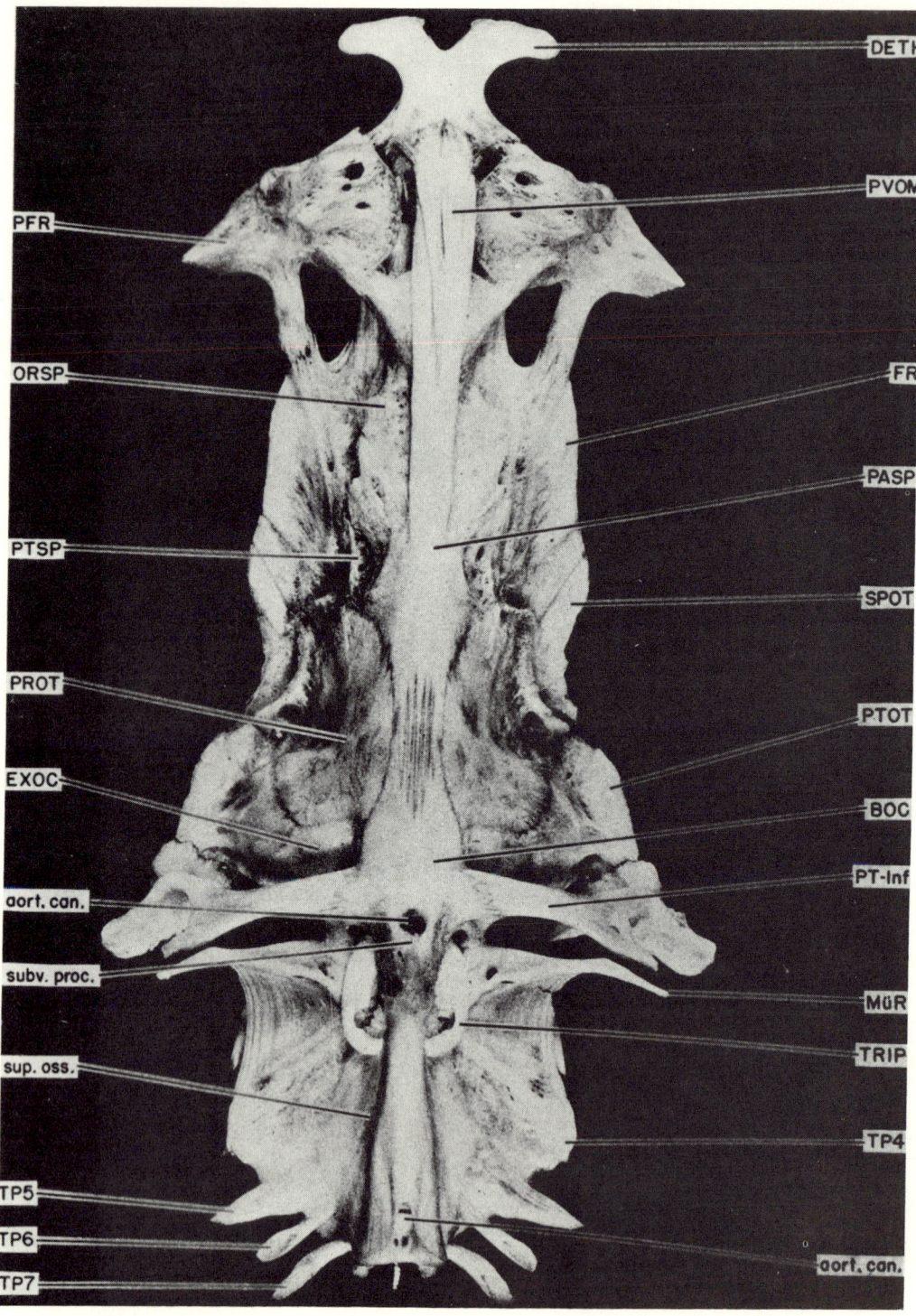

Ventral view of neurocranium and anterior vertebral complex of *Galeichthys felis*. × 2

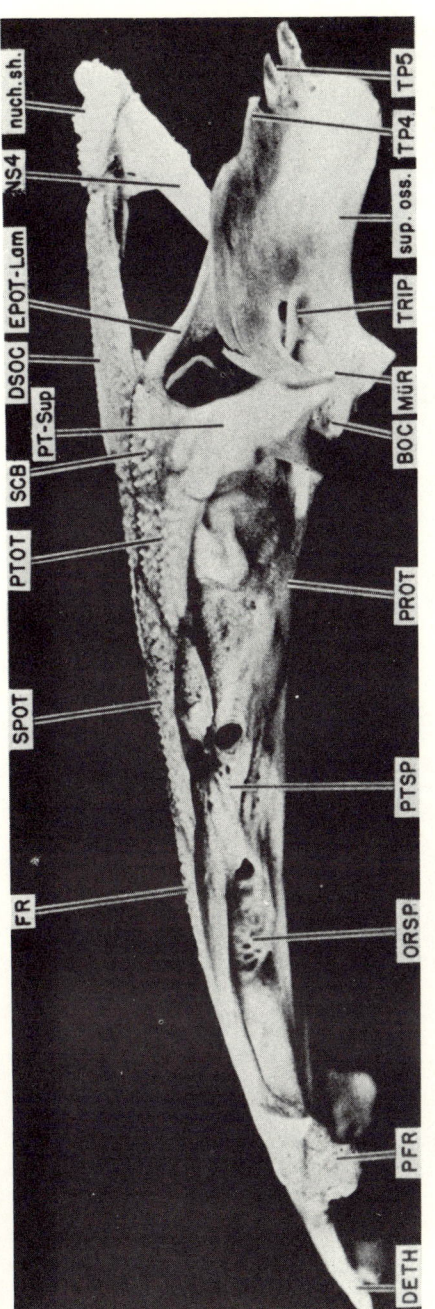
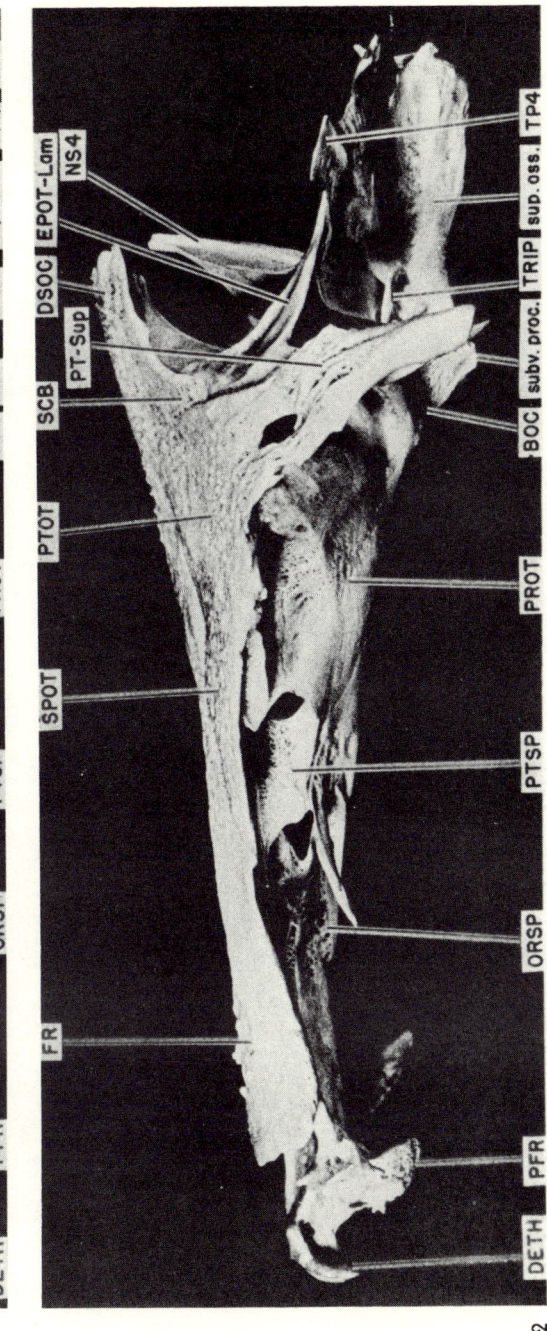

1. Left lateral view of neurocranium and anterior vertebral complex of *Galeichthys felis*. ×2
2. Left lateral view of neurocranium and anterior vertebral complex of *Bagre marinus*. ×1

243

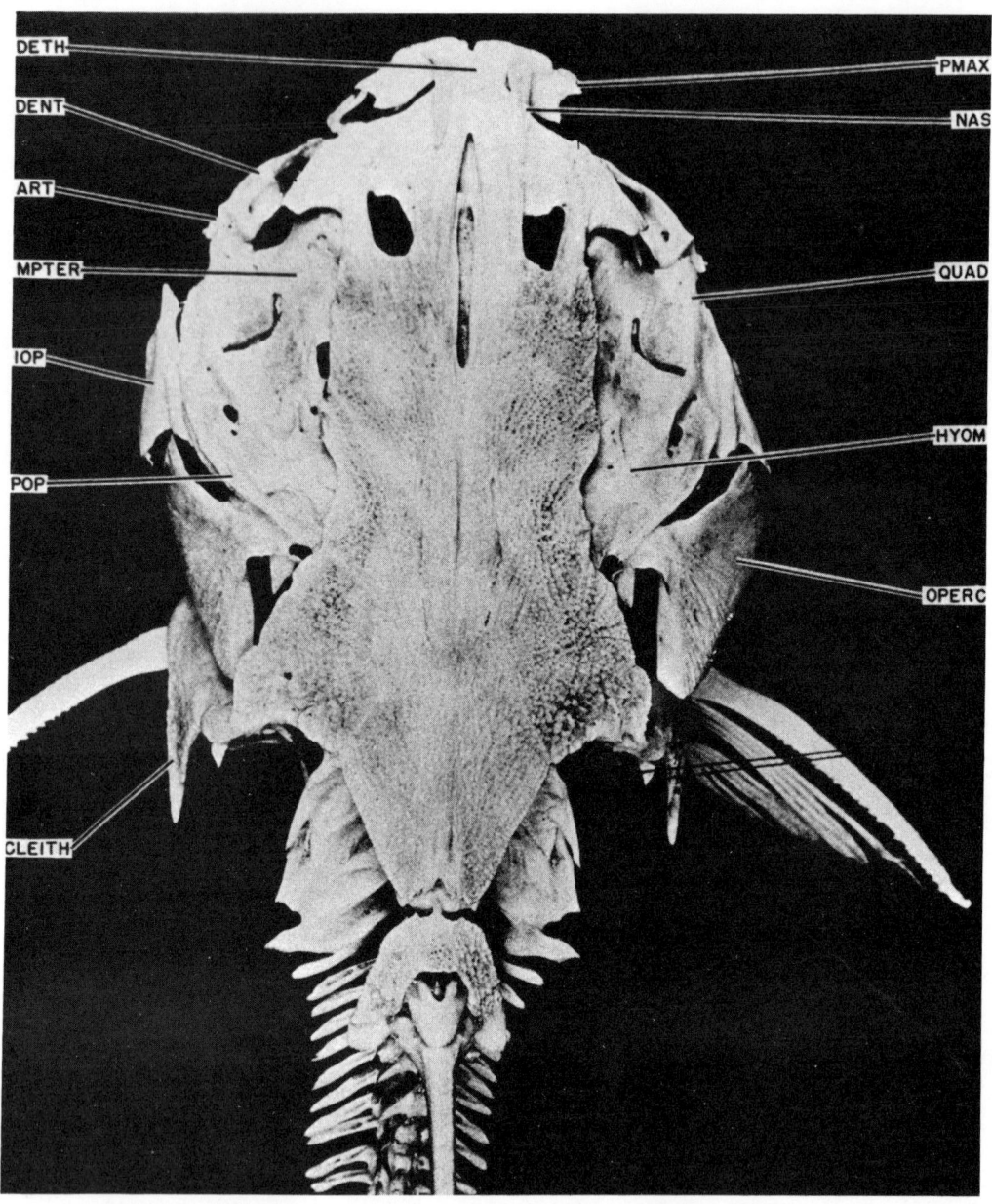

Dorsal view of cranial and pectoral skeleton of *Galeichthys felis*. × 1

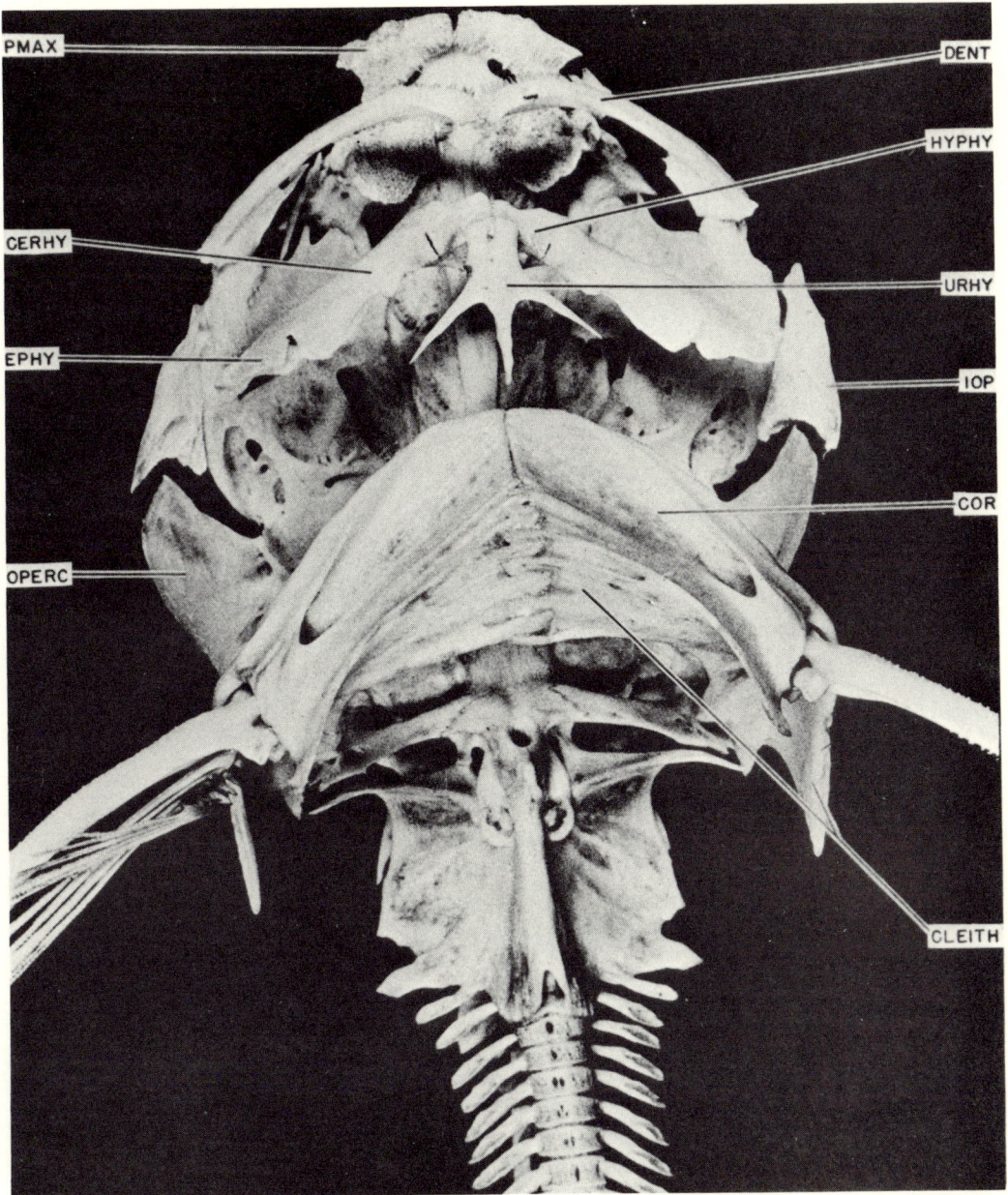

Ventral view of cranial and pectoral skeleton of *Galeichthys felis*. ×1

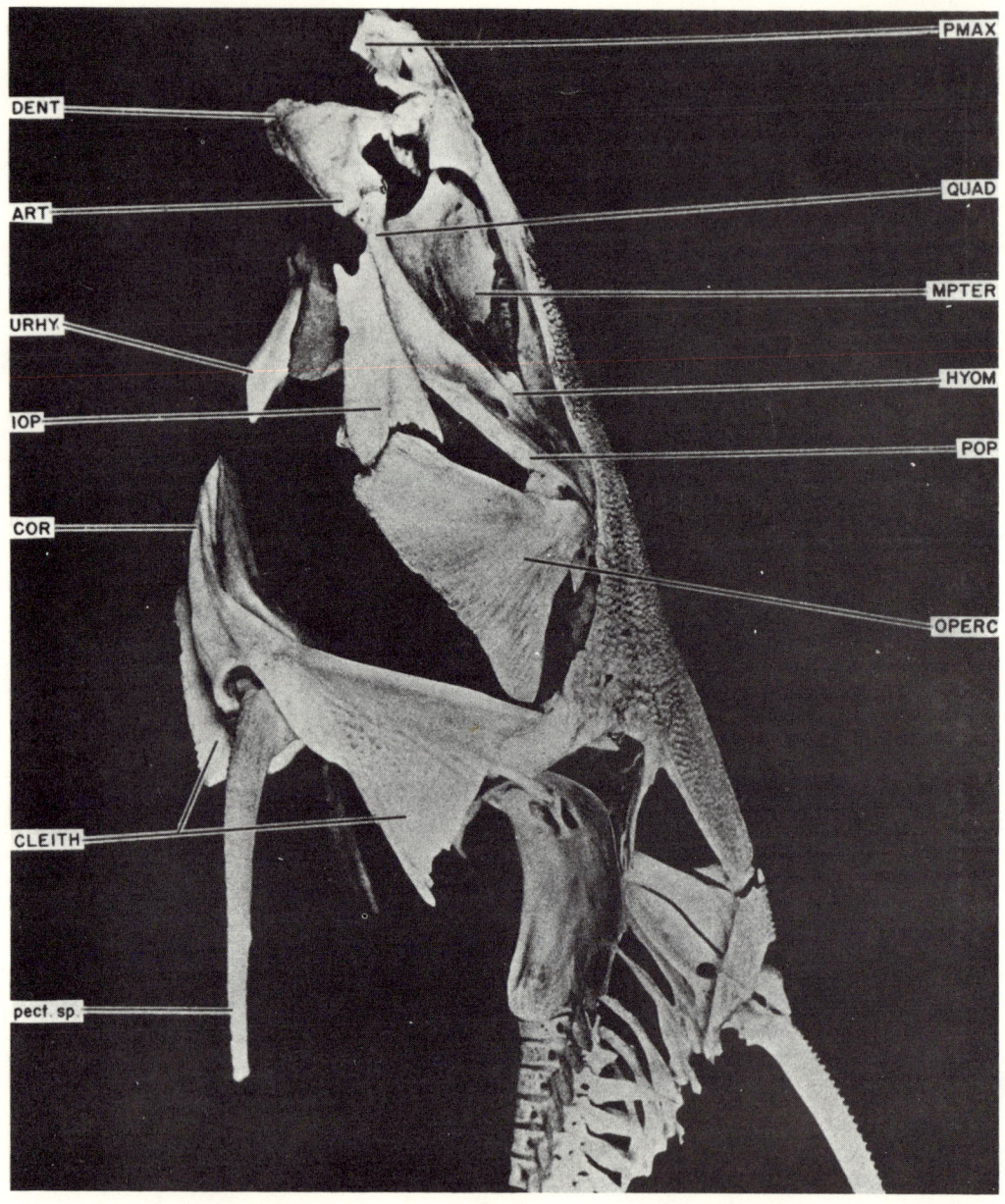

Left lateral view of cranial and pectoral skeleton of *Galeichthys felis*. ×1

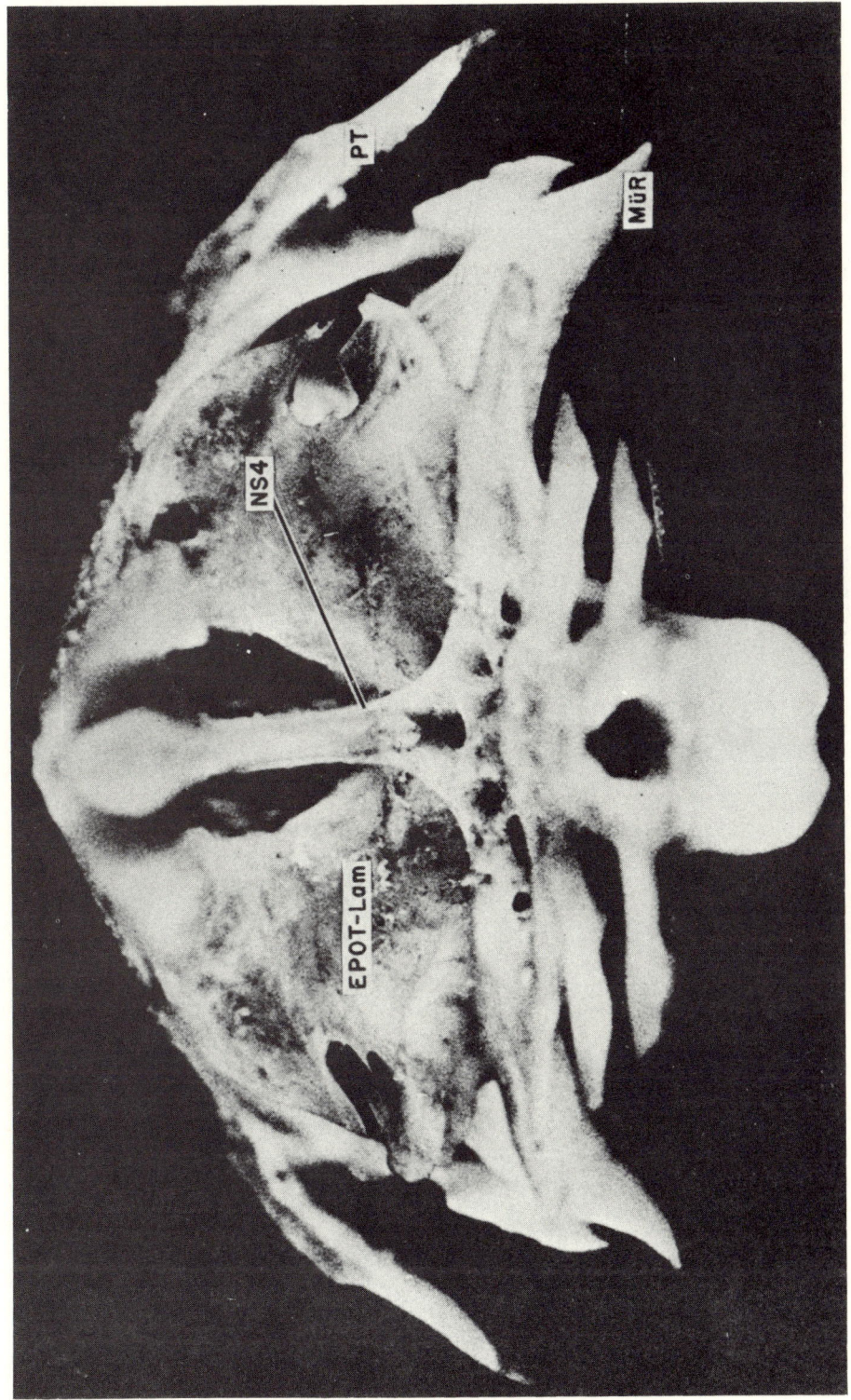

Occipital view of neurocranium and anterior vertebral complex of *Galeichthys felis*. × 4

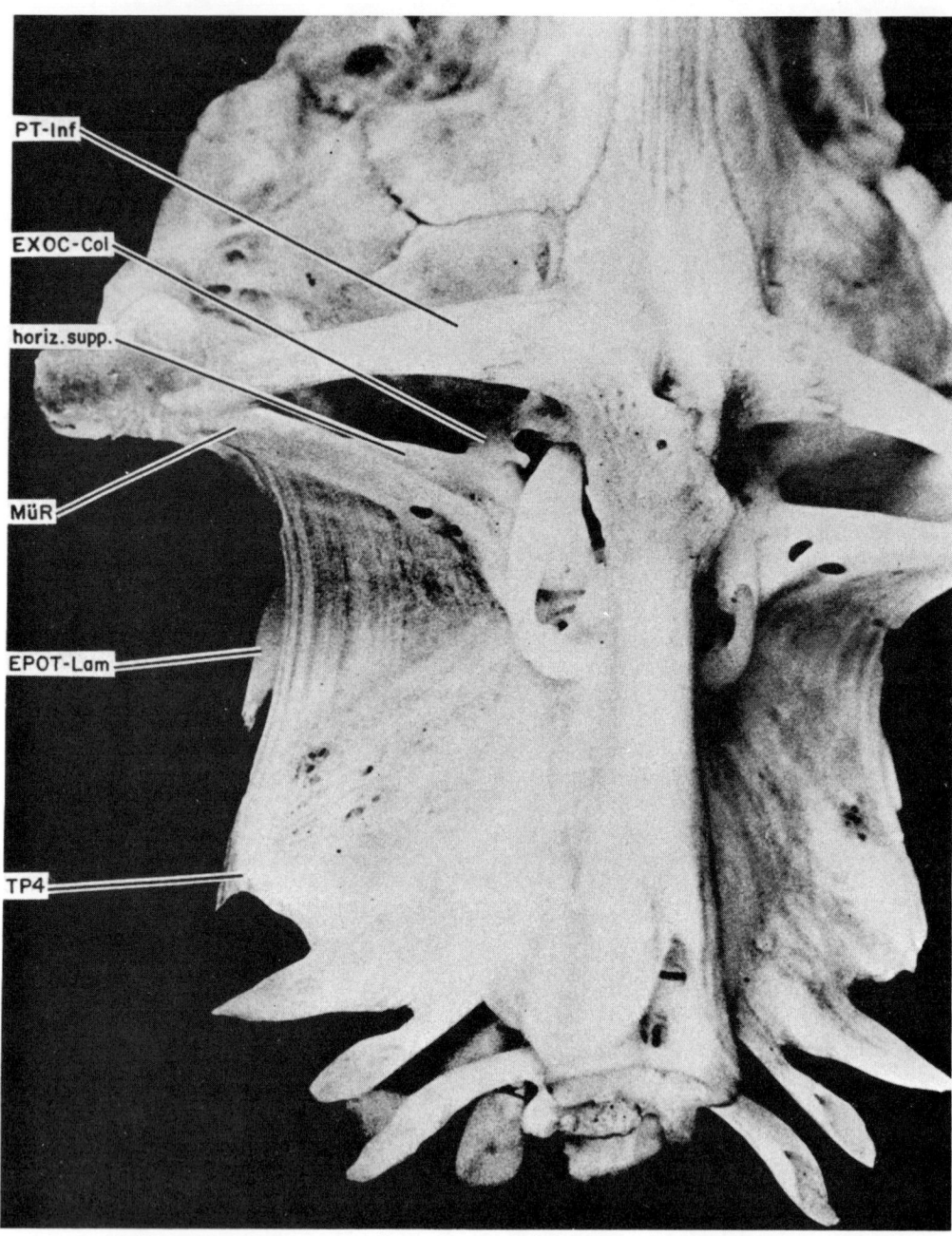

Ventral view of *Springfederapparat* and associated structures of *Galeichthys felis*. × 4

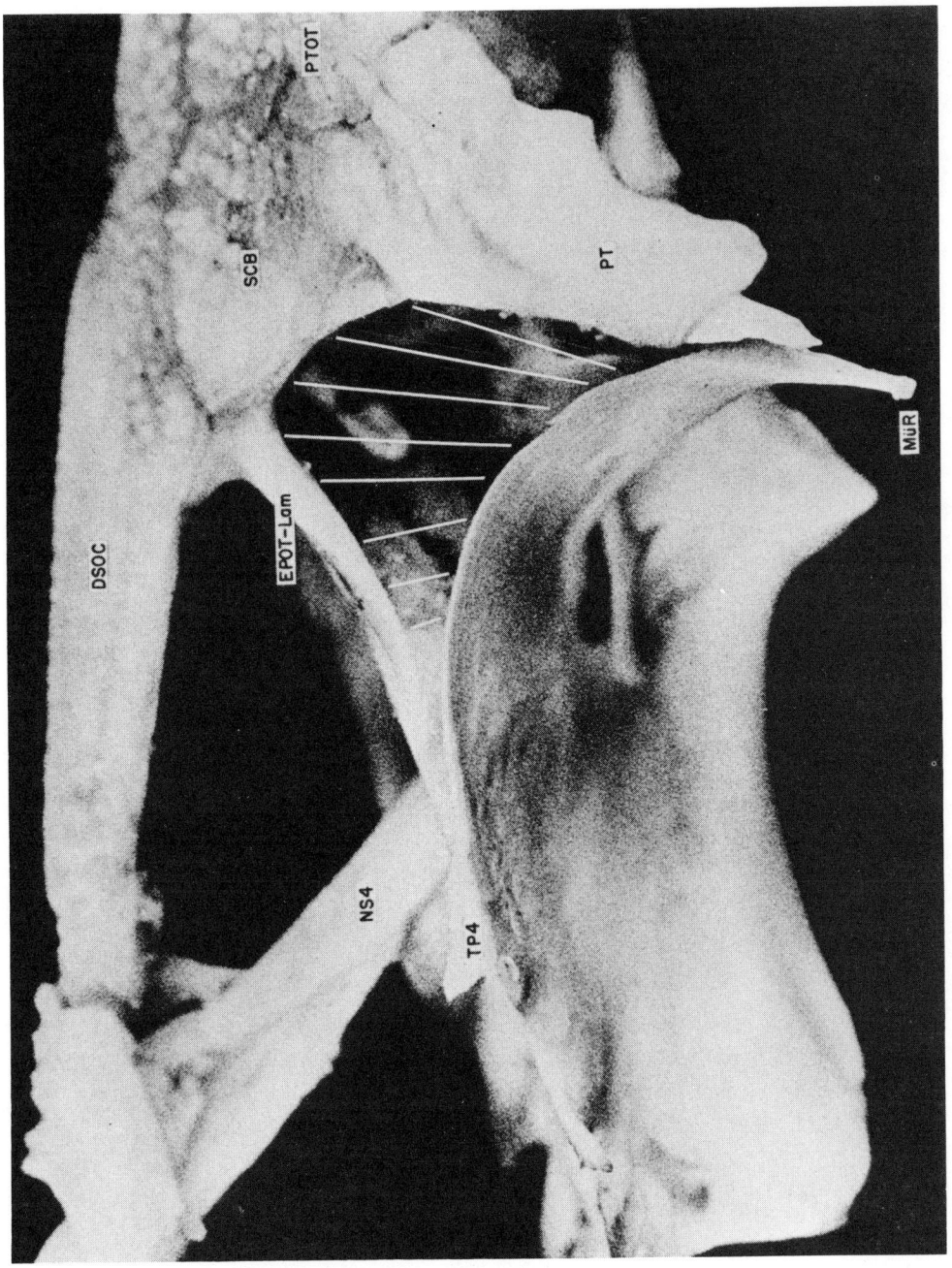

Right lateral view of *Springfederapparat* and associated structures of *Galeichthys felis*. White lines represent position of protractor muscle. × 6

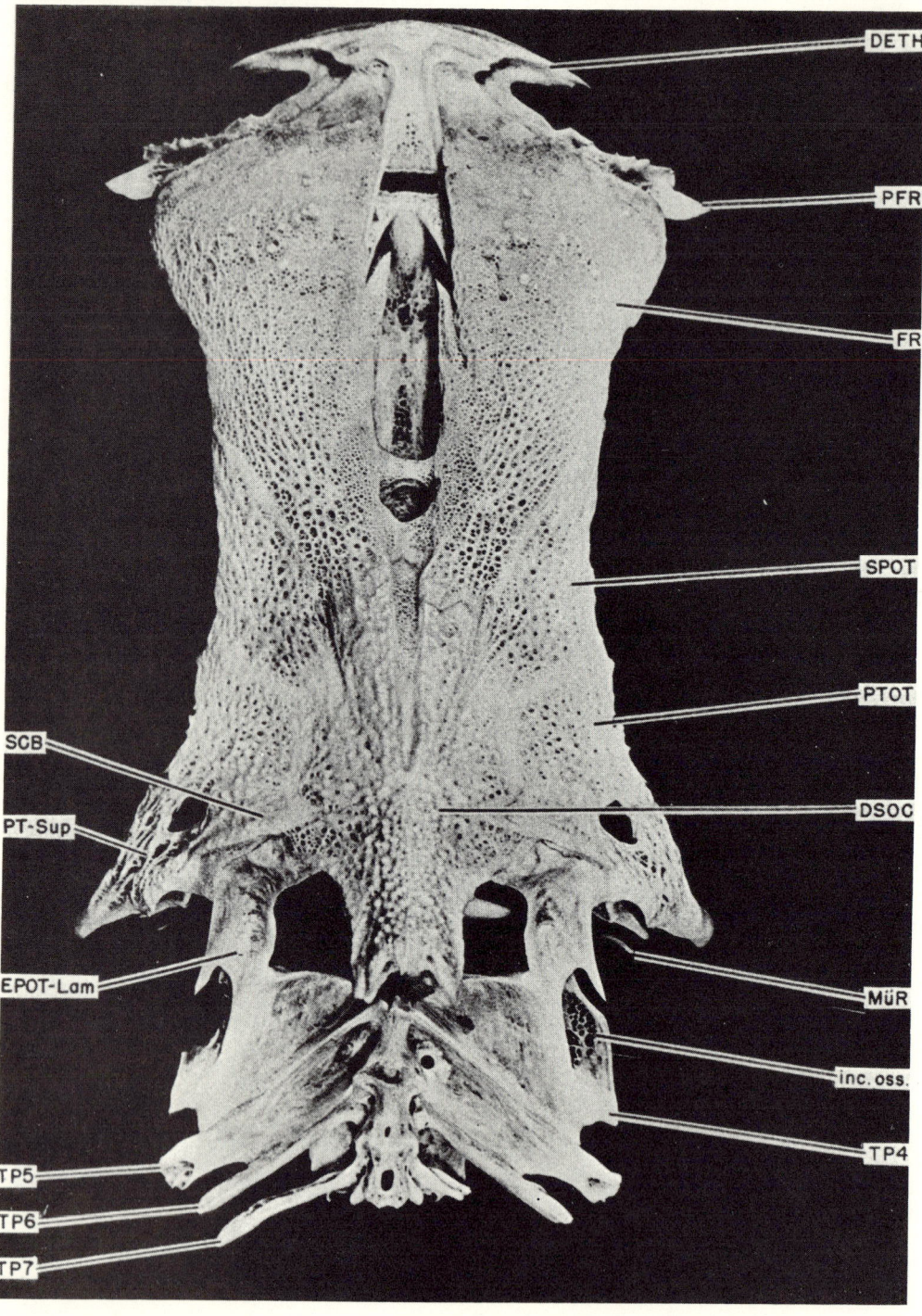

Dorsal view of neurocranium and anterior vertebral complex of *Bagre marinus*. ×1

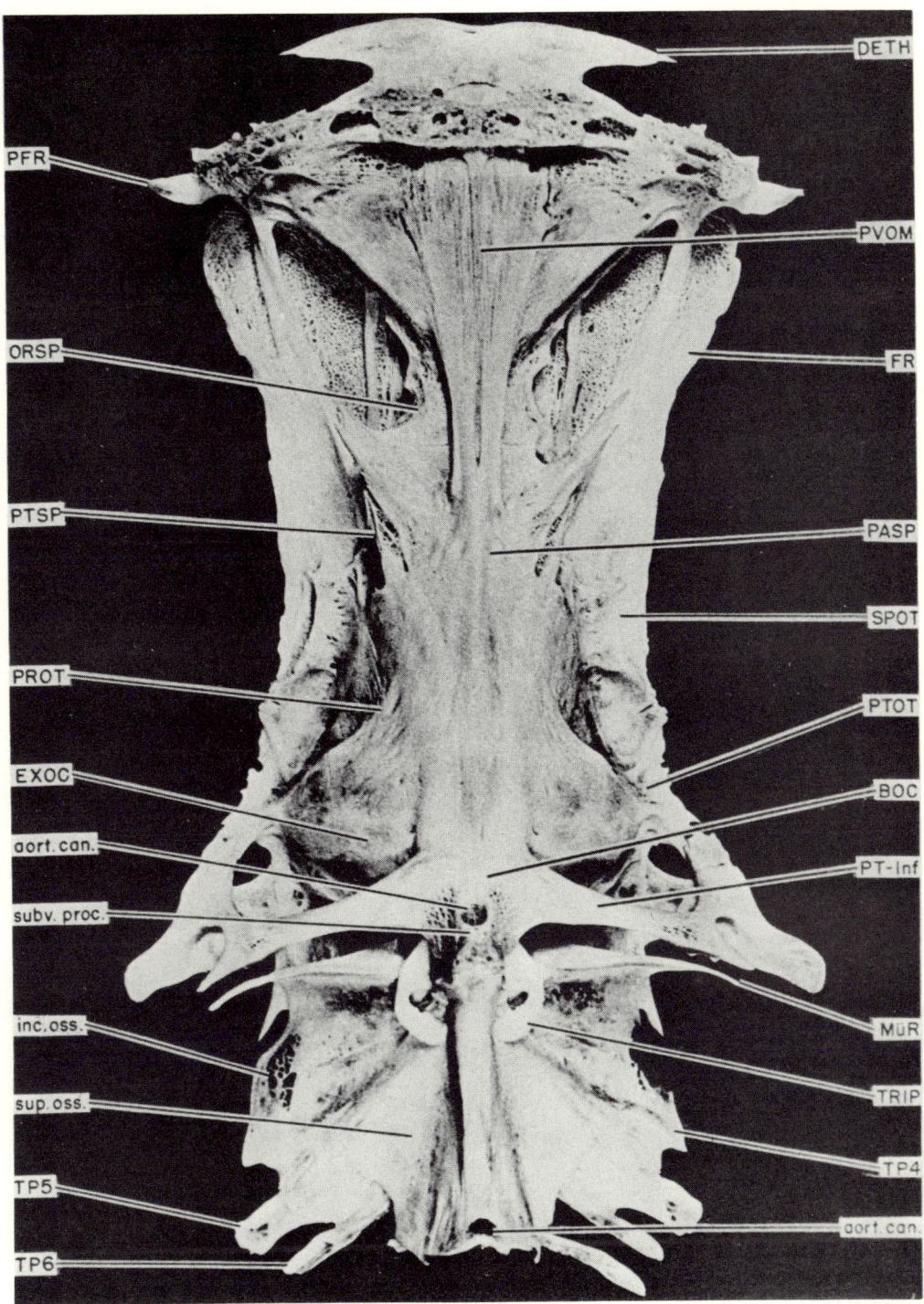

Ventral view of neurocranium and anterior vertebral complex of *Bagre marinus*. × 1

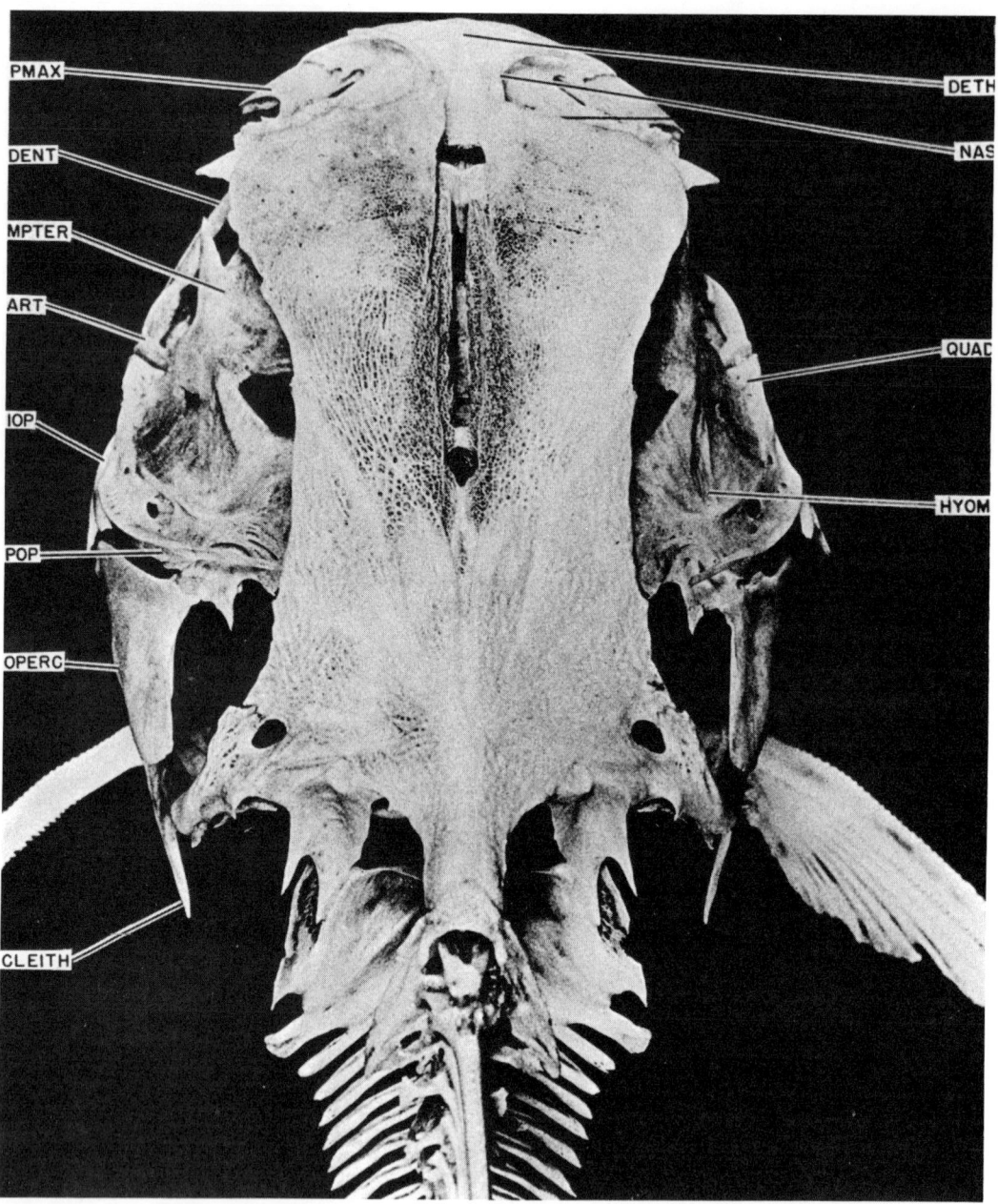

Dorsal view of cranial and pectoral skeleton of *Bagre marinus*. ×1

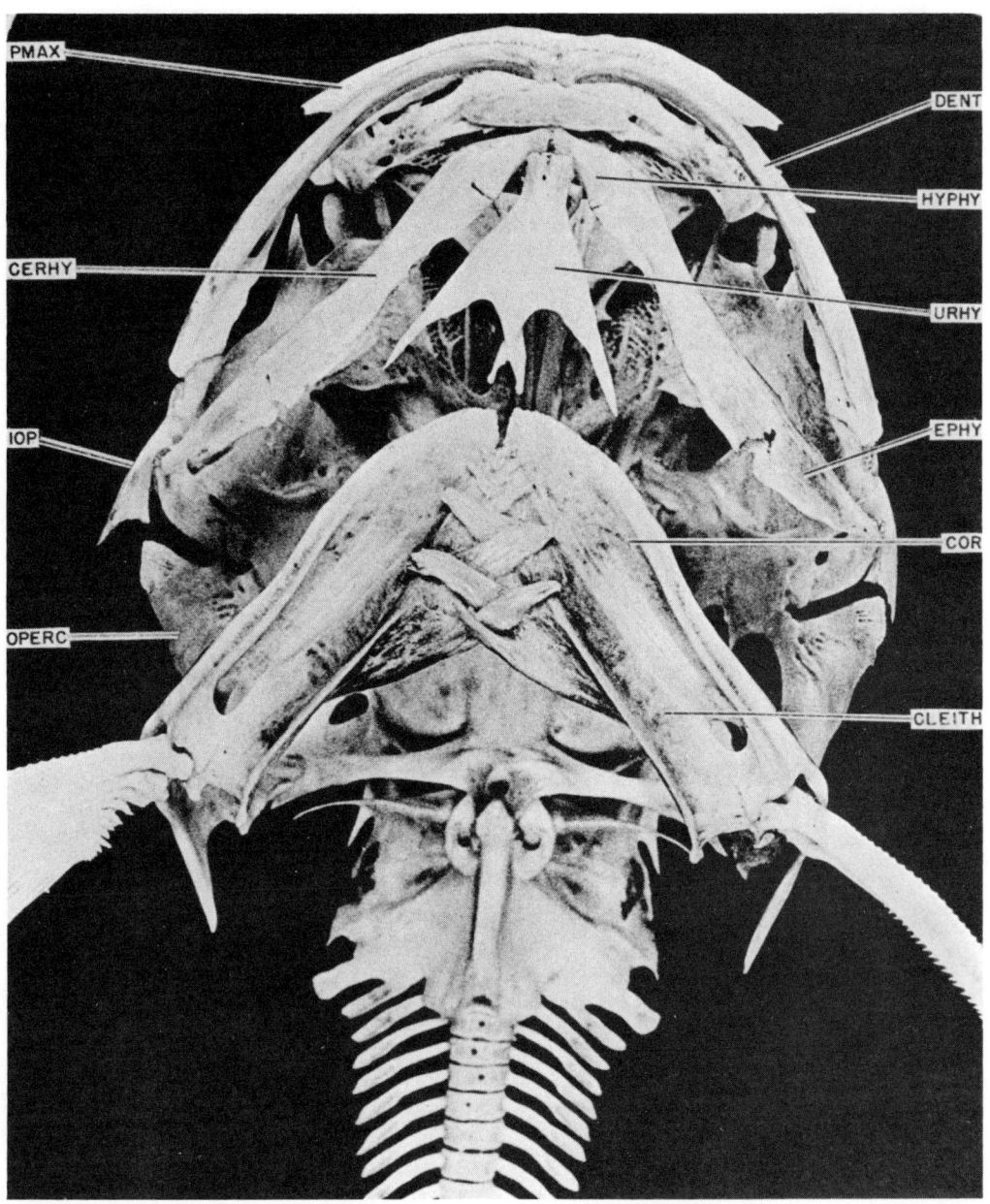

Ventral view of cranial and pectoral skeleton of *Bagre marinus*. ×1

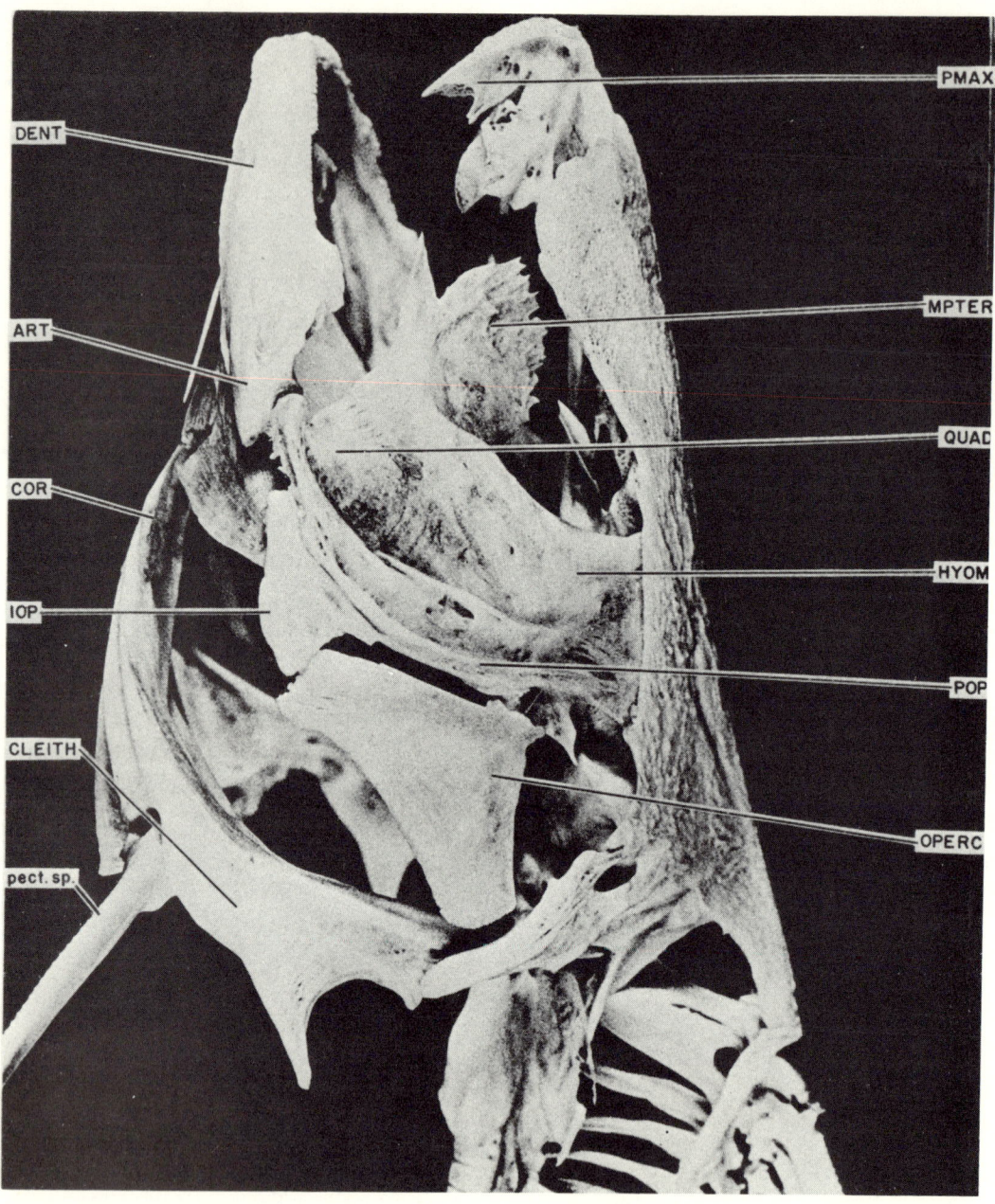

Left lateral view of cranial and pectoral skeleton of *Bagre marinus*. ×1

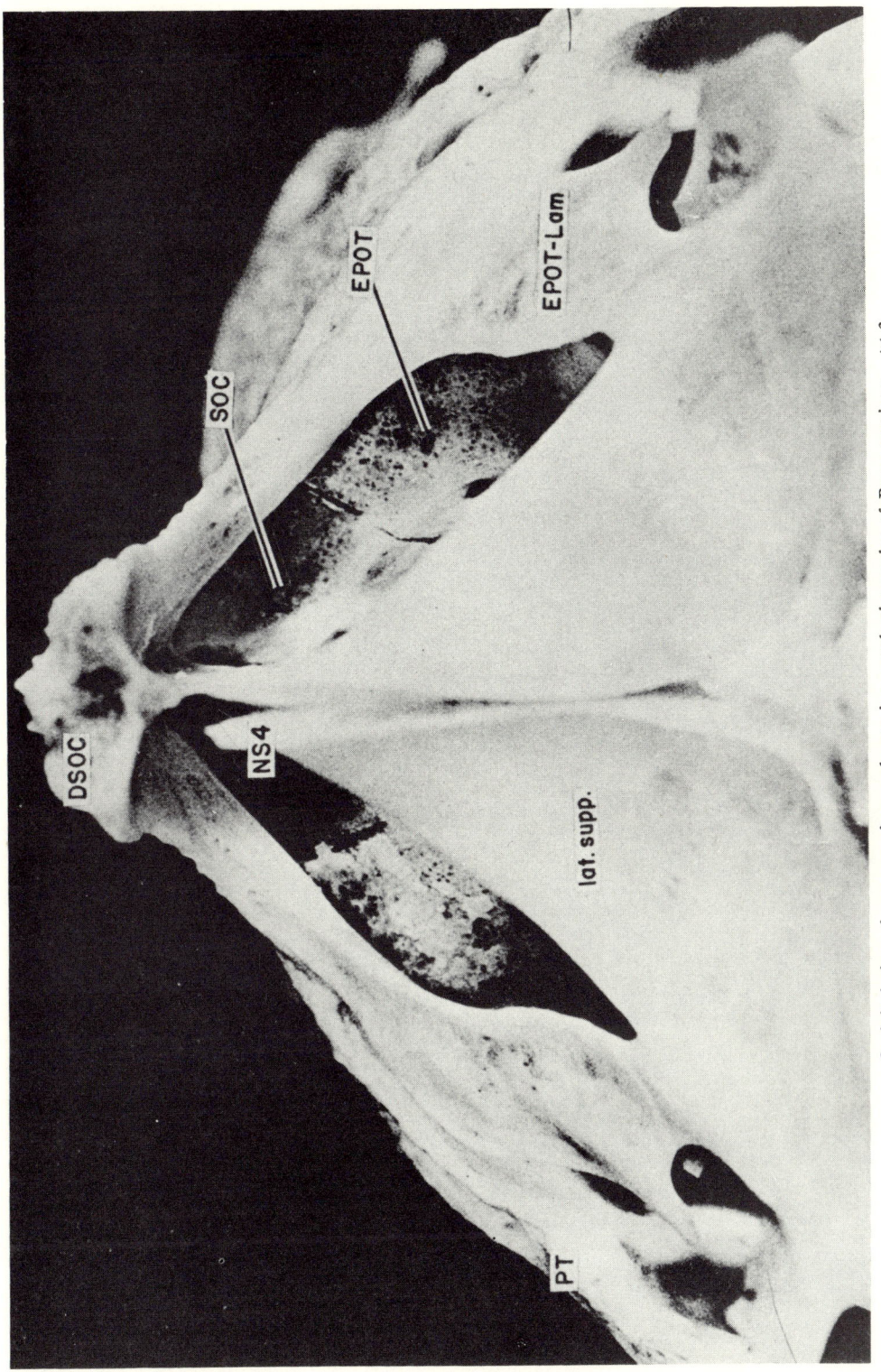

Occipital view of neurocranium and anterior vertebral complex of *Bagre marinus*. × 3

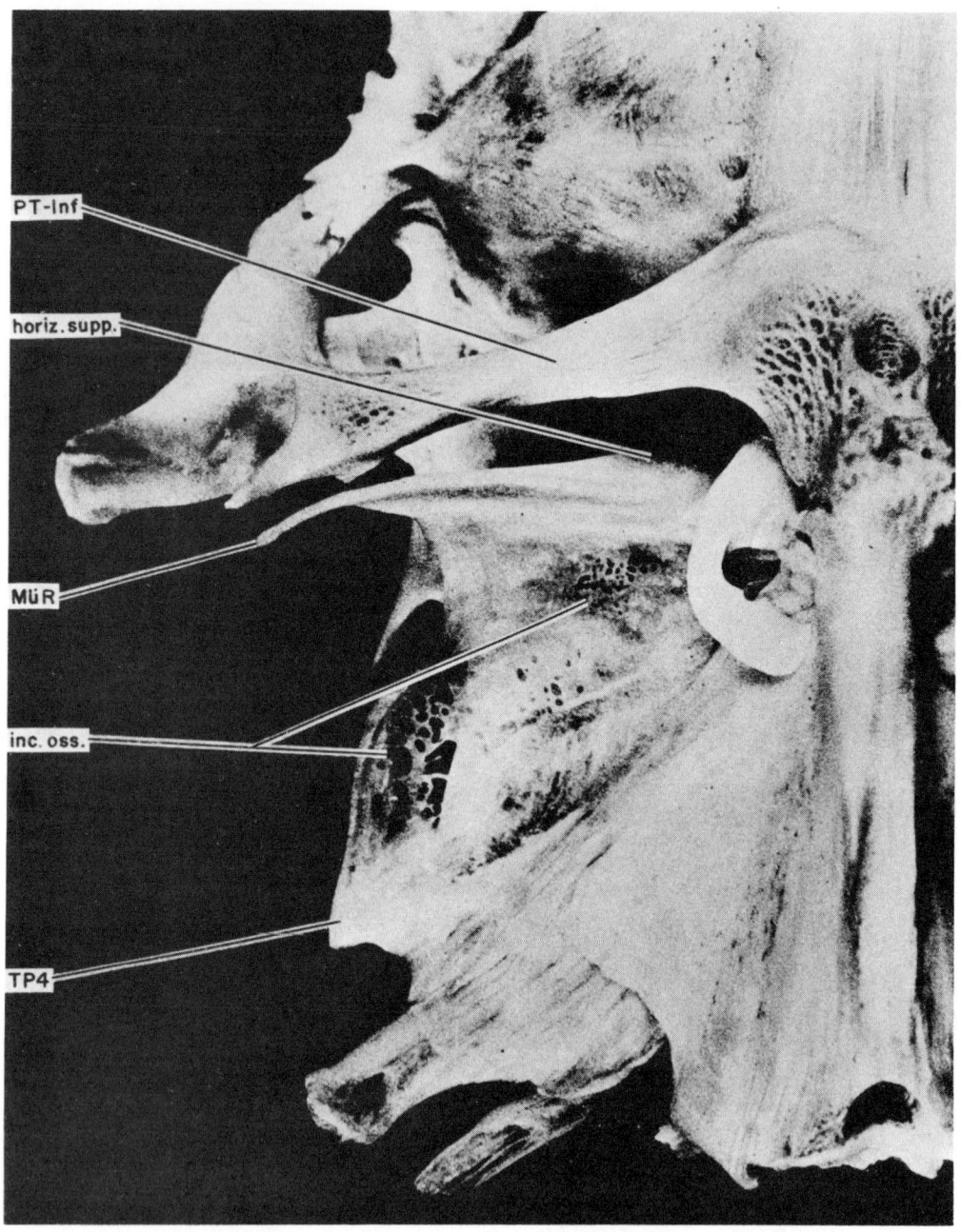

Ventral view of *Springfederapparat* and associated structures of *Bagre marinus*. ×3

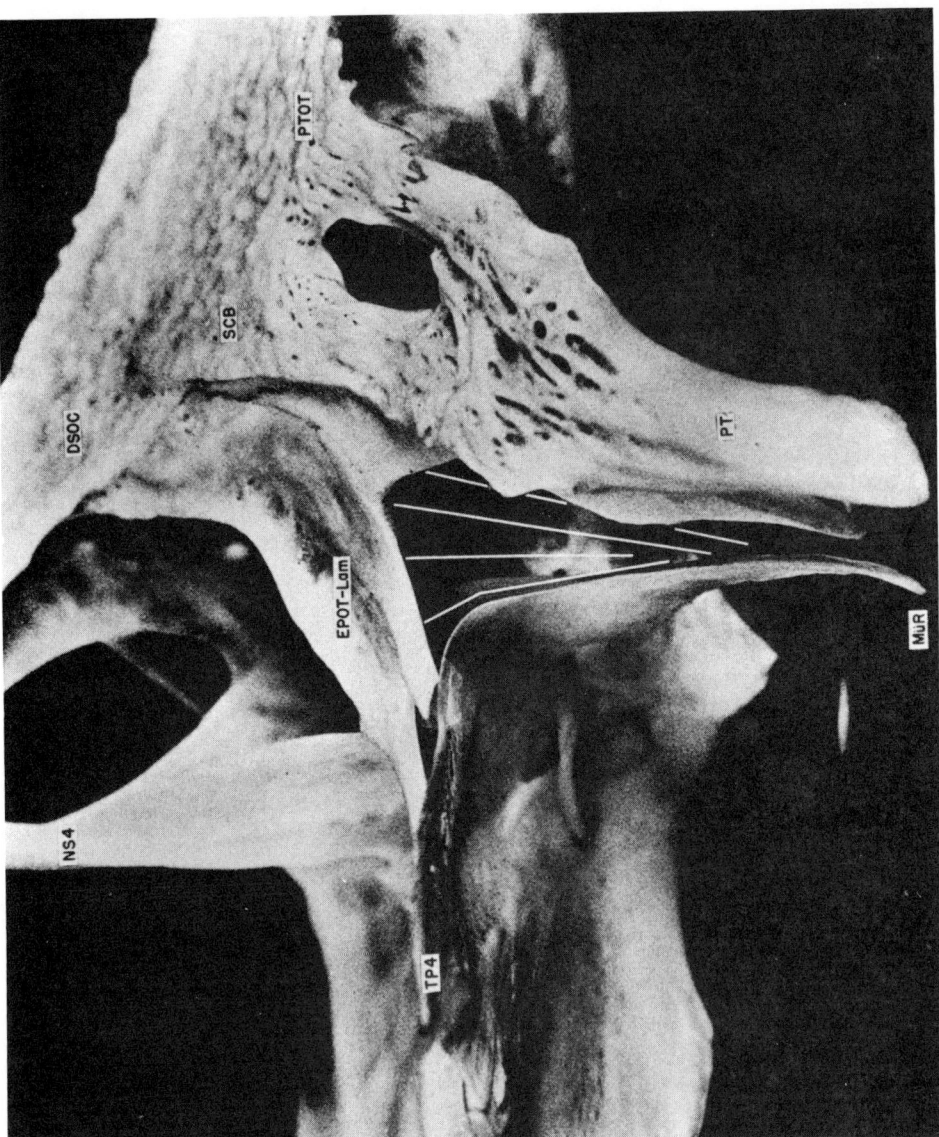

Right lateral view of *Springfederapparat* and associated structures of *Bagre marinus*. White lines represent position of protractor muscle. ×5

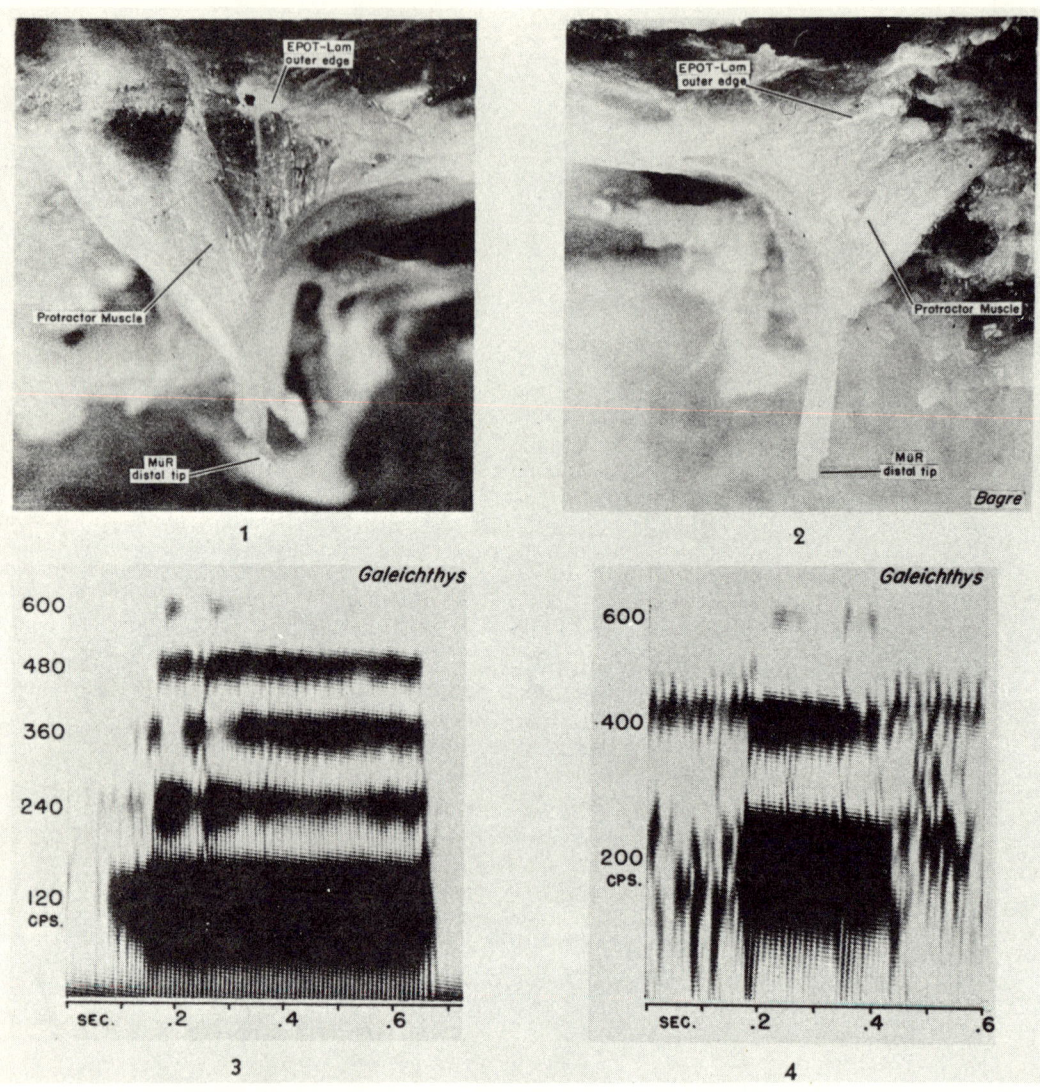

1, 2. Dissections of preserved specimens to show structure of sound-producing protractor muscle. Posttemporal bone removed; origin of muscle along line indicated as outer edge of epiotic lamina. 1. *Galeichthys felis*, left lateral view. 2. *Bagre marinus*, right lateral view. Both × 4

3, 4. Sonagrams of sounds produced by stimulation of occipital nerve in *Galeichthys*. 3. At 120 pulses per second. 4. At 200 pulses per second

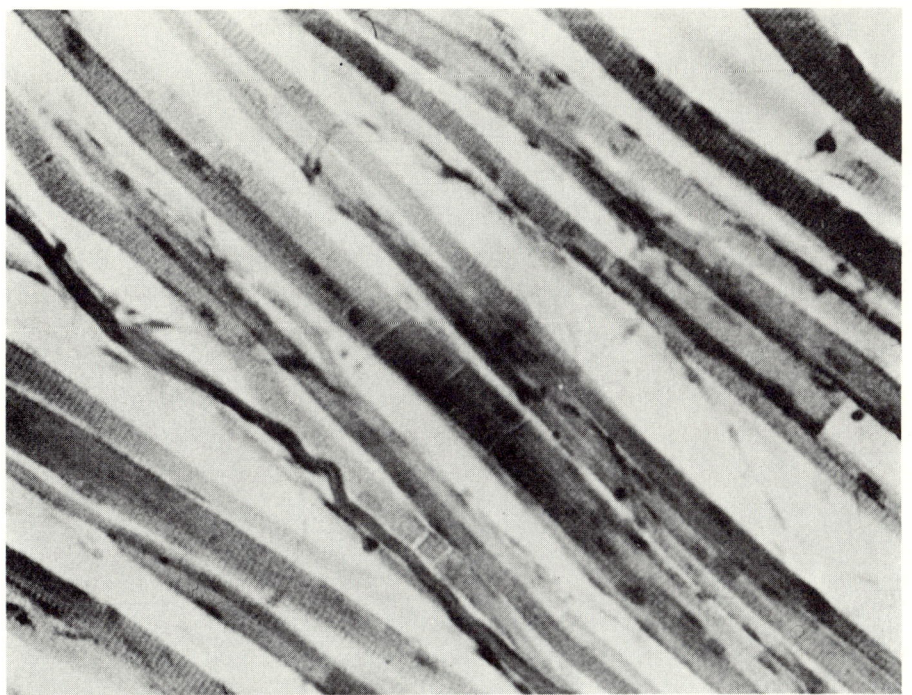
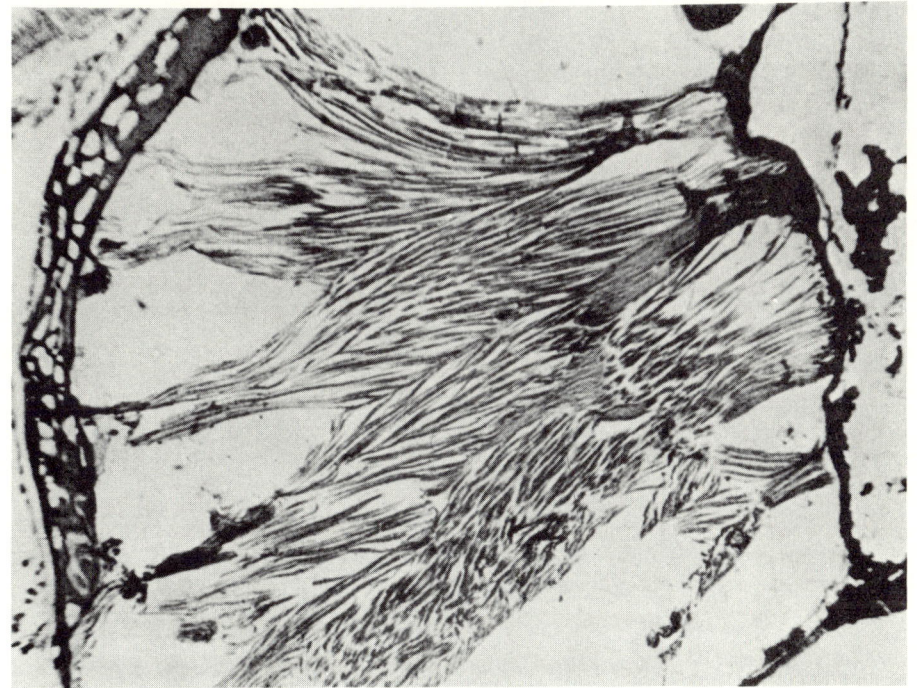

1. Transverse section of juvenile *Galeichthys* in region of protractor muscle. Thick cancellous bone in upper part of picture is a section through epiotic lamina, and dark line where muscle fibers insert is thin shelf formed by expanded fourth transverse process. ×50
2. High-power view of protractor muscle fibers. ×500

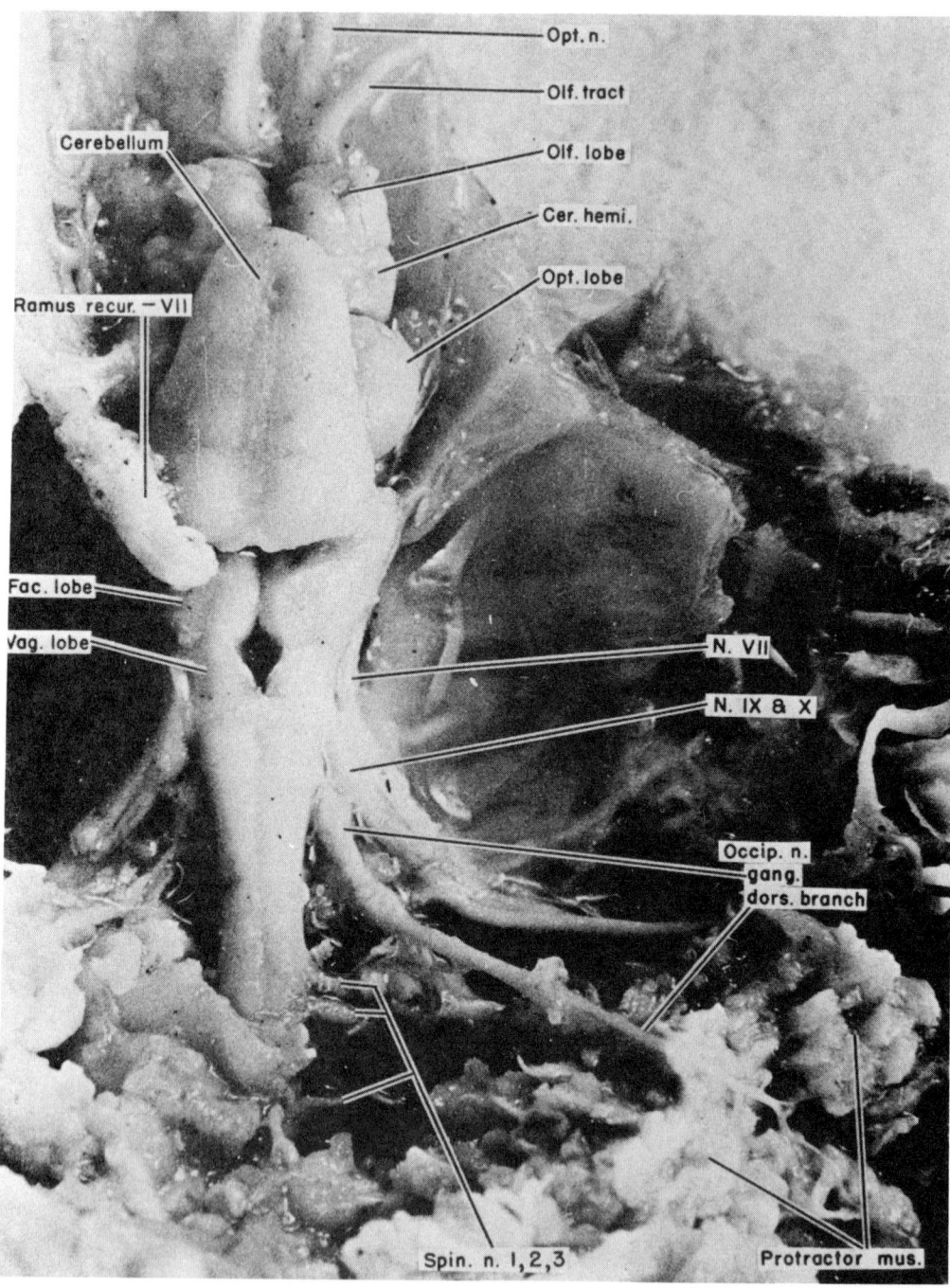

Dissection of brain and cranial nerves of formalin-preserved *Galeichthys felis*. Right epiotic lamina removed to expose dorsal surface of protractor muscle and its innervation by dorsal branch of occipital nerve. Large utricular otolith (lapillus) removed. × 3

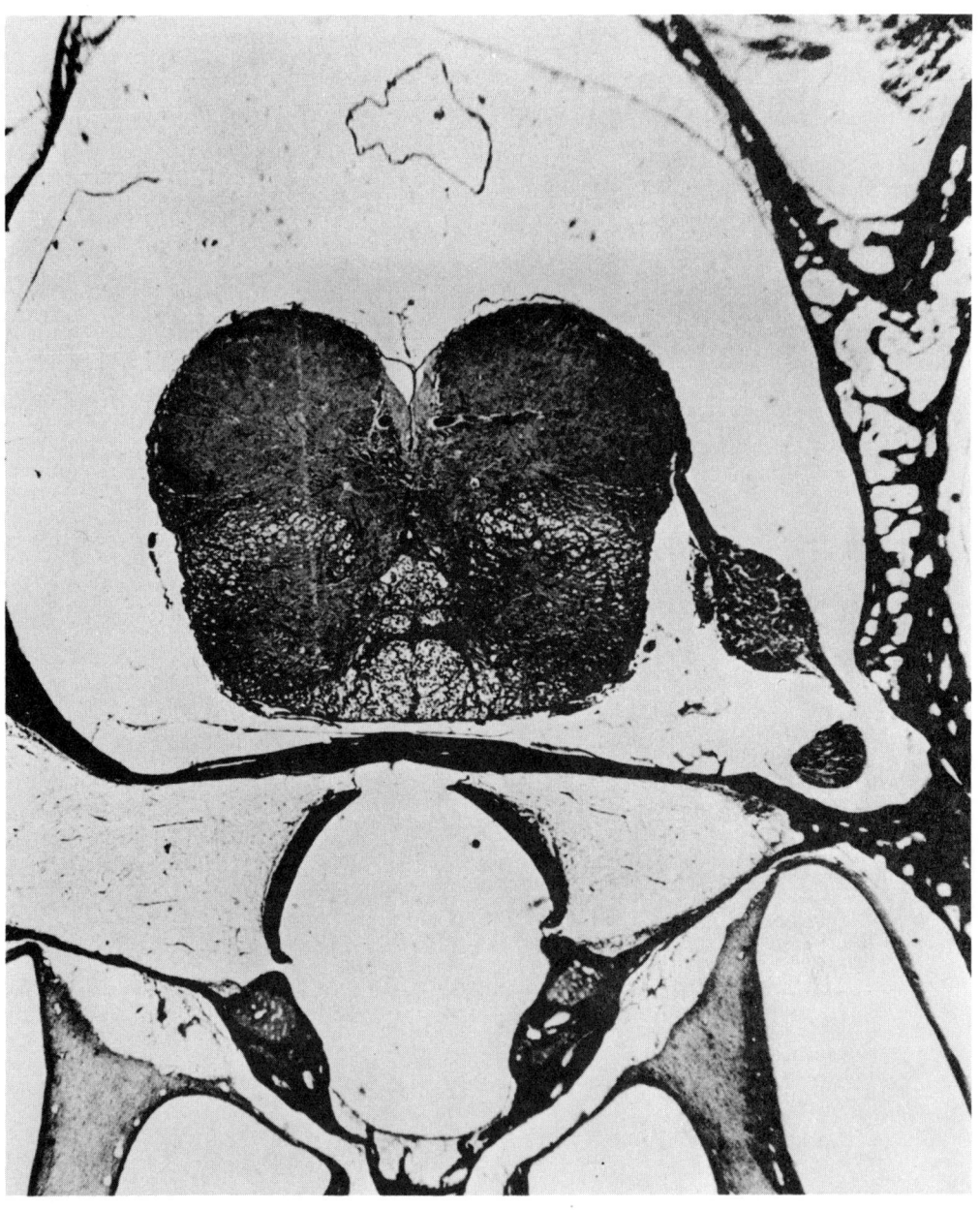

Transverse section of juvenile *Galeichthys felis* at level of occipital nerve, of which dorsal root ganglion and ventral root can be seen on right side of nerve cord. Ventral to nerve cord is portion of perilymphatic cavity (cavum sinus imparis). × 75

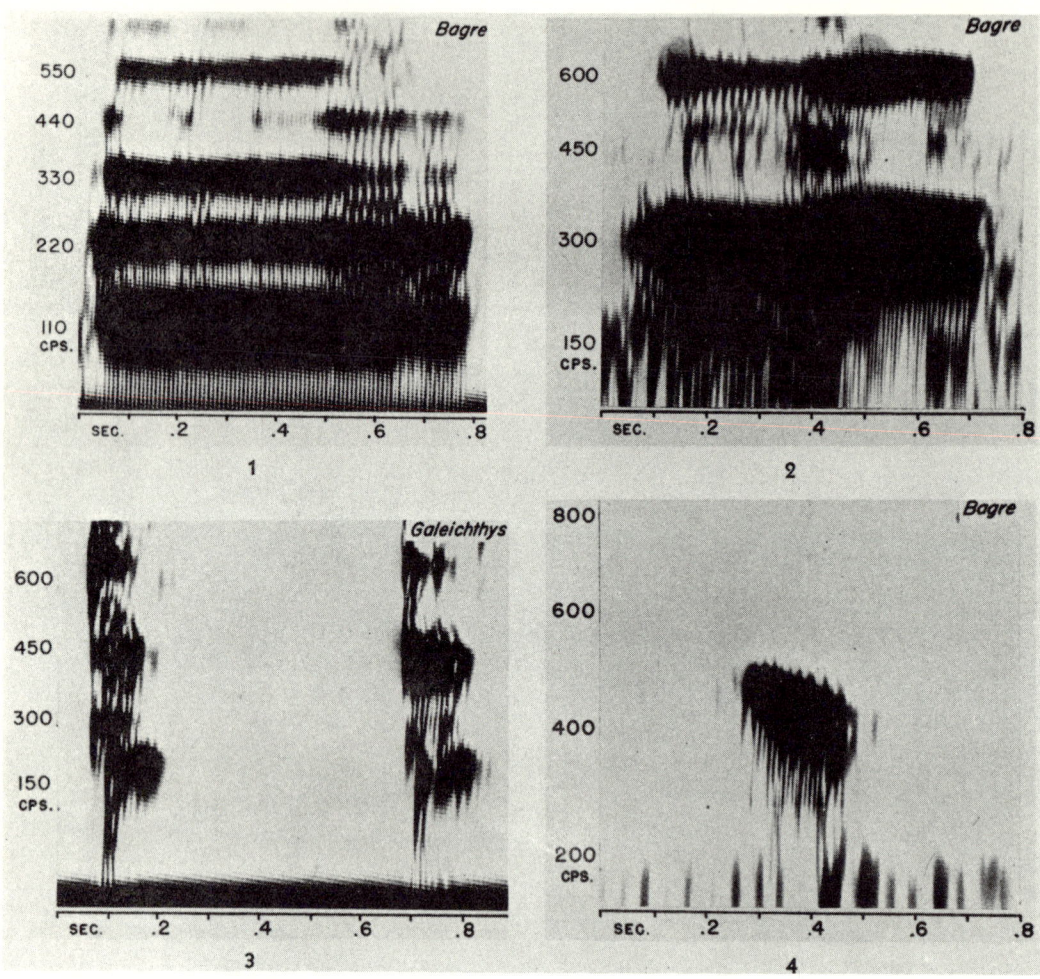

1, 2. Sonagrams of sounds produced by stimulation of occipital nerve in *Bagre*. 1. At 110 pulses per second. 2. At 300 pulses per second. Note partial response at 150 cycles per second
3. Sonagram of distress sounds produced by a captive *Galeichthys*
4. Sonagram of a "sob-like" sound produced by a captive *Bagre*

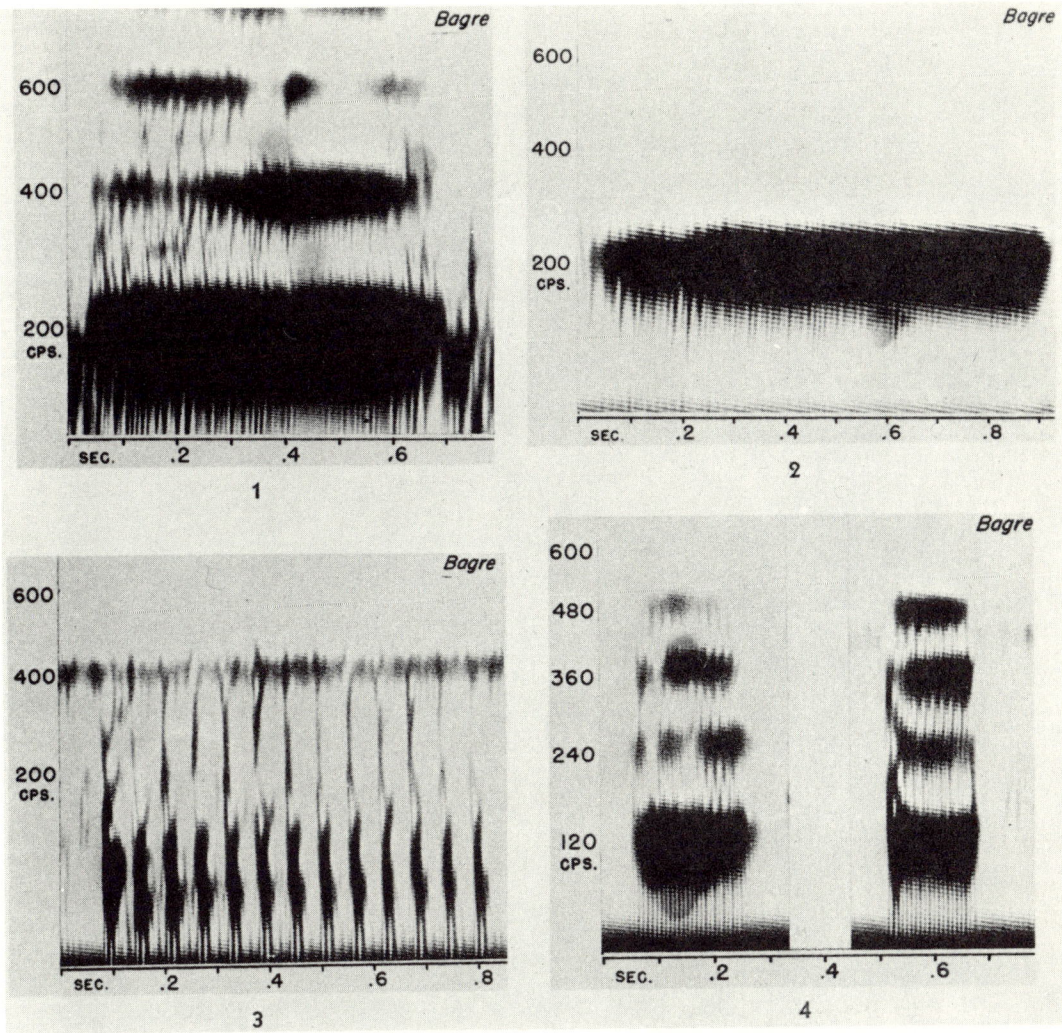

1, 2. Sonagrams of sounds produced by stimulation of protractor muscle in *Bagre*. 1. At 200 pulses per second. 2. After entire preparation was covered with fiberglass batting

3. Sonagram of sound produced by stimulation of occipital nerve at approximately 20 pulses per second in *Bagre*. Trace at 400 cycles per second is of external origin

4. Sonagram at left resulted from stimulation of left occipital nerve at 120 pulses per second in *Bagre*. Sonagram at right was produced by same specimen when both occipital nerves were simultaneously stimulated

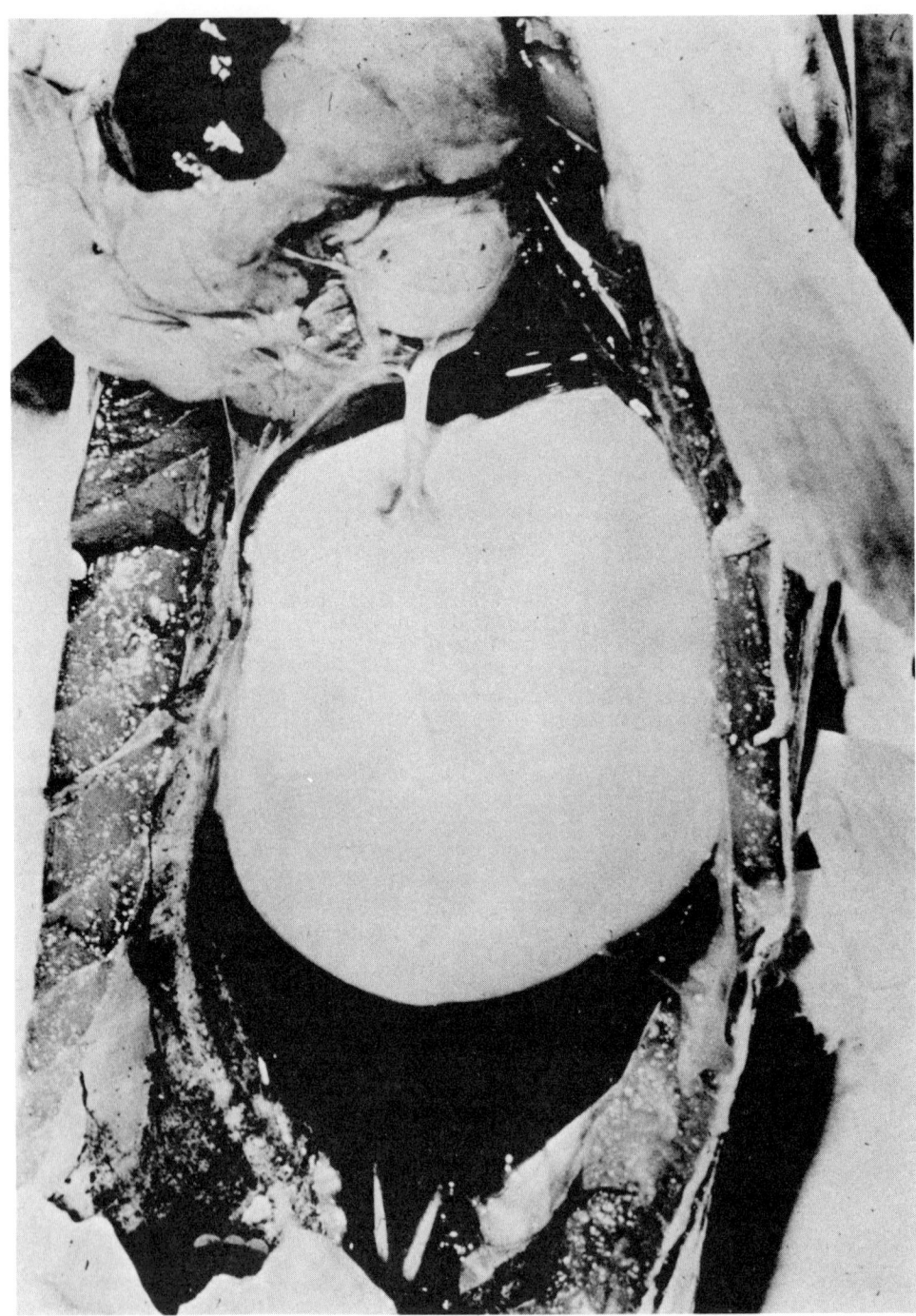

Dissection showing intact swim bladder of *Galeichthys felis*. Viscera reflected anteriorly, and pneumatic duct is visible. × 1

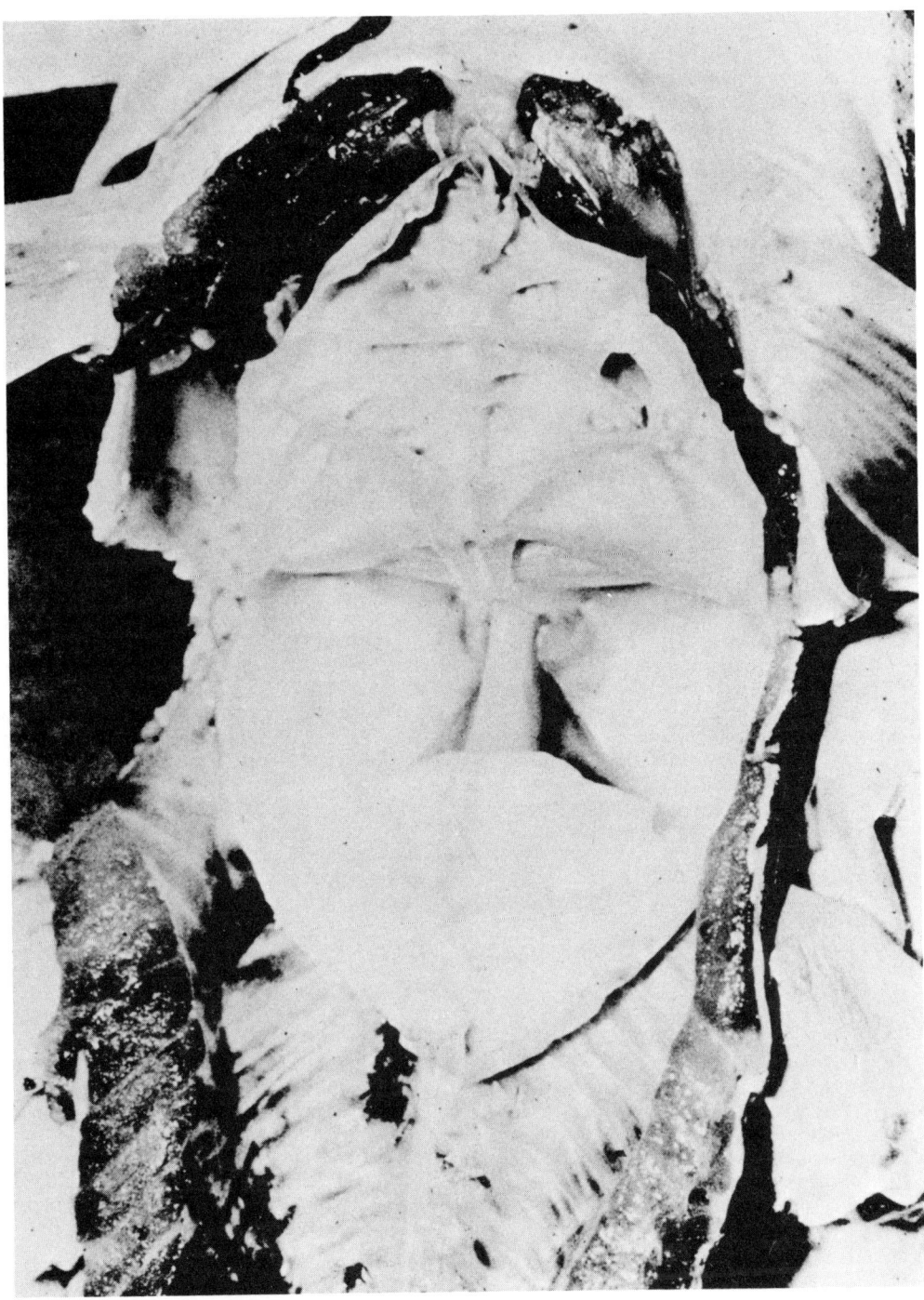

Dissection of swim bladder of *Galeichthys felis*. Bladder has been split open along frontal plane and its ventral surface reflected anteriorly. × 1

to the anterior ceratobranchials. Although in *Galeichthys* and *Bagre* these muscles appear to be reduced, probably they correspond to the branchial depressors and retractors (pharyngo-clavicularis of McMurrich, 1884, or cleido-branchialis of Fürbringer, 1897). The medial branch continues craniad and innervates a large, conical muscle, the origin of which is the anterodorsal surface of the ventral arm of the cleithrum and the insertion of which is on the posterior surface of the urohyal (see pls. 8, 9, 16, 17 for skeletal parts). This muscle has been variously called "sternohyoid" (Allis, 1897), "hyopectoralis" (McMurrich, 1884), and "coraco-hyoideus" and "cleido-hyoideus" (Fürbringer, 1897). Its function is to draw the floor of the branchial basket ventrad and caudad, expanding the branchial cavity.

The nomenclature and homology of the postvagal nerves in fishes were the subject of a large monograph by Fürbringer (1897). The spino-occipital nerves as a group are characterized by being small and in some instances lacking in dorsal ganglia. The precise distinction between the last spino-occipital and the first true spinal nerve has never been made entirely clear and unambiguous, except on the basis of their emergence from the skull. Fürbringer divided the spino-occipitals into two groups, with reference to their relationships to the segmentation of the occipital region. The nerves immediately behind the vagus he called the "occipital" nerves. These were named "w," "x," "y," and "z," with the posteriormost and most persistent being "z." The occipital nerves emerge through foramina in the exoccipitals, and they were considered as representing the earliest metameres to fuse with the paleocranium to form the neocranium. The second group were called the "occipito-spinal" nerves which emerge through the foramen magnum and out between the occipital ring and the first neural arch. Fürbringer called these "a," "b," and "c" and considered that they represented vestiges of vertebral metameres that secondarily fused with the neocranium. The first true spinal nerve, therefore, was derived from metamere number 4. Teleosts generally lack any vestige of the occipital nerves, e.g., x, y, z, and occipitospinal nerve a is also absent. In the Ostariophysi, siluroids included, nerve c is absent, leaving only the occipitospinal b.

Allis (1903) described three occipitospinal nerves in *Scomber*, the first one, presumably Fürbringer's a, lacking a dorsal root. In addition, Allis (1897, 1898) disagreed with Fürbringer on the homology and terminology of these nerves. Based upon his work on *Amia*, he stated that the first true spinal nerve actually represented metamere 5 and, in the Ostariophysi, it is nerve c that is present, not b. Some authors, including Black (1917), have followed Fürbringer's (1897) interpretations. More recently, however, Addens (1933) recommended the use of the term "occipital nerves" for all these structures of uncertain homology between the vagus and the first spinal nerve. The metamerism of the hind part of the skull is subject to a number of different interpretations (Goodrich, 1930), and De Beer (1937) has severely criticized Fürbringer's concepts and reviewed the entire subject of the segmentation of the head.

The possible homology of these occipital nerves with the hypoglossal of higher vertebrates was suggested by Haller (1895) and Fürbringer (1897). Herrick (1899), and several other authors, did not agree with such an interpretation. Most later authors, however, tend to concur with the original theory. Beccari (1922) stated that nerves z, a, b, and c combined to form the hypoglossal, and Addens (1933) unequivocally averred that the occipital nerves gave rise to the twelfth.

Wright (1884) described the fist spinal nerve (= occipital nerve) in *Ameiurus* as emerging through the exoccipital in two branches. The medial branch turns ventrad and supplies the pharyngo-claviculare (= ceratobranchial retractor) musculature. The lateral branch joins with the first true spinal nerve and supplies some of the pectoral fin muscles. Wright also described a branch innervating the trapezius muscle. Although this was thought to be in error by Herrick (1901), the present report confirms both observations—at least in *Galeichthys* and *Bagre*.

It appears generally true that the spino-occipital nerves are primarily motor and supply some of the hypobranchial musculature. In *Amia* (Allis, 1897) two muscles are innervated: the branchiomandibularis and the sternohyoid. The former is the protractor of the tongue and is not present in teleosts.

The sternohyoid originates from the ventral limb of the cleithrum and inserts on the urohyal. Its action depresses and expands the floor of the branchial cavity. The term "sternohyoid" probably originated with Cuvier, from a superficial resemblance to the sternohyoid in human anatomy. In fishes, obviously there is no sternum as such, and therefore other terms have been used. McMurrich (1884) called it the "hypopectoralis," and Fürbringer (1897) named it the "coracohyoideus" or "cleido-hyoideus." Edgeworth (1911) states that in ontogeny the sternohyoid is derived from the hypobranchial spinal musculature.

In teleosts, the spino-occipital nerves also innervate the pharyngo-clavicularis muscles (Wright, 1884; Fürbringer, 1897). Fürbringer called them "cleido-branchialia." They are present in two pairs (internus and externus) arising from the dorsal surface of the ventral arm of the cleithrum and inserting on the ventral ends of the fifth ceratobranchials. They retract and depress the floor of the branchial cavity. In *Amia* (Allis, 1897), the same muscles are supplied by a branch of the fifth branchial (vagus) nerve. This fact, plus some ontogenetic observations, led Edgeworth (1911) to conclude that the pharyngoclavicularis is derived from the fifth branchial myotome.

In most instances, the spino-occipital nerves in teleosts have been found to contribute to the brachial plexus, the anterior portion of which is often referred to as the "ramus cervicalis." In *Ameiurus* (Wright, 1884), there are branches to the abductor and deep adductor of the pectoral fin spine.

The innervation of the protractor muscle of the Müllerian ramus has not been hitherto described, but it seems evident that, in the Ariidae at least, the occipital nerve is the sole supply. From these data it can be inferred that the protractor muscle is most probably derived from the hypobranchial moiety, although its position suggests an origin from epaxial trunk musculature.

PHYSIOLOGICAL ASPECTS OF THE SOUND-PRODUCING MECHANISM

In brief, the technique consisted of the stimulation of the nerves or muscles with a spike form of repetitive potential. The response of the protractor muscle was an audible sound, the pitch of which corresponded to the frequency of stimulus repetition. The muscle reponse was detected with a microphone or hydrophone, recorded on magnetic tape, and monitored by earphones and an oscilloscope.

Equipment and Methods

The stimulus source was a variable-frequency, square-wave generator, the output of which was controlled by a telegraph key. By passage through a 500-micromicrofarad capacitor and a pair of silicon diodes, the wave form was changed to a train of spike-form potentials. These were amplified, and the stimulus was applied to the tissues of a decapitated or anesthetized animal by means of a pair of stainless steel wire electrodes. The electrodes were insulated in plastic and glass except for the terminal millimeter. The output at the electrodes was measured with a vacuum tube voltmeter and monitored visually on an oscillosope. At frequencies of up to 700 cycles per second, the spike potentials had a rise time of about one microsecond and an exponential decay time of one and eight-tenths to two milliseconds. (See fig. 1.)

The audible response was detected either with a hydrophone, if the preparation was under water, or with a crystal microphone, if in air. The hydrophone was a small barium-titanate crystal unit. In either case, the signal was passed through a voltage pre-amplifier and thence to a magnetic tape recorder. The tape recorder output was monitored with earphones, and the signal was also made visible on an oscilloscope screen.

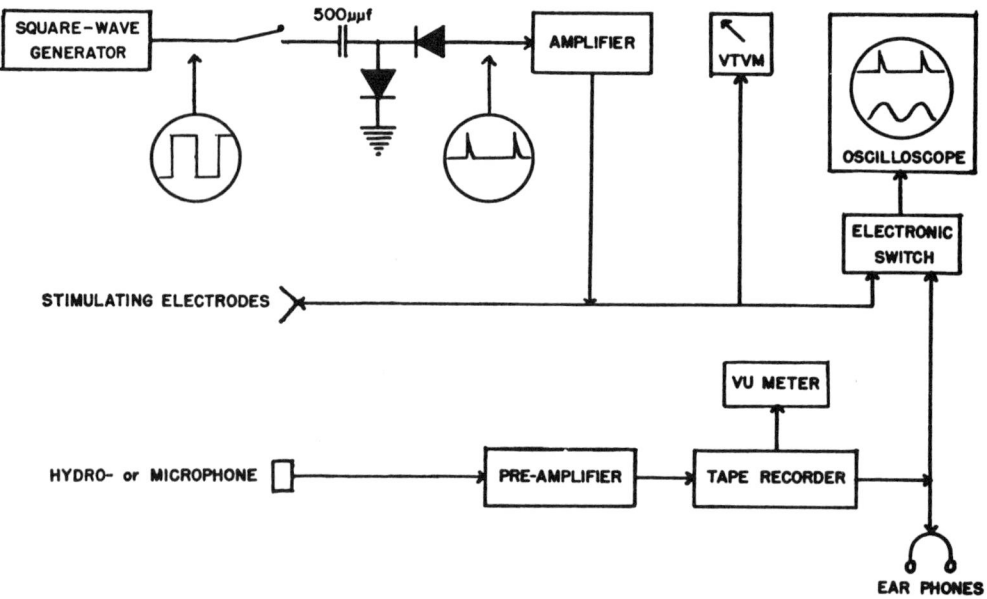

Fig. 1. Block diagram of equipment used in stimulating and recording sound production in nerve-muscle preparations of catfishes.

By means of an electronic switch, both the stimulator output and the response could be observed on the oscilloscope at the same time. The relative response intensity was measured on a VU meter across the output of the tape recorder.

The tape recordings were subsequently analyzed by means of a sound spectrograph (Sona-Graph, manufactured by the Kay Electric Co.). Recording levels and the "reproduce" levels of the Sona-Graph were always monitored with a VU meter to prevent harmonic distortion from overloaded amplifiers. Plates 21, figures 3 and 4, and plates 25 and 26 represent a sample of the records used as data in this section.

The animals used were *Galeichthys felis* and *Bagre marinus* of all sizes from about 4 inches to 15 inches in standard length. The results are based on more than 50 specimens. The fish were immobilized in three ways. One method was to tie them down in a form-fitting basket of stiff wire screening. This was feasible for the smallest specimens. A second method was to cut the head off at a point just anterior to the medulla. The third method was to use a solution of one part in 20,000 of MS-222 in sea water flushed over the gills. The responses to stimulations were the same in all cases.

Stimulation was applied, in some cases, directly to the protractor muscle, by the insertion of the electrodes through the skin at a point just below the lateral edge of the epiotic lamina. Stimulus voltages of 5 volts or more were needed to obtain detectable responses. The voltages given here are as read from a voltmeter calibrated to give root-mean-square values for sine waves, so that the actual peak potentials are somewhat higher. In most cases, the stimulus was applied directly to the dorsal branch of the occipital nerve by the removal of the roof of the skull and dissection along the path of the nerve. Here a stimulus intensity of 0.2 volt was adequate for the obtaining of a clear response from the muscle. The electrodes were about 1 mm. apart and placed directly across the exposed nerve trunk at various points along its length. Except when tetanizing rates and fatigue determinations were desired, the stimulus duration was kept short, i.e., about half a second.

In the majority of the tests, the dissected animal was kept moist with wet paper towels, and a crystal microphone in a thin plastic bag was placed underneath the specimen in the region of its swim bladder. In some cases, the specimen was partially immersed in sea water in a shallow tray, and the hydrophone was placed a few inches away. In a few tests, stimulating electrodes were inserted into both protractor muscles, and the anesthetized animal was immersed in a 15-gallon aquarium, with a hydrophone several inches away, i.e., in an effort to duplicate conditions under which some spontaneous sounds were recorded.

Sonic Properties of Protractor Muscle

The establishment of the protractor muscle as the sound producer and the occipital nerve as its innervation was accomplished by a trial-and-error method. Various nerves and muscles were stimulated with a spike-potential train of 100 pulses per second. At first, a stimulus of about 10 volts was used across the entire pectoral region of an unanesthetized animal. This evoked an audible response of 100 cycles per second. Gradually, with the use of lower voltages and more localized stimuli, the exact source of the sonic response was determined. This was the protractor muscle (described above), and its response could not only be heard but felt. In an anesthetized specimen with its viscera and swim bladder removed, the vibration in response to electrical stimulation could be felt by placing the fingers against the inner surfaces of the *Springfederapparat*. This observation was also reported by Sörensen (1894) by which he demonstrated the sonic function of the "elastic spring" in *Doras*. In his experiment the fish was not anesthetized and was producing sounds spontaneously. I was able to duplicate Sörensen's observations in both *Galeichthys* and *Bagre*. Cutting the fibers of the protractor muscle resulted in a complete loss of audible response, whereas considerable damage can be done to neighboring pectoral musculature without destroying sound production. Damage to the epiotic lamina or Müllerian ramus also destroyed sound production.

Similarly, the application of a stimulus of

100 pulses per second to various parts of the nervous system showed that the dorsal branch of the occipital nerve was the sole motor innervation of the protractor muscle. A spectrogram of a sound produced by a stimulus of 120 pulses per second is shown in figure 3 of plate 21.

The fact that the protractor muscle and its nerve could respond to a stimulus of 100 pulses per second with a sound of 100 cycles per second without immediately going into tetany was itself considered remarkable. All other muscles that could be tested became tetanized immediately, and, indeed, they became tetanic at stimulus frequencies of more than 10 pulses per second. It was of interest, therefore, to determine the limits of frequency response of the protractor muscle before it became tetanized. In all these tests, the occipital nerve was stimulated near its base at the medulla with the minimum voltage (0.2 to 0.4 volt) necessary to evoke a readily detectable and measurable response. Lower voltages or greater distances between electrodes produced proportionately lower intensity responses. Such a decremental response was undoubtedly the result of the stimulation of part of the nerve fibers. Above a given stimulus voltage, there was no increase in response intensity, which indicated that a maximum of nerve fibers were firing. Stimulus tests were spaced about one minute apart, and after each higher-frequency stimulus a stimulus of 100 pulses per second was used as the next test. In this manner the nerve-muscle preparation was given a period of rest, and a constant check on possible effects of fatigue on response intensity could be kept.

FREQUENCY RESPONSE: The graphs (fig. 2) summarize the data on frequency response in both species. Each point represents an average of five or 10 observations. A total of 15 animals (10 *Galeichthys* and five *Bagre*) of mature size were used. In some individuals, a complete curve was obtained in a single series of stimulations, but in most cases relative values of only two or three frequencies were possible and reliable. Relative intensity measurements were extremely variable, because the slightest change in the position of the animal, electrodes, or microphone between tests altered the VU-meter reading

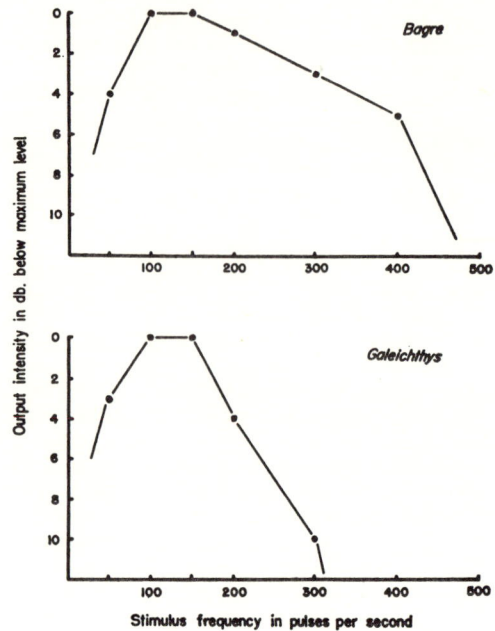

FIG. 2. Graphs showing differences in frequency response of the sonic apparatus in *Bagre* and *Galeichthys*, as determined from artificial stimulation experiments.

drastically. Thus a large number of determinations were dismissed as unreliable, and only those that could be repeated several times in a single preparation are included here.

Despite the few data, the results are quite indicative. In both species, the best response was to a stimulus of from 100 to 150 pulses per second. The response intensity dropped sharply at 50 pulses per second. At stimulus frequencies of over 150, there was a clear difference between the two species. In *Galeichthys*, there was a 4-decibel drop at 200 pulses per second. At 300 pulses per second, the response intensity was approximated at about 10 decibels below that at 150, and the muscle became tetanized in less than two-tenths of a second. At 400 pulses per second, the tetanization was immediate, and there was no sonic response. Figures 3 and 4 of plate 21 show sound spectrograms resulting from stimulations at 120 and 200 pulses per second.

TABLE 1
Time (in Seconds) in Which Protractor Muscles Reached Tetany

Stimulus Frequency in Pulses per Second	Galeichthys (Average of 10)	Bagre (Average of 5)
20	15	20
50	12	18
100	3	12
150	3	12
200	0.3	3
300	<0.2	1.5
400	Immediate	0.4
500	Immediate	<0.1

In *Bagre*, there was only a 1-decibel drop at 200 pulses per second, a 3-decibel drop at 300 pulses per second, and a 5-decibel drop at 400 pulses per second. At 500 pulses per second, tetanization occurred in less than one-tenth of a second, and the response intensity could not be measured, but a brief sound was detectable at a pitch of 500 cycles per second. Sound spectograms (pl. 25, figs. 1, 2) show responses to stimulations of 110 and 300 pulses per second.

In short, the sound-producing system in *Bagre* was capable of a frequency response almost an octave higher than that of *Galeichthys*.

TETANIZATION: The rate at which the protractor muscle became tetanized at various frequencies is obviously a factor in the responsiveness of the system. In this respect also there was a difference between the species. Table 1 shows the tetanizing times at various frequencies. These were measured as the length of time from initiation of the stimulus to a point at which the sonic response could no longer be detected. The time was determined by the measuring of the magnetic tape on which the sound was recorded.

In order to test the effects of fatigue after a tetanizing stimulus, the preparation was given a 30-second rest period and then stimulated for half of a second at 100 pulses per second. In all cases, the response was at least 4 decibels lower than at a previous stimulus of 100 pulses per second. Following a two-minute rest, tetanization took place more rapidly, but after a half-hour rest, the original determinations could be replicated. Probably the abnormally strong and prolonged stimulation produced fatigue in the muscle, and possibly accumulation of the by-products of fatigue reduced the resistance of the muscle to tetanization.

The physiology of a muscle capable of a response to such high rates of stimulus repetition is of interest. Fast-acting muscles of such a nature are not common but are widely distributed among animals. The best known are the flight muscles of insects (Gilmour, 1953; Chadwick, 1953). Some species exhibit a contraction and relaxation rate of up to 1000 per second, but 200 to 300 is the more common range. It is not known, however, if such muscles are actually capable of responding *in toto* to a stimulus administered with such a frequency. Among vertebrates, the muscle with the highest known fusion frequency is the internal rectus of the cat, in which complete tetanization is achieved at a stimulus repeated at 350 pulses per second (Cooper and Eccles, 1930). The swim-bladder muscle of the toadfish (*Opsanus*) is a fast-acting tissue comparable to the protractor muscle in the catfishes. Skoglund (1959) showed that it had a contraction-relaxation cycle of 10 to 15 milliseconds, and Fawcett and Revel (1961) studied it with techniques of the electron microscope. Their study revealed a highly developed sarcoplasmic reticulum which is presumed to be an important factor in the conduction of the impulse from the motor end plate to the contractile elements within the muscle fiber. Cytological investigations of this type on the catfish protractor muscle would seem to be highly desirable, with a view toward an investigation of both its fast-acting and quick-recovery properties.

CHARACTER OF NORMAL AND ARTIFICIALLY PRODUCED SOUNDS

Spontaneously produced sounds in marine catfish have been variously described as "grunts," "pops," and so on by Dobrin (1947), Knudson, Alford, and Emling (1948), and Kellogg (1955). Tavolga (1960) used spectrographic analysis on these sounds, and the following description is based mainly on that report.

GALEICHTHYS FELIS: Sounds of animals both under captive conditions and in the field are all similar in possessing a fundamental frequency at about 150 cycles per second and several harmonics at intervals of 150 cycles per second (pl. 25, fig. 3). The durations vary from 20 to 40 milliseconds, with distress sounds tending to be longer (over 100 milliseconds) than those uttered during the formation of nocturnal schools. The harmonics were found to be variable in strength and number, and under some conditions the fundamental of 150 cycles per second could not even be detected, although the spacing of the harmonics indicated that it must be present.

BAGRE MARINUS: Distress sounds from these animals were essentially the same as those of *Galeichthys*, with a fundamental at 150 cycles per second. Sounds during night schooling were, however, quite different. These were long, sob-like cries, with a distinct fundamental at about 400 cycles per second, and a duration of almost two-tenths of a second (pl. 25, fig. 4).

Although the nocturnal schooling sounds could not be elicited at the experimenter's will, the distress sounds of both species were easily evoked by prodding or electric shock (D.C.). Thus the harmonic content of these sounds could be analyzed under various conditions. It was found that the surrounding environment and the distance of the microphone or hydrophone made a considerable difference in the strength and occurrence of harmonics in these distress sounds.

In a small, 2-gallon aquarium, with the hydrophone within an inch of the animal, the sounds were almost pure tones at a frequency of 150 cycles per second. Using large wooden tanks 6 feet or more in the longest dimension resulted in an almost complete loss of the fundamental and an emphasized harmonic at 300 cycles per second. If the animal was held in the air, with it and a microphone wrapped in acoustical padding (fiberglass batting), the sound output was a pure tone at 150 cycles per second. Actually, the harmonic content of the catfish grunts could be altered easily by changing the conditions under which the recording was made. If a sound speed in sea water of about 5000 feet per second is assumed, the wave length at 150 cycles per second would be about 33.3 feet; and at 300 cycles per second, about 16.6 feet. Thus a large aquarium could function as a half-wave resonating chamber. Standing waves created under these and similar circumstances could conceivably reënforce or dampen certain frequencies and even create harmonics that did not exist in the original sound source. Recording under water always presents such problems more acutely than that in air, because it is virtually impossible to do away with the ever-present, reflecting, air-water interface. As much as 99.9 per cent of sound propagated in water is reflected back from the surface layer (Vigoureux, 1960; Horton, 1959). It can be concluded that, unless sound reflections are rigidly controlled or eliminated, only the fundamental frequency of these fish sounds is of any significance with regard to interpretations of mechanisms of sound production.

The artificially produced sounds, by stimulations of the occipital nerves, were with few exceptions at a pitch that exactly matched the frequency of the stimulus repetition rate. The few exceptions occurred sporadically with specimens of *Bagre*. In some cases a stimulus of 400 pulses per second produced a fleeting response at a fundamental pitch of 200 cycles per second, and in one instance a stimulus of 300 pulses per second produced a response of 150 cycles per second (pl. 25, fig. 2). In each case the stimulus level was a low one, and unfortunately the situation could not be duplicated except by chance. Very probably the nerve was responding to alternate pulses.

Spectrographic analysis of the responses to pulsed stimuli showed the presence of harmonics. However, the occurrence and strengths of the harmonics were extremely variable, and, as with the spontaneous sounds, the harmonic content could be varied with the environmental conditions. Stimulation at 150 pulses per second in a large aquarium would almost duplicate the harmonic content of spontaneous sounds uttered in the same container. Plate 26 shows sound spectrograms of an artificially stimulated (200 pulses per second) specimen before (fig. 1) and after (fig. 2) the preparation was covered with fiberglass batting. Furthermore, stimulations at 20 pulses per second or less produced completely non-harmonic sounds (pl. 26, fig. 3). This is, then, a second indication that harmonics are not intrinsic in the sonic mechanism.

In most of the tests, only a single occipital nerve was stimulated. In a few cases, a bilateral stimulation was accomplished. Two pairs of electrodes were connected in parallel, one pair across each of the occipital nerves. The distances between the electrodes, their position across the nerves, and the stimulus levels were all adjusted so that the response from each side alone was measured, and both were equivalent. When both nerves were thus stimulated at 100, 150, and 200 pulses per second, the response intensity was increased by about 1 decibel over that of a unilateral response. Harmonically there was no significant difference between a unilaterally and a bilaterally stimulated response (pl. 26, fig. 4).

EFFECTS OF SWIM-BLADDER DAMAGE ON SOUND OUTPUT

The anatomy of the swim bladder in *Arius* and many other siluroids has been described in detail by Bridge and Haddon (1893). The following is summarized from descriptions by Tavolga (1960) for *Galeichthys felis* and *Bagre marinus* (pls. 27, 28).

The swim bladder is composed of a soft, fibrous, slightly elastic tissue. It is ovoid in shape, with a straight, flattened, anterior margin, and comes to a rounded apex posteriorly. A thin pneumatic duct connects the ventral surface to the anterodorsal wall of the cardiac stomach (pl. 27). The bladder is never strongly distended, and puncture of the wall causes little or no visible collapse. There are no muscles inside, outside, or connected to the bladder. No red glands are present. Internally there are a large anterior chamber and a smaller posterior chamber (pl. 28). The latter is divided by a sagittal septum and two secondary transverse septa. All chambers are connected by dorsolateral passageways through the transverse septa. The internal architecture is strongly reminiscent of a series of sound-absorbing and directing baffles within a loud-speaker enclosure, particularly in *Bagre*, in which the walls of the posterior chamber possess numerous small pockets and other irregularities. The posterior chamber is attached loosely to the dorsal body wall, but the anterior chamber is firmly knitted dorsally to the thin shelf of bone formed by the fourth and fifth pairs of parapophyses, including the entire inner surface of the *Springfederapparat*. A pair of tough, flat tendons extend along the anterior margin of the bladder from the distal tips of the Müllerian rami to the protruding subvertebral process.

In most cases, an unanesthetized animal ceased sound production when immobilized and dissected. A single specimen of *Galeichthys*, however, continued a steady stream of grunts, even after complete evisceration and destruction of its swim bladder. The sounds were recorded but not measured at that time, but the relative intensities and harmonic contents were analyzed from the tape recordings. Evisceration had no effect on the intensity or quality of the sounds, nor did a small incision into the ventral surface of the swim bladder. As the swim bladder was opened up, the sound output became greatly reduced, to a level at least 10 to 15 decibels below the original. Although the recording was made in air, with a crystal microphone and considerable background noise, weak harmonics at 300 and 400 cycles per second were still detectable even after complete excision of the swim bladder except for the portion attached to the *Springfederapparat*.

The same observations were duplicated with anesthetized, artificially stimulated specimens. In air, the destruction of the swim bladder reduced the sound output at the frequencies tested (100 and 150 pulses per second) by at least 10 decibels, but did not effect the quality of the sound. Similarly, under water, if the swim bladder were ruptured and permitted to fill with water, the sound output dropped at least 20 decibels, almost to inaudibility. Filling the swim bladder with water, while the specimen was in air, gave essentially the same results. With the attachment of a large hypodermic syringe to the pneumatic duct, the air pressure in the swim bladder could be changed. An increase of the pressure almost to the bursting point had no detectable effect on either the intensity or the quality of the artificially produced sounds. A reduction of pressure gradually reduced the sound output, but again the quality of the sound remained unaffected.

From the above it can be concluded that the air bladder serves as a means of transmitting and amplifying the sound produced at the *Springfederapparat*.

MECHANICS OF SOUND-PRODUCTION: CONCLUSIONS

It has always been assumed that the structure of the sonic organs determined the timbre and pitch of the sound emitted. The acoustical characteristics of the swim bladder were considered as fixing the character of the sounds. In the marine catfishes at least, the evidence points strongly to the fact that the fundamental frequency of the sounds is the direct equivalent of the frequency of contraction of the protractor muscles. The stiff, spring-like, Müllerian ramus acts as the antagonist of the muscle. The ability of the protractor muscle to contract at such frequencies, i.e., its remarkably rapid contraction-recovery cycle, resides in four factors.

First is its high vascularization, as evidenced by its color and spongy consistency. A rich blood supply and drainage would be essential. The relatively small diameter of the muscle fibers increases surface area and enhances the efficiency of exchange of materials between the blood and the tissue.

Second is the intrinsic cellular structure and physiology of the protractor muscle fibers. The small size of the fibers, the closely packed myofibrils, and small amount of sarcoplasm are only indicative of fast-acting properties. It will be necessary to do some fine cytological and histochemical work to establish further the relation between cytology and cytophysiology. In addition, it would be of interest to know what the contractile properties of isolated fibers and fiber bundles would be when removed from the restretching effect of the stiff Müllerian ramus.

A third factor is the ability of the occipital nerve fibers to fire with such a rapid recovery rate. This is under the control of some unknown central medullary mechanism with the unique ability to trigger and regulate the train of stimuli. Here is a most intriguing problem for the neurophysiologist. Even if it be assumed that in natural sound production all fibers do not fire simultaneously, there still must be an unusually rapid recovery rate in this neural mechanism.

Fourth is the structure of the Müllerian ramus. Its very stiffness is a considerable advantage, because it produces a quick return with the minimum of overshoot. It might be postulated that the spring has a natural oscillating frequency of its own and that the muscle simply "twangs" it. If this were so, then the pitch of the natural sounds would be quite high, at least over 1000 cycles per second, because the spring is short and extremely stiff. In acoustical and electronic terms, the system has a low compliance and a high damping factor. Although this property usually indicates a poor low-frequency response, it does increase the speed with which the muscle fibers are returned to "rest" position. Furthermore, the system would tend to be extremely resistant to the effects of reverberations and, as a result, to the production of harmonics.

Possibly the differences in sonic properties between *Bagre* and *Galeichthys* are the result of differences in the compliance of the Müllerian ramus. The extent to which the ramus is bent during the contraction phase of the muscle is not known, except that it is a very small, grossly imperceptible distance. The distal tip of the Müllerian ramus in a freshly dissected specimen or a dried skeleton can be moved dorsally only. In *Galeichthys*, it takes a weight of 10 to 20 grams to deflect the tip a distance of 0.5 mm. Larger specimens have stiffer, less compliant springs. In *Bagre*, the limit of deflection is only 0.25 mm., and a force of 50 grams is necessary to accomplish this. The structure of the supporting bones also indicates that *Bagre* has the system of lower compliance. This factor may account for the higher limits of frequency response in *Bagre*.

The role of the swim bladder in sound production appears to be relatively minor. The evidence indicates that the timbre, i.e., harmonic content, of the sounds is not a function of the bladder, but is extrinsic. It is clear, however, that destruction, severe damage, or removal of air from the swim bladder causes a large decrease in the amplitude of sounds produced by the *Springfederapparat*. In effect, then, the bladder amplifies the sound, but I can suggest no mechanism by which amplification in the true sense can be achieved, if by amplification we mean the increase in amplitude, i.e., amount of sound energy. Vigoureux (1960) pointed out that water has an extremely high impedance, i.e., resistance to initial propagation, as compared to air. Energy transmission, on the other hand, is more efficient in water than in air. The problems of under-water sound production and the physics of water-borne sound are reviewed by Horton (1959) and Richardson (1957). Comparatively, the area of the vibrating *Springfederapparat* is much smaller than that of the entire swim-bladder wall. If we assume that the function of the air within the bladder is simply to transmit and distribute the sound energy from the *Springfederapparat* to the bladder wall, then the bladder acts as an impedance matching device and not as an amplifier. An analogy can be made here with function of the paper cone of a loud-speaker which transmits the vibrations of the voice coil to the air. We can probably ignore the rest of the fish, because, in the words of Griffin (1955), it is essentially transparent to water-borne sound. This contention has been made by many other authors, including Marshall (1951). Jones and Pearce (1958) showed that with a 30-kilocycle sonar signal about 50 per cent of the sound energy of the echo was caused by the swim bladder.

What, then, remains as the function of the swim bladder in these catfishes? Its vascular supply is small, and it lacks red glands, so that its role in gas secretion and storage is probably negligible. Very likely its structure is important in sound reception, as is evidenced by the insertion of the tripus into its dorsal wall. The sensitivity of these ostariophysine fishes to frequencies of more than 1000 cycles per second could well be enhanced by the swim bladder which could act as an acoustical discontinuity and, again, as an impedance matching device. The elaborate septa and irregularities in the inner wall might have two functions. One would be to prevent the internal echoes that would be possible at higher frequencies. In air, a half wave length at 2000 cycles per second would be in the order of 3 inches, a magnitude within the dimensions of the average catfish swim bladder. Among the catfishes, *Ameiurus* has been shown to have an extremely high auditory acuity dependent on its swim bladder (Kleerekoper and Roggenkamp, 1959). Another function might be suggested, that of sound localization. Conceivably a sound-receiving device with transverse baffles as present in the catfish swim bladder would tend to have an axis of optimum response. Indications are that at least one ostariophysine fish (*Semotilus*) possesses localization abilities (Kleerekoper and Chagnon, 1954). To carry this hypothesis one step farther, is it possible for the entire sound-producing and receiving complex to act as an echo-ranging mechanism? Considering the absence of any direct evidence for such an hypothesis, I advance it here simply as an intriguing speculation. Griffin (1955) suggested the possibility of an echo-ranging method in fishes based on the detection of modal intensity changes in standing waves, in which case the emission of sounds without intrinsic harmonics would be advantageous.

SUMMARY

The skeletal basis of the sound producing mechanism in the ariid catfishes *Galeichthys felis* and *Bagre marinus* consists of a thin shelf of bone which is firmly attached to the anterior dorsal wall of the swim bladder. This so-called "elastic spring," or "*Springfederapparat*," is formed from the parapophysis of the fourth vertebra. The anterior ramus of this parapophysis, herein named the "Müllerian ramus" (after Johannes Müller who, in 1842, first described the structure), is the main vibrating element in sound production.

The skulls and anterior vertebral complexes of *Galeichthys* and *Bagre* are figured and described, with comparisons with a non-sonic silurid form, *Wallago*.

The "protractor muscle" (so named by Bridge and Haddon in 1893) which activates the Müllerian ramus is a highly vascularized conical muscle, the origin of which is on the under side of the epiotic lamina and insertion on the Müllerian ramus. It is innervated by the dorsal branch of the occipital nerve. The relationship of this nerve to the cranial nerves of higher vertebrates is controversial, but it is probably homologous with the hypoglossal (XII).

Stimulation of the protractor muscle or its nerve supply with repetitive spike-form potentials results in an audible sound output from the *Springfederapparat*. This response can be recorded and analyzed, and its fundamental pitch is equivalent to the pulse frequency of the stimulus. The protractor muscle is remarkable in that it can withstand stimulations of 300 or more pulses per second without going into immediate tetany. In *Bagre*, the frequency response of the sonic system is about an octave higher than that in *Galeichthys*.

Spontaneous sounds from these species consist of low-pitched grunts (fundamental pitch of about 150 cycles per second) and, in *Bagre*, higher, sob-like sounds (400 cycles per second or over).

The harmonic components of both artificially induced and spontaneous sounds can be influenced by the amount of sound reflectance in the external environment. Even under partially anechoic conditions, the sound output is virtually a pure tone. Damage to the swim bladder reduces the amplitude of the sound but not its timbre (i.e., harmonic content). It was concluded that the swim bladder does not serve as a resonating chamber for these sounds, nor is it a true amplifier. Rather, it transfers the energy from the small area of the vibrating Müllerian ramus to the larger area of its entire outer surface, thus making the propagation of the sound from its source to the water more efficient. In acoustical and electronic terms, the swim bladder is an impedance matching device.

LITERATURE CITED

Addens, J. L.
1933. The motor nuclei and roots of the cranial and first spinal nerves of vertebrates. Zeitschr. Anat. Entwick., I. Zeitschr. Ges. Anat., vol. 101, pp. 307–410.

Agassiz, J. L.
1850. [Manner of producing sounds in catfish and drumfish.] Proc. Amer. Acad. Arts Sci., vol. 2, p. 238.

Allis, E. P., Jr.
1897. The cranial muscles and cranial and first spinal nerves in *Amia calva*. Jour. Morph., vol. 12, pp. 487–808.
1898. The homologies of the occipital and first spinal nerves of *Amia* and teleosts. Zool. Bull. [Biol. Bull.], vol. 2, pp. 83–97.

1903. The skull, and the cranial and first spinal muscles and nerves in *Scomber scomber*. Jour. Morph., vol. 18, pp. 45–328.

Beccari, N.
1922. Studi comparative sulla struttura del rhombencefalo. I. Nervi spino-occipitali e nervo ipoglosso. II. Centri tegmentali. Arch. Italiano Anat. Embriol., vol. 19, pp. 123–291.

Berg, L. S.
1947. Classification of fishes both recent and fossil. [Reprint in Russian and English.] Ann Arbor, Michigan, Edwards Bros., Inc., pp. 85–517.

Berkelbach van der Sprenkel, H.
1915. The central relations of the cranial

nerves in *Silurius glanis* and *Mormyrus caschiva*. Jour. Comp. Neurol., vol. 25, pp. 5–63.

BHIMACHER, B. S.
1933. On the morphology of the skull of certain Indian catfishes. Half-yearly Jour. Mysore Univ., vol. 7, pp. 1–35.

BLACK, D.
1917. The motor nuclei of the cerebral nerves in phylogeny: A study of the phenomena of neurobiotaxis. Part I. Cyclostomi and Pisces. Jour. Comp. Neurol., vol. 27, pp. 467–564.

BRIDGE, T. W., AND A. C. HADDON
1889. Contribution to the anatomy of fishes. I. The airbladder and Weberian ossicles in the Siluroidea. Proc. Roy. Soc. London, vol. 46, pp. 309–328.
1893. Contributions to the anatomy of fishes. II. The air-bladder and Weberian ossicles in the siluroid fishes. Phil. Trans. Roy. Soc. London, ser. B, vol. 184, pp. 65–333.
1894. Notes on the production of sounds by the air bladder of certain siluroid fishes. Proc. Roy. Soc. London, vol. 55, pp. 439–441.

BURKENROAD, M. D.
1931. Notes on the sound-producing marine fishes of Louisiana. Copeia, pp. 20–28.

CHADWICK, L. E.
1953. The motion of wings. *In* Roeder, K. D. (ed.), Insect physiology. New York, John Wiley and Sons, Inc., pp. 577–614.

CHRANILOV, N. S.
1929. Beiträge zur Kenntnis des Weber'schen Apparates der Ostariophysi. 2. Der Weber'sche Apparat bei Siluroidea. Zool. Jahrb. Anat., vol. 51, pp. 323–462.

COOPER, S., AND J. C. ECCLES
1930. Isometric responses of mammalian muscles. Jour. Physiol., vol. 69, pp. 377–385.

DE BEER, G. R.
1937. The development of the vertebrate skull. Oxford, Oxford University Press.

DIJKGRAAF, S.
1949. Untersuchungen über die Funktionen der Ohrlabyrinths bei Meeresfischen. Physiol. Comp. Oecol., vol. 2, pp. 81–106.
1952. Bau und Funktionen der Seitenorgane und des Ohrlabyrinths bei Fischen. Experientia, vol. 8, pp. 205–216.
1960. Hearing in bony fishes. Proc. Roy. Soc. London, ser. B, vol. 152, pp. 51–54.

DOBRIN, M. B.
1947. Measurements of underwater noise produced by marine life. Science, vol. 105, pp. 19–23.

DUFOSSÉ, M.
1874. Recherches sur les sons expresifs que font entendre les poissons d'Europe. Ann. Sci. Nat., Paris, ser. 5, vol. 19, pp. 1–53; vol. 20, pp. 1–134.

EDGEWORTH, F. H.
1911. On the morphology of the cranial muscles in some vertebrates. Quart. Jour. Micros. Sci., vol. 56, pp. 167–316.

EVANS, H. M.
1925. A contribution to the anatomy and physiology of the airbladder and Weberian ossicles in Cyprinidae. Proc. Roy. Soc. London, vol. 97, pp. 545–576.
1935. Hearing in fishes. Jour. Ipswich Nat. Hist. Soc., vol. 1, pp. 217–230.

FAWCETT, D. W., AND J. P. REVEL
1961. The sarcoplasmic reticulum of a fast-acting fish muscle. Jour. Biophys. Biochem. Cytol., vol. 10, suppl., pp. 89–109.

FISH, M. P.
1954. The character and significance of sound production among fishes of the western North Atlantic. Bull. Bingham Oceanogr. Coll., vol. 14, pp. 1–109.

FRISCH, K. VON
1923. Ein Zwergels der dommt, wenn man ihm pfeift. Biol. Zentralbl., vol. 43, pp. 439–446.
1936. Über den Gehörsinn der Fische. Biol. Rev. Cambridge Phil. Soc., vol. 11, pp. 210–246.

FRISCH, K. VON, AND H. STETTER
1932. Untersuchungen über den Sitz der Gehörsinnes bei der Elritze. Zeitschr. Vergleich. Physiol., vol. 17, pp. 686–801.

FÜRBRINGER, M.
1897. Ueber die spino-occipitalen Nerven der Selachier und Holocephalen und ihre vergleichende Morphologie. *In* Festschrift zum siebenzigsten Geburtstage von Carl Gegenbaur. Leipzig, vol. 3, pp. 349–788.

GEOFFROY ST.-HILAIRE, I.
1829. Histoire naturelle des poissons du Nil. *In* Description de l'Égypte. Seconde édition. Paris, vol. 24, pp. 141–338.

GILMOUR, D.
1953. The biochemistry of muscle. *In* Roeder, K. D. (ed.), Insect physiology. New York, John Wiley and Sons, Inc., pp. 404–422.

GOODRICH, E. S.
1930. Studies on the structure and develop-

ment of vertebrates. London, the Macmillan Co., 2 vols.

GREGORY, W. K.
1933. Fish skulls. A study of the evolution of natural mechanisms. Trans. Amer. Phil. Soc., vol. 23, pp. 75–481.

GRIFFIN, D. R.
1955. Hearing and acoustical orientation in marine animals. Deep Sea Res., Papers Marine Biol. Oceanogr., vol. 3, suppl., pp. 406–417.

HALLER, B.
1895. Untersuchungen über des Rückenmark der Teleostier. Morph. Jahrb., vol. 23, pp. 21–122.

HARRINGTON, R. W., JR.
1955. The osteocranium of the American cyprinid fish, *Notropis bifrenatus*, with an annotated synonymy of teleost skull bones. Copeia, pp. 267–290.

HERRICK, C. J.
1899. The cranial and first spinal nerves of Menidia: A contribution upon the nerve components of the bony fishes. Jour. Comp. Neurol., vol. 9, pp. 153–455.
1900. A contribution upon the cranial nerves of the codfish. *Ibid.*, vol. 10, pp. 205–316.
1901. Cranial nerves and cutaneous sense organs of North America siluroid fishes. *Ibid.*, vol. 11, pp. 177–249.

HORTON, J. W.
1959. Fundamentals of sonar. Second edition. Annapolis, United States Naval Institute.

HUBBS, C. L.
1936. Fishes of the Yucatan peninsula. *In* Pearse, A. S., E. P. Creaser, and F. G. Hall (eds.). The canotes of Yucatan. Washington, Carnegie Institution, pp. 157–288.

JAYARAM, K. C.
1956. Nomenclatorial status of the names *Bagre* Cuvier (Oken), *Bagrus* Valenciennes and *Porcus* Geoffroy St. Hilaire. Copeia, pp. 248–249.

JONES, F. R. H.
1957. The swimbladder. *In* Brown, M. E., Physiology of fishes. New York, Academic Press, Inc., vol. 2, pp. 305–322.

JONES, F. R. H., AND N. B. MARSHALL
1953. The structure and functions of the teleostean swimbladder. Biol. Rev. Cambridge Phil. Soc., vol. 28, pp. 16–83.

JONES, F. R. H., AND G. PEARCE
1958. Acoustic reflexion experiments with perch (*Perca fluviatilis* Linn.) to determine the proportion of the echo returned by the swimbladder. Jour. Exp. Biol., vol. 35, pp. 437–450.

JORDAN, D. S.
1917. The genera of fishes. California, Stanford University.
1923. A classification of fishes, including families and genera as far as known. Stanford Univ. Publ., Univ. Ser. Biol. Sci., vol. 3, pp. 79–243.

JOSEPH, N. I.
1960. Osteology of *Wallago attu* Bloch and Schneider. Part I. Osteology of the head. Proc. Natl. Inst. Sci. India, ser. B, vol. 26, pp. 205–233.

KARANDIKAR, K. R., AND V. B. MASUREKAR
1954. Weberian ossicles and other related structures of *Arius platystomus* Day. Jour. Univ. Bombay, new ser., vol. 22, pp. 1–28.

KELLOGG, W. N.
1955. Sounds of sea animals. Vol. II. Florida. Folkways Record Album No. FPX 125; Science Series. New York, Folkways Record and Service Corp.

KINDRED, J. E.
1919. The skull of *Amiurus*. Illinois Biol. Monogr., vol. 5, pp. 1–120.

KLEEREKOPER, H., AND E. C. CHAGNON
1954. Hearing in fish, with special reference to *Semotilus atromaculatus atromaculatus* (Mitchill). Jour. Fish. Res. Board Canada, vol. 11, pp. 130–152.

KLEEREKOPER, H., AND P. A. ROGGENKAMP
1959. An experimental study on the effect of the swimbladder on hearing sensitivity in *Ameiurus nebulosus nebulosus* (Lesueur). Canadian Jour. Zool., vol. 37, pp. 1–8.

KNUDSEN, V. O., R. S. ALFORD, AND J. W. EMLING
1948. Underwater ambient noise. Jour. Marine Res., vol. 7, pp. 410–429.

KOSCHKAROFF, D. N.
1905. Beiträge zur Morphologie des Skelets der Teleostier. Das Skelet der Siluroidei. Bull. Soc. Imp. Nat. Moscou, new ser., vol. 19, pp. 209–307.

KRUMHOLZ, L. A.
1943. A comparative study of the Weberian ossicles in North American ostariophysine fishes. Copeia, pp. 33–40.

LANGE, J. W.
1953. The singing fish of the Batticoloa Lagoon. Jour. Brit. Roy. Asiatic Soc., Ceylon, new ser., vol. 3, pp. 12–24.

MCMURRICH, J. P.
1884. The myology of *Amiurus catus* (L.) Gill.

Proc. Canadian Inst., ser. 3, vol. 2, pp. 311–351.

MARSHALL, N. B.
1951. Bathypelagic fish as sound scatterers in the ocean. Jour. Marine Res., vol. 10, pp. 1–17.

MERRIMAN, D.
1940. Morphological and embryological studies on two species of marine catfish, *Bagre marinus* and *Galeichthys felis*. Zoologica, vol. 25, pp. 221–248.

MÜLLER, J.
1842. Beobachtungen über die Schwimmblase der Fische, mit Bezug auf einige neue Fischgattungen. Arch. Anat. Physiol. (Meckel's Arch.), pp. 307–329.
1843. Untersuchungen über die Eingeweide der Fische. Abhandl. K. Akad. Wiss. Berlin, pp. 109–170.

POGGENDORF, D.
1952. Die absoluten Hörschwellen des Zwergwelses (*Amiurus nebulosus*) und Beiträge zur Physik des Weberschen Apparatus der Ostariophysen. Zeitschr. Vergleich. Physiol., vol. 34, pp. 222–257.

PROSSER, C. L.
1950. Muscle and electric organs. *In* Prosser, C. L. (ed.), Comparative animal physiology. Philadelphia W. B. Saunders Co., pp. 576–629.

REGAN, C. T.
1911. The classification of the teleostean fishes of the order Ostariophysi. 2. Silurioidea. Ann. Mag. Nat. Hist., ser. 8, vol. 8, pp. 553–577.

RICHARDSON, E. G.
1957. Propagation of sound in the atmosphere and the sea. *In* Richardson, E. G. (ed.), Technical aspects of sound. Amsterdam, Elsevier, vol. 2, pp. 1–30.

SKOGLUND, C. R.
1959. Neuromuscular mechanisms of sound production in *Opsanus tau*. (Abstract.) Biol. Bull., vol. 117, p. 438.

SMITH, H. M.
1927. The so-called musical sole of Siam. Jour. Siam Soc. Nat. Hist., vol. 7, suppl., pp. 49–54.

SÖRENSEN, W.
1884. Om Lydorganer hos Fiske: En physiologisk og comparative-anatomisk Undersögelse. Copenhagen. (Not seen.)
1894. Are the extrinsic muscles of the air bladder in some Siluroidea and the "elastic spring" apparatus of others subordinate to the voluntary production of sound? Jour. Anat. Physiol., ser 9, vol. 29, pp. 109–139, 205–229, 399–423, 518–552.

STETTER, H.
1929. Untersuchungen über den Gehörsinn der Fische, besonders von *Phoxinus laevis* L. und *Amiurus nebulosus* Raf. Zeitschr. Vergleich. Physiol., vol. 9, pp. 339–477.

TAVOLGA, W. N.
1960. Sound production and underwater communication in fishes. *In* Lanyon, W. E., and W. N. Tavolga (eds.), Animal sounds and communication. Publ. Amer. Inst. Biol. Sci., no. 7, pp. 93–136.

VIGOUREUX, P.
1960. Underwater sound. Proc. Roy. Soc. London, ser. B, vol. 152, pp. 49–51.

WEBER, E. H.
1820. De aure et auditu hominis et animalium. Pars I. De aure animalium aquatilium. Leipzig.

WRIGHT, R. R.
1884. On the nervous system and sense organs of *Amiurus*. Proc. Canadian Inst., ser. 3, vol. 2, pp. 352–386.

15

Copyright © 1964 by Pergamon Press, Inc.

Reprinted from *Marine Bio-Acoustics*, W. N. Tavolga, ed., Pergamon Press, Inc., Oxford, 1964, pp. 233–247

CONSIDERATIONS ON THE PHYSICS OF SOUND PRODUCTION BY FISHES*

G. G. HARRIS

Bell Telephone Laboratories, Inc.
Murray Hill, New Jersey

ALTHOUGH there has been much written about underwater sound there is not much material available which is appropriate to the study of sound production and detection by fishes. The frequency range of hearing in fish varies from 100–3000 cps while the frequency range important to the lateral line organ is around 20–500 cps. In addition, the distances involved in experimental studies, and perhaps also those which are important in the life of fish, may be only a few wavelengths. The range of variables involved in the hearing of fish is associated with acoustical effects which are not often considered (Harris and van Bergeijk, 1962). Because of this it is relevant to consider here certain features of sound production which are important in the study of the acoustical and lateral line organs of fishes.

In the main we wish to draw qualitative conclusions since it is difficult to control the boundary conditions when experimenting with fishes and their sense organs. Nevertheless, it will be useful to start by considering an idealized case with some rigor. Such a study will yield insight as to the variables which are important. The idealized case which will be considered will be the sound radiated from a sphere.

RADIATION FROM A SPHERE

We will consider radiation from a sphere whose surface undergoes an arbitrary radial oscillation which is constrained to be axially symmetric. This means that the motion of the sphere at any point on the surface is a function only of the polar angle θ. θ is measured as the angle between the radius terminating on the point in question and the diameter which defines the axis of symmetry.

The radial velocity of the sphere will then be given by $U(\theta, t) = U(\theta)e^{-i\omega t}$. (See the explanation of symbols in the Appendix.) Our task is to relate this

---

* Editor's note: This paper was not originally scheduled in the Symposium and it represents an extension of Dr. Harris' remarks during the general discussion period of this session.

to a pressure wave $p(r, \theta, t)$ which is radiated from the sphere. The pressure will be a solution to the spherical wave equation

$$\frac{1}{r^2}\frac{\partial}{\partial r}\left(r^2\frac{\partial p}{\partial r}\right) + \frac{1}{r^2 \sin\theta}\frac{\partial}{\partial \theta}\left(\sin\theta\frac{\partial p}{\partial \theta}\right) = \frac{1}{c^2}\frac{\partial^2 p}{\partial t^2} \qquad (1)$$

The reader is referred to standard texts such as Morse (1948) for details of the solution of Eq. (1). When the spherical wave equation is solved the pressure can be represented as a sum of the series

$$p = \sum_{m=0}^{\infty} A'_m [j_m(kr) + i n_m(kr)] P_m(\cos\theta) e^{-i\omega t} \qquad (2)$$

A'_m is a constant which will be determined; $j_m(kr)$ and $n_m(kr)$ are the spherical Bessel and Neumann functions of order m; and $P_m(\cos\theta)$ is a Legendre function of order m; $k = \omega/c = 2\pi/\lambda$. For the present discussion we will use only the first three terms, i.e. for $m = 0, 1, 2$ which are given in Table 1.

TABLE 1. FORMULAS FOR THE SPHERICAL BESSEL AND NEUMANN FUNCTIONS j_m AND n_m, AND THE LEGENDRE FUNCTIONS OF ORDER m, $P_m(\cos\theta)$ FOR $m = 0, 1, 2$. THE FORMULAS ARE GIVEN AS FUNCTIONS OF θ THE POLAR ANGLE AND kr A CONVENIENT DIMENSIONLESS CONSTANT, $k = 2\pi/\lambda = \omega/c$.

m	j_m	n_m	$P_m(\cos\theta)$
0	$\dfrac{\sin kr}{kr}$	$-\dfrac{\cos kr}{kr}$	1
1	$\dfrac{\sin kr}{(kr)^2} - \dfrac{\cos kr}{kr}$	$-\dfrac{\sin kr}{kr} - \dfrac{\cos kr}{(kr)^2}$	$\cos\theta$
2	$\left(\dfrac{3}{(kr)^3} - \dfrac{1}{kr}\right)\sin kr - \dfrac{3}{(kr)^2}\cos kr$	$-\dfrac{3}{(kr)^2}\sin kr - \left(\dfrac{3}{(kr)^3} - \dfrac{1}{kr}\right)\cos kr$	$\tfrac{1}{4}(3\cos 2\theta + 1)$

From Table 1 we can construct explicitly the first three terms of Eq. (2)

$$p = p_0 + p_1 + p_2$$

$$= A_0 \frac{e^{ik(r-ct)}}{kr} + A_1\left(\frac{1}{kr} + \frac{i}{(kr)^2}\right)\cos\theta\, e^{ik(r-ct)}$$

$$+ A_2\left(\frac{1}{kr} + \frac{3i}{(kr)^2} - \frac{3}{(kr)^3}\right)\tfrac{1}{4}(3\cos 2\theta + 1) e^{ik(r-ct)} \qquad (3)$$

Only the real part of Eq. (3) is important. The imaginary unit, i, is retained in order to express the phase relations of the different components in a convenient form. Equation (3) represents an outgoing spherical pressure wave. kr is a convenient dimensionless quantity.

The radial particle velocity u_r, is related to the pressure by Newton's equation,

$$\rho \frac{\partial u_r}{\partial t} = -\frac{\partial p}{\partial r} \qquad (4)$$

Since the time function for u_r is of the form $e^{-i\omega t}$, we have

$$u_r = \frac{1}{ik\rho c} \frac{\partial p}{\partial r} \qquad (5)$$

In a similar manner the angular particle velocity can be found from the equation $u_\theta = (1/ik\rho cr)(\partial p/\partial \theta)$.

Once the pressure is known, u_r can be found from Eq. (5). The value of u_r at the surface of the sphere is u_a, and can be equated to the radial velocity of the sphere. The constants A_m can then be found in terms of this radial velocity. In order to do this the radial velocity of the sphere is expanded in a series of Legendre functions, $P_m(\cos \theta)$. Then it can be shown that each term in the series for the radial velocity of the sphere can be equated to the corresponding term in the series for the radial velocity of the water at the surface of the sphere. Thus if

$$U(\theta, t) = U(\theta)e^{-i\omega t} = \sum_{m=0}^{\infty} U_m P_m(\cos \theta)e^{-i\omega t} \qquad (6)$$

where

$$U_m = (m + \tfrac{1}{2}) \int_0^\pi U(\theta) P_m(\cos \theta) \sin \theta \, d\theta$$

we can set

$$U_m P_m(\cos \theta) = \frac{1}{ik\rho c} \frac{\partial p_m}{\partial r} \quad \text{at} \quad r = a$$

and solve for A_m in terms of U_m.

Table 2 lists the explicit solutions for A_m for the first three terms $m = 0, 1, 2$, of the series expansion for the pressure with the condition that $ka \ll 1$, where a is the radius of the sphere. It also lists other properties of these solutions which may prove useful. Some insight into the nature of these solutions may be obtained by considering the type of motion the sphere undergoes for each of these first three terms.

TABLE 2. FORMULAS WHICH ARE APPROPRIATE FOR MONOPOLE, DIPOLE AND QUADRAPOLE RADIATION

Term, $m =$	General Symbol	0
Description		Pulsating Sphere Monopole Source
Radial Velocity of Source	$U_m P_m(\cos\theta)$	U_0
Constant[1,2]	A_m	$-i\rho c(ka)^2 U_0 e^{-ika}$
Pressure Amplitude[3]	$\dfrac{p_m}{e^{ik(r-a-ct)}}$	$-\dfrac{i\rho c(ka)^2 U_0}{kr}$
Particle Velocity Amplitude Radial Component	$\dfrac{u_{mr}}{e^{ik(r-a-ct)}}$	$-i(ka)^2 U_0 \left(\dfrac{1}{kr} + \dfrac{i}{(kr)^2}\right)$
Particle Velocity Amplitude Angular Component	$\dfrac{u_{m\theta}}{e^{ik(r-a-ct)}}$	0
Near-Field Approximation, Radial Particle Velocity Amplitude	$\dfrac{u_{mr}}{e^{ik(r-a-ct)}}; kr \ll 1$	$U_0 \dfrac{a^2}{r^2}$
Far-Field Approximation, Pressure Amplitude	$\dfrac{p_m}{e^{ik(r-a-ct)}}; kr \gg 1$	$-\dfrac{i\rho c k a^2 U_0}{r}$
Energy Radiated per Sec per cm²	I	$\dfrac{\rho c k^2 a^4 U_0^2}{2r^2}$
Total Energy Radiated per Sec	E	$2\pi\rho c U_0^2 k^2 a^4$
Radiation Impedance	Z_r	$4\pi\rho c k^2 a^4 - (i\omega)4\pi\rho a^3$

NOTES:
1. The symbols used in the formulas are: $k = 2\pi/\lambda = 2\pi v/c$, $a =$ radius of sphere in cm, $\rho =$ density of water in g/cm³, $c =$ velocity of sound in cm/sec, $p =$ pressure in dynes/cm², $u =$ particle velocity in cm/sec. Only real parts are to be considered, imaginary parts are shown to indicate phase. Energy is in ergs.

FROM A SPHERE UNDER THE CONDITION $ka \ll 1$ WHERE a IS THE RADIUS OF THE SPHERE AND $k = 2\pi/\lambda$.

1	2
Vibrating Sphere Dipole Source	Deforming Sphere Quadrapole Source
$U_1 \cos \theta$	$U_2 \tfrac{1}{4}(3 \cos 2\theta + 1)$
$-\dfrac{\rho c (ka)^3}{2} U_1 e^{ika}$	$\dfrac{i\rho c(ka)^4 U_2 e^{ika}}{9}$
$-\dfrac{\rho c(ka)^3 U_1}{2}\left(\dfrac{1}{kr} + \dfrac{i}{(kr)^2}\right)\cos\theta$	$\dfrac{i}{36}\rho c(ka)^4 U_2\left(\dfrac{1}{kr} + \dfrac{i3}{(kr)^2} - \dfrac{3}{(kr)^3}\right)(3\cos 2\theta + 1)$
$-\dfrac{(kr)^3}{2} U_1 \left(\dfrac{1}{kr} + \dfrac{2i}{(kr)^2} - \dfrac{2}{(kr)^3}\right)\cos\theta$	$\dfrac{i}{36}(ka)^4 U_2\left(\dfrac{1}{kr} + \dfrac{i4}{(kr)^2} - \dfrac{9}{(kr)^3} - \dfrac{i9}{(kr)^4}\right)(3\cos 2\theta + 1)$
$-\dfrac{(ka)^3}{2} U_1 \left(\dfrac{i}{(kr)^2} - \dfrac{1}{(kr)^3}\right)\sin\theta$	$i\dfrac{(ka)^4}{6} U_2\left(\dfrac{i}{(kr)^2} - \dfrac{3}{(kr)^3} - \dfrac{i3}{(kr)^4}\right)\sin 2\theta$
$U_1 \dfrac{a^3}{r^3} \cos \theta$	$U_2 \dfrac{a^4}{r^4} \dfrac{(3\cos 2\theta + 1)}{4}$
$-\dfrac{\rho c k^2 a^3}{2r} U_1 \cos\theta$	$\dfrac{i\rho c k^3 a^4 U_2}{9r} \dfrac{(3\cos 2\theta + 1)}{4}$
$\dfrac{\rho c k^4 a^6 U_1^2 \cos^2\theta}{8r^2}$	$\dfrac{\rho c k^6 a^8 U_2^2}{162 r^2} \dfrac{(3\cos 2\theta + 1)^2}{16}$
$2\pi \rho c U_1^2 \dfrac{k^4 a^6}{12}$	$2\pi \rho c U_2^2 \dfrac{k^6 a^8}{405}$
$\dfrac{\pi \rho c k^4 a^6}{3} - (i\omega)\tfrac{2}{3}\pi \rho a^3$	—

2. Defined by the equation
$$p = \sum_{m=0}^{\infty} A_m(j_m + in_m)P_m(\cos\theta)e^{i\omega t}.$$
3. For convenience the common phase factor $e^{ik(r-a-ct)}$ has been divided out of many of the formulas thus leaving only the complex amplitude.

$m=0$, RADIATION FROM A PULSATING SPHERE—MONOPOLE SOURCE

The radial velocity of the sphere is given by $U = U_0 e^{-i\omega t}$. This is the equation for a pulsating sphere which expands and contracts, i.e. the volume (zeroth moment) of the sphere changes. This motion is also called the monopole motion of the sphere (Fig. 1).

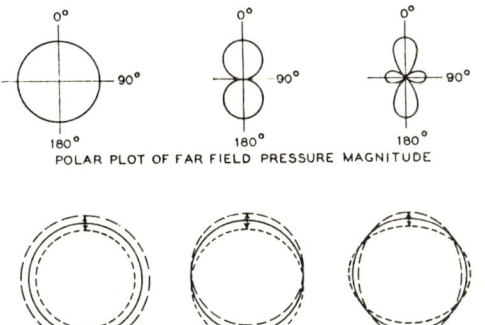

Fig. 1. Radial displacement and polar plots of far-field pressure for monopole, dipole and quadrupole radiation from a sphere.

From Table 2 the near-field velocity (neglecting phase factors) is given by $u_{0r} \approx U_0 a^2/r^2$ and decreases as the inverse square of the distance. (The displacement magnitude d may be found from the equation $\partial d/\partial t = u$ which for sinusoidal time dependence becomes $d = (i/\omega)u$.) The far-field velocity is given by $u_{0r} \approx -(ika^2 U_0/r)$. The factor i means that the far-field velocity is 90° out of phase with the near-field velocity. The far-field pressure is given by $p_0 = \rho c u_{0r} = -i\rho c(ka^2 U_0/r)$. Both the far-field pressure and velocity decrease as $1/r$ as they must for a radiated spherical wave.

The total energy E radiated per second is given by

$$E_0 = 2\pi\rho c U_0^2 k^2 a^4 \qquad (7)$$

Since $k = \omega/c = 2\pi\nu/c$, the energy radiated is proportional to the square of the frequency.

The radiation impedance contains two terms,

$$4\pi\rho c k^2 a^4 - (i\omega)4\pi\rho a^3 \qquad (8)$$

The first is a resistance term. The second is reactive and represents the effect of the mass of the water which has to be moved, $4\pi\rho a^3$. It is equivalent to moving a volume of water equal to three times the volume of the sphere.

Some additional equations which may be of use are those relating the velocity, pressure and displacement in the field of a pulsating sphere. The radial velocity in terms of the pressure is given by

$$u_{0r} = \frac{p_0}{\rho c}\left(1 + \frac{i}{kr}\right) \qquad (9)$$

The radial displacement d in terms of the pressure is given by

$$d_{0r} = \frac{p_0}{\omega\rho c}\left(i - \frac{1}{kr}\right)$$

Or, if one considers only magnitudes,

$$d_{0r} = \frac{p_0}{\omega\rho c}\left(1 + \frac{1}{(kr)^2}\right)^{1/2} \qquad (10)$$

The displacement magnitude D_0 of the source with radius a can be calculated once the pressure is known at a distance r from the source. This is found from Eq. (11);

$$D_0 = \frac{p_0 r}{\omega\rho c a}\left(1 + \frac{1}{(ka)^2}\right)^{1/2} \qquad (11)$$

$m = 1$, RADIATION FROM A VIBRATING SPHERE—DIPOLE SOURCE

The radial velocity of the sphere is given by $U = U_1 \cos\theta e^{-i\omega t}$. This equation is for a sphere displacing along an axis with velocity $U_1 e^{-i\omega t}$. It is called the vibrating sphere and also the dipole motion of the sphere. Figure 1 gives a representation of the motion. The volume (zeroth moment) remains constant but the center of gravity (first moment) displaces.

The near-field radial velocity (neglecting phase factors) is given by $u_{1r} \approx (a^3/r^3)U_1 \cos\theta$ and is inversely proportional to the cube of the distance. The θ component of the near-field velocity is given by $u_{1\theta} \approx \frac{1}{2}(a^3/r^3)U_1 \sin\theta$ and also is inversely proportional to the cube of the distance.

The far-field radial velocity is given by

$$u_{1r} \approx -\frac{k^2 a^3 U_1 \cos\theta}{2r}$$

and is 180° out of phase with the near-field velocity. Only a radial velocity remains in the far field. The far field is angular dependent. A plot of pressure magnitude vs. angle shows two cosine lobes (see Fig. 1).

The total energy E radiated per second is given by

$$E_1 = 2\pi\rho c U_1^2 \frac{k^4 a^6}{12} \qquad (12)$$

This is a factor $k^2a^2/12$ smaller than the energy radiated by the pulsating sphere of the same velocity magnitude, i.e. $U_0 = U_1$. The energy radiated is seen to depend on the frequency to the fourth power.

As with monopole radiation, the radiation impedance contains two terms. The first is a resistance term. The second is reactive and represents the effect of the external mass of water. It is equivalent to moving a volume of water equal to one-half the volume of the sphere.

$m=2$, RADIATION FROM A SPHERE UNDERGOING DEFORMATION—QUADRUPOLE SOURCE

The radial velocity of the sphere is given by $U = U_2 \tfrac{1}{4}(3\cos 2\theta + 1)e^{-i\omega t}$. This is the equation for a sphere which deforms by contraction and expansion along one axis. The motion at the extremity of the axis is given by $U_2 e^{-i\omega t}$. We call it the deforming sphere because the volume (zero moment) and the center of gravity (first moment) do not change but the sphere undergoes a deformation and the moment of inertia (second moment) does change (Fig. 1). It is also called the quadrupole motion of the sphere.

The near-field radial velocity (neglecting phase factors) is given by $u_{2r} \approx (a^4/r^4)U_2 \tfrac{1}{4}(3\cos 2\theta + 1)$ which is inversely proportional to the fourth power of the distance. The far-field velocity is given by

$$\frac{ik^3 a^4 U_2}{9r} \tfrac{1}{4}(3\cos 2\theta + 1)$$

and is 90° out of phase with the near-field velocity. Again both the far-field velocity and pressure vary as $1/r$. The far field is a function of the angle θ. Figure 1 shows a polar plot of the pressure magnitude showing four directional lobes.

The total energy E radiated per second is given by

$$E_2 = 2\pi\rho c U_2^2 \frac{k^6 a^8}{405} \tag{13}$$

This is a factor of $k^4 a^4/405$ smaller than the energy radiating from an equivalent pulsating sphere and a factor $4/135 \cdot k^2 a^2$ smaller than from an equivalent vibrating sphere. The energy radiated varies as the sixth power of the frequency.

The higher-order motions of the sphere, i.e. when $m = 3, 4$, etc., have correspondingly less energy radiated and are therefore less important. (Our assumption is that we are working in conditions where $ka \ll 1$.)

APPLICATIONS

The motivation behind the method of multipole expansion can be simply stated. Any axially symmetric motion of the sphere can be represented by a series of Legendre functions, Eq. (6). We can calculate the radiation produced by each term of the series, i.e. monopole, dipole, quadrapole, etc. The energy radiated is the sum of the energies radiated by each separate term. For equal U_m the energy radiated by each term is smaller than the energy radiated by the preceding term by a factor smaller than k^2a^2. Since we are investigating a region where $ka \ll 1$, only the lowest-order terms for any type of motion need to be considered.

As an example of this method consider an underwater sound source as described by Tavolga and Wodinsky (1963) in their experiments on the threshold of hearing in fish. This source, consisting of a rubber bulb attached to a loudspeaker, can be fairly accurately described as a pulsating sphere

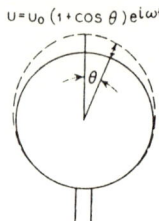

FIG. 2. Schematic representation of the motion of a pulsating rubber bulb when held fixed at a point on its surface.

which is fixed to the loudspeaker at its lowest point (see Fig. 2). The fixed point at the base of the pulsating sphere makes it act like a combination of a pulsating sphere and a dipole source. The radial velocity can be approximated by $U = U_0(1 + \cos\theta)e^{-i\omega t}$. In this approximation we can say by inspection from Eq. (6) that only the first two terms of the expansion are nonzero and that $U_0 = U_1$. From Table 2 we find for the pressure

$$\frac{p}{e^{ik(r-a-ct)}} = -\frac{ic\rho ka^2 U_0}{r}\left(1 - \frac{ika}{2}\cos\theta + \frac{a}{2r}\cos\theta\right) \quad (14)$$

The term in brackets in Eq. (14) is the amount the pressure is altered from that produced by a pure pulsating sphere. In Tavolga's case $a = 2$ cm and $ka \approx 1/10$ for $f = 1150$ cps. Thus the second term can be neglected for frequencies below about 1000 cps. The distance r varied between 10 cm and 20 cm, and θ was around 60° so that the third factor represents about a 5 per cent (0.5 dB) correction over the distances important in the experiment. A pure pulsating sphere is consequently a good approximation to this source.

These formulas can also be used in a qualitative discussion of sound production in fishes. Tavolga (1962) has described the mechanism of sound production in certain catfishes. Figure 3 shows a schematic representation of the sound producing mechanism consisting of the swim bladder and the

"*Springfederapparat*". The swim bladder is attached to a spring-like bone which moves something like a wing when driven by a special muscle. When producing sound the muscle contracts about 150 times per sec displacing the anterior part of the swim bladder with sharp movements. The sharp movements will produce harmonics of the fundamental. The harmonics will be

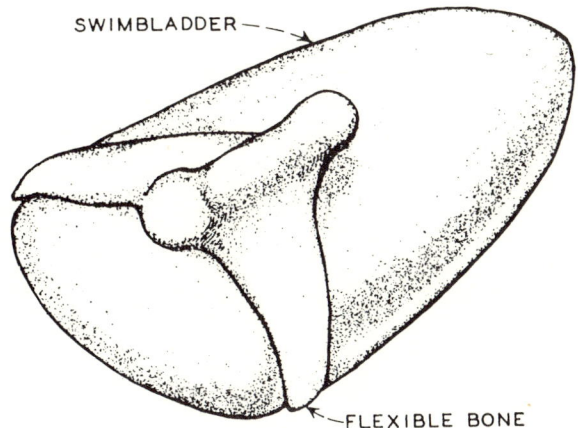

FIG. 3. Schematic representation of the swim bladder and *Springfederapparat* of the *Bagre*.

accentuated as sources of sound because of the way the radiated energy increases as a function of frequency. However, due to a swim bladder resonance which is described below, the main source of energy will probably still be at the fundamental frequency. The swim bladder is about 10 cm long and for a frequency of 150 cps, ka is about 0.15 and k^2a^2 is about 0.02. Consequently only the lowest terms in the radiation expansion should be important. The main question is whether it should be regarded as a pulsating sphere, a vibrating sphere or a mixture of both.

Tavolga (1962) has described the bladder as not strongly distended and that puncture of the wall causes little or no visible collapse. He also says that "there are no muscles inside or outside connected to the bladder. The posterior part of the bladder is attached loosely to the dorsal body wall, but the anterior part is firmly knitted dorsally to the thin shelf of bone which includes the entire inner surface of the *Springfederapparat*."

If the bladder were strongly distended it would tend to maintain a constant volume when deformed by the *Springfederapparat* and dipole radiation would be the most important term. Because the bladder is not strongly distended there will be a compromise between the compressional force of the air in the bladder and the inertial forces of the water which would have to be moved. Therefore there probably will be changes in volume as well as changes in position. It is seen from Table 2 that monopole radiation or change in volume is the most efficient way to radiate energy for $ka \ll 1$.

There is little quantitative data to test the assumption that volume change is important but we may make some qualitative calculations. Tavolga has reported that when the swim bladder is opened up, there is about a 16 dB loss in intensity in the sound energy radiated. When the swim bladder is intact the pressure magnitude p will be given by the equation

$$p \approx \frac{\rho c k a^2 U_0}{r}$$

With the swim bladder cut the pressure p' will be given by

$$p' \approx \frac{\rho c k a'^3 U_1}{2r^2}$$

(assuming we are measuring in the near-field region and $\cos \theta \sim 1$). The ratio of the two pressures is given by

$$\frac{p}{p'} \approx \frac{2r a^2 U_0}{a'^3 U_1}$$

In order to calculate this pressure ratio we make the following estimates: r, the hydrophone distance from source ~ 16 cm; a, the radius of equivalent pulsating sphere ~ 4 cm; a', the radius of equivalent vibrating sphere ~ 2 cm; U_0, the amplitude of equivalent pulsating sphere; U_1, the amplitude of equivalent vibrating sphere $\sim 8 U_0$. With the above qualitative estimates of the magnitude we have $p/p' \approx 8$ which is equivalent to an 18 dB change. Although it must be remembered that these estimates are highly qualitative the answer is at least of the right order of magnitude.

There is another and stronger argument in favor of monopole radiation which has to do with the swim bladder resonance. If there is a change in volume the interaction between the elastic forces of the air and swim bladder walls on the one hand and the inertial forces of the water and fish tissue on the other will produce a resonance. We can estimate the resonant frequency by examining the equation of motion of a bubble of air in water (Minnaert, 1933; Meyer, 1957).

$$F = 4\pi a^3 \rho \frac{d^2 a}{dt^2} + R \frac{da}{dt} - 4\pi a^2 \frac{\partial p'}{\partial a} \Delta a \qquad (15)$$

F is the forcing function which we shall assume is sinusoidal with time, $F = F_0 e^{-i\omega t}$. The term $4\pi a^3 \rho$ is the effective mass of the water moved (see Eq. (8)). R is a resistive term which will include the radiative resistance of the sphere, thermal damping effects and, for a fish, R will also include viscous damping terms of the swim bladder wall and surrounding tissue. The term $4\pi a^2 (\partial p'/\partial a)\Delta a$ is a force due to the change in volume of the sphere.

We can assume that the change in radius Δa is small compared to the radius and is of the form $\Delta a = De^{-i\omega t}$; thus $da/dt = -i\omega \Delta a$ and $d^2a/dt^2 = -\omega^2 \Delta a$. Before substituting this into Eq. (14) we must determine $\partial p'/\partial a$, the change in pressure as a function of radius. For a perfect gas undergoing isothermal compression $\partial p'/\partial a = -3p'/a$ where p' is the equilibrium pressure. Usually adiabatic compressions are considered in acoustics. If this were done it would increase $\partial p'/\partial a$ by a factor γ, which is 1.4 for air. Pfriem (1940) has shown that the adiabatic condition holds only partially for low frequencies and so in our rough calculations we shall neglect it. The value $3p'/a$ is a good approximation for a swim bladder without distension. When there is distension p' is more complicated but we can conceive of it as a generalized pressure. The term $\partial p'/\partial a$ can then be determined experimentally by increasing the static pressure and noting the change in radius which is produced. Such measurements have been made by Alexander (1959, 1961) and the results can be expressed by the equation

$$\frac{\partial p'}{\partial a} = -\frac{3p'}{a} \cdot b \tag{16}$$

where b represents the factor of increase over the no wall condition. Alexander (1959, 1961) finds that in Cypriniformes $b \approx 2$, b should be close to unity when there is little excess pressure.

Equation (14) can now be solved for D the amplitude of the forced pulsations.

$$D = \frac{F_0}{(-4\pi a^3 \rho \omega^2 - iR\omega + 12\pi abp')} \tag{17}$$

The denominator is a complex number since the second or resistive term is 90° out of phase with the first and third terms. Equation (17) is considerably simplified at the resonance frequency which is defined as that frequency at which the first and third terms cancel. This frequency, ω', is therefore given by the equation

$$\omega' = 2\pi f = \left(\frac{3bp'}{\rho a^2}\right)^{1/2} \tag{18}$$

If $a \approx 3$ cm, $b \approx 1$, $p' \approx 2$ atmospheres or 2×10^6 dynes/cm^2 (values which are not inappropriate for the catfish) then f would be about 130 cps. This resonant frequency is close enough to the actual frequency produced by the catfish to strongly suggest that the catfish makes use of the resonant condition and that the sound is produced by volume changes.

A special case of the forcing term would be that produced by water borne sound. As long as the swim bladder is small compared to a wavelength, $F_0 = s \cdot p$, where s is the surface area of the swim bladder and p is the sound pressure. The amplitude of the swim bladder would be greatest at its own resonant frequency. Van Bergeijk, this volume, points out that if a fish both produces sound and receives sound via the swim bladder the source and receiver are automatically matched and that the fish would be especially tuned to its own species.

It is interesting to conjecture whether the swim bladder resonance has an importance for hearing in fish. In particular is the frequency of the minimum threshold of hearing the same as the resonance frequency in fish which have their ears connected to their swim bladder? Rough calculations indicate that the resonance condition should vary between 100–1000 cps for different species. Because of the large number of factors which determine the resonant condition a detailed consideration will be necessary for each species.

The very intense, 20 cps pulses, which may be produced by whales, could be an interesting example of underwater sound production. The figures quoted by Walker (this volume) and Patterson (this volume) of source strengths from 1 W up to 25 W are difficult to explain by the usual methods of sound production via vocal chords. If these sounds were produced under water, use could be made of the bladder-like resonance of the lung. For a sphere at a depth of 20 fathoms, the radius would have to be about 30 cm. This figure may be considerably altered by the fact that the lung is not a free sphere, but is severely constricted. Moreover, the resonance of the lung may be broad and the 20 cps pulses would then have to be a forced type of motion from a 20 cycle power input. The velocity amplitude which would produce 1 W output from a pulsating sphere whose radius is 30 cm is $U_0 = 3$ cm/sec. This is equivalent to a displacement amplitude of about 0.05 cm. For a 25 W output the amplitude would be 0.25 cm. These amplitudes of oscillation do not seem unreasonable. Other mechanisms of sound production could be considered but it is difficult to see how so much energy at so low a frequency could be radiated except by utilizing a volume change. Underwater volume changes of this magnitude can be achieved most easily with an air-water interface.

CONCLUSIONS

We have undertaken an analysis of sound production by a spherical source under the conditions where the wavelength of the produced sound is much greater than the size of the source. This is done in the hope that the qualitative conclusions if not the quantitative conclusions, will be of use in the understanding of sound production and sound reception in fishes. In experimental conditions as well as in the actual life conditions of the fish, the receiver is often separated from the source by less than a wavelength. The analysis shows when this is so both near-field and far-field effects must be considered for the organs of hearing as well as the organs of the lateral line. Two illustrative examples are discussed. A particularly interesting result of the analyses is that the bubble resonance of a volume of air such as the swim bladder or a lung is probably quite important in the production of sound by the catfish, and may also be important in the production of 20 cps tones by whales. This resonance may also be important for the hearing of fish. For a more precise understanding of sound production and reception by the swim bladder each species must be considered separately.

APPENDIX I

Definition of Symbols

- a radius of sphere in cm
- b inverse sensitivity
- c velocity of sound in water, approximately 150,000 cm/sec
- f frequency in cycles per second
- i imaginary unit $=\sqrt{-1}$, represents a 90° phase change from 1
- j_m spherical Bessel's function of order m
- k wave number $=2\pi/\lambda=2\pi f/c=\omega/c$
- λ wavelength in cm
- n_m spherical Neumann's function of order m
- ω angular frequency in radians per second
- p sound pressure in dynes/cm^2
- p' pressure in dynes/cm^2
- P_m Legendre's function of order m
- r radial distance in cm
- R resistance in dyne sec/cm
- ρ density in water in g/cm^2
- t time in seconds
- θ polar angle in degrees, angle between axis of symmetry and radius to point in question
- U radial velocity of sphere
- u_r radial component of particle velocity
- u_θ angular component of particle velocity

APPENDIX II

Complex notation is used for convenience in expressing phase relationships. These phase relationships can be visualized by noting that $+1, i, -1, -i, +1$ represent points 90° apart on the circumference of a circle. The amplitude of any function is determined by the real part of any formula. This is de-determined by noting that $e^{ik(r-a-ct)} = \cos k(r-a-ct) + i \sin k(r-a-ct)$. When this phase factor is multiplied into the amplitude factor only the real parts are retained. Thus,

$$u_{0r} = -i(ka)^2 U_0 \left(\frac{1}{kr} + \frac{i}{(kr)^2}\right) e^{ik(r-a-ct)}$$

and

$$\text{Real}(U_{0r}) = (ka)^2 U_0 \left(\frac{\sin k(r-a-ct)}{kr} + \frac{\cos k(r-a-ct)}{(kr)^2}\right)$$

REFERENCES

ALEXANDER, R. M. (1959) The physical properties of the swim bladder in intact *Cypriniformes*. *J. Exp. Biol.* **36**, 315–332.

ALEXANDER, R. M. (1961) The physical properties of the swim bladders of some South American *Cypriniformes*. *J. Exp. Biol.* **38**, 403–410.

HARRIS, G. G. and W. A. VAN BERGEIJK (1962) Evidence that the lateral-line organ responds to near-field displacements of sound sources in water. *J. Acoust. Soc. Amer.* **34**, 1831–1841.

MEYER, E. (1957) Air bubbles in water. In E. G. Richardson (ed.)., *Technical Aspects of Sound*. Elsevier, Amsterdam, vol. 2, pp. 222–239.

MINNAERT, F. M. (1933) On musical air bubbles and the sounds of running water. *Phil. Mag.* **16**, 235–248

MORSE, P. M. (1948) *Vibration and Sound*. McGraw-Hill, New York.

PFRIEM. H. (1940) Zur thermischen Dämpfung in kugelsymmetrisch schwingenden Gasblasen. *Akust. 2.* **5**, 202.

TAVOLGA, W. S. (1962) Mechanisms of sound production in the ariid catfishes *Galeichthys* and *Bagre*. *Bull. Amer. Mus. Nat. Hist.* **124**, 1–30.

TAVOLGA, W. S. and J. WODINSKY (1963) Auditory capacities in fishes. Pure tone thresholds in nine species of marine teleosts. *Bull. Amer. Mus. Nat. Hist.* (in press).

Part V
BIOLOGY AND COMMUNICATION

Editor's Comments on Papers 16 Through 19

16 TAVOLGA
The Significance of Underwater Sounds Produced by Males of the Gobiid Fish, Bathygobius soporator

17 GRAY and WINN
Reproductive Ecology and Sound Production of the Toadfish, Opsanus tau

18 WINN, MARSHALL, and HAZLETT
Behavior, Diel Activities, and Stimuli That Elicit Sound Production and Reactions to Sounds in the Longspine Squirrelfish

19 WINN
The Biological Significance of Fish Sounds

Probably the most trenchant and most difficult research area in fish sonics is the question of the specific function of the sounds in the behavior of the species and the value of the sounds in social communication. Combined laboratory and field investigations are required and, inevitably, many of the conclusions are based on partial or inadequately controlled experimental data. In spite of the fact that the goby, a benthic species, appears to have no specialized sonic mechanisms, the sounds produced by males have a very specific function in the process of reproductive behavior. One of the significant points made by the Tavolga article reprinted here is the fact that the sounds are only part of a complex of stimuli involved in courtship. Redundancy of signals combined with their low level of specificity gives the entire courtship interaction its ultimate effectiveness.

Howard E. Winn has been one of the leaders in this field for many years, and three of his major studies are reprinted here. His paper with Grace-Ann Gray was the result of the first of several field experiments on the toadfish, a species whose courtship includes the emission of powerful sound blasts characterized as "boat-whistles." Winn's later students have added significantly to this study, but it remains as a fine example of a design for a field experiment with the minimum disruption to the subject animals. In another paper with his students J. A. Marshall and B. A. Hazlett, Winn provided a model of how sound production in the crepuscular squirrelfish can function in territorial behavior, and how the behavior operates in that particular ecological niche.

The last article by Winn is taken from a volume now out of print. His major theoretical contribution here was to show how the simple aspect of temporal patterning is the critical factor in sonic communications in fishes. In most cases, such properties as fundamental frequency and harmonic structure are of little importance, but the repetition rate and regularity are essential. This can be considered as further evidence of the relatively low level of organization of communication systems in most fishes (see Paper 2 in this volume).

THE SIGNIFICANCE OF UNDERWATER SOUNDS PRODUCED BY MALES OF THE GOBIID FISH, BATHYGOBIUS SOPORATOR

WILLIAM N. TAVOLGA[1]

Department of Animal Behavior, American Museum of Natural History, New York

During the early phase of investigations on the production of underwater sound by fishes, Griffin (1950) stated: "The discovery that a wide variety of underwater sounds are produced by fish and other marine animals has raised many unsolved problems concerning the biological significance of these sounds. Are they purely accidental by-products of other activities, are they used for communication from one animal to another, or do they serve in any way for orientation?" This, of course, does not exclude the possibility that more than one function can be served by a given sound under different circumstances. As one of the primary steps in the study of these problems, Fish and other workers (1952, 1954) began to catalogue the various sounds produced by fishes and, wherever possible, correlate the sound production with observations on behavior.

Many of these fish sounds were found to be in some way related to reproductive and territorial behavior. This conclusion was based upon the correlation of sound production with the occurrence of animals in spawning or prespawning condition. This is particularly true of the Sciaenidae, in which actual choruses of drums, croakers, grunts, etc., can be heard during spawning seasons. The "boat-whistle" sound of the toadfish (*Opsanus tau*), described by Fish (1954), and the "staccato" call of the sea robin (*Prionotus*), studied by Moulton (1956), seem to be closely related to territorial behavior and occur during spawning periods. It is not known, however, whether these sounds are "mating calls," in the sense of frog and toad calls, or whether they serve some other function, as, for example, species discrimination or territorial dominance.

Sounds made by the goby (*Bathygobius soporator*) are uttered only at a very specific stage in its reproductive process, that is, they are produced only by the male and only while he is courting a female (Tavolga, 1956a). The further

[1] These studies were aided by a contract between the Office of Naval Research, Department of the Navy, and the American Museum of Natural History. Contract No. NR 163-322.

The author is indebted to Dr. Eugenie Clark, director of the Cape Haze Marine Laboratory, for making the facilities of the laboratory available. Mr. Robert Laupheimer, of the Institute of Mathematical Sciences, Mr. Philip Brown and Mr. Nathaniel Tillman, of the Electrical Engineering Department of the City College, were extremely helpful in the instrumentation, as well as in their advice on the electronic problems involved in this project. The author is also grateful to Dr. Lester R. Aronson, of the American Museum of Natural History, for his comments on the manuscript.

study and analysis of these sounds was undertaken in order to determine their precise role in sex discrimination, territory establishment, and orientation.

Bathygobius soporator is abundant in tropical and subtropical coastal waters. It is hardy and adaptable in captivity, spawns readily, and its behavior patterns are not easily disrupted by handling or other laboratory procedures. The author has been investigating various aspects of the behavior of this species for some time (1954, 1956b), and, with the earlier work as a basis, the interpretations of the sounds could be made within the context of the total reproductive behavior pattern.

This work was undertaken during the summer of 1957 at the Cape Haze Marine Laboratory, located on Gasparilla Sound, a few miles north of Placida, Florida. The animals were collected in the vicinity of the laboratory and were maintained in aquaria supplied with running sea water.

EQUIPMENT

The hydrophone that was used for both pickup and playback of the underwater sounds was a Barium Titanate transducer, Model LF-400, manufactured by the Chesapeake Instrument Corporation, Shadyside, Maryland. The unit is cylindrical, and the sensitive portion measures $5\frac{1}{2}$ inches in height and $3\frac{1}{4}$ inches in diameter. Calibration curves for this instrument were supplied by the manufacturer. This type of transducer was used in preference to the standard U.S. Navy QBG equipment because its smaller size made it possible to use in small aquaria.

A low-noise voltage amplifier was designed for the above transducer by Mr. Robert Laupheimer and constructed by Mr. Nathaniel Tillman. The power supply of this unit provided highly filtered B+ voltages and a filtered, rectified filament current. The amplifier had a balanced input and push-pull output, using 12AD7 and 12AY7 tubes. The voltage gain of this instrument was approximately $1,000\times$.

The signal from this amplifier was fed into a magnetic tape recorder, Crestwood (Daystrom) model 303. Recordings were made at $7\frac{1}{2}$ inches per second, and they were monitored with either earphones or an external speaker. This tape recorder was also used as a signal source in the playback of sounds.

Sounds were played back through the LF-400 transducer, and the amplifier used to power the transducer was a Heathkit model A-9B. This unit, using two 6L6 tubes in its output stage, provided sufficient power to activate the LF-400, although there was some distortion at higher intensities, especially at frequencies below about 500 cps. Oscilloscope observations of these distortions showed a clipping of the sine waves into square waves, and the quality of the sound to the ear became a buzz rather than a tone. At intensities up to ten times those of normal outputs of the fish, distortions became detectable only below 100 cps.

In addition to the tape recorder, an audio generator (Heathkit model AG-9), was used as a signal source. The switching of this generator was controlled by a gating circuit designed and constructed by Mr. Philip Brown and Mr. Nathaniel Tillman. The circuit was battery-powered and employed three transistors and two crystal diodes. The timing was controlled by the selection of any one of ten capacitors and the closing of a triggering microswitch. This unit was inserted between the output of the generator and the input of the power amplifier. The open-gate durations ranged from 5 to 1,500 milliseconds.

Another hydrophone was constructed by Mr. Nathaniel Tillman which was essentially similar to model BM-110 (discontinued) of the Brush Electronics Company. A piezoelectric ceramic B tubular element (No. 55347B) was capped with bakelite with the leads brought through the top end-cap. This unit was smaller than the LF-400, measuring $1\frac{1}{2}$ inches in diameter and about 5 inches in over-all height, but it proved to be insufficiently sensitive at low frequencies, and its output was badly distorted at frequencies below 1,000 cps. It was used only as a monitor of the output of the LF-400.

The analysis of the sounds was made with a sound spectrograph (Sona-graph, Kay Electric Company). Some measurements and observations were made using a cathode-ray oscilloscope (Heathkit model O-11).

CONDITIONS UNDER WHICH SOUNDS ARE DETECTED

The low-pitched grunts which were described in an earlier report are produced only by the *Bathygobius* males during the portion of prespawning behavior described as courtship. Courtship by the *Bathygobius* male consists of an approach to the female accompanied by a distinctive color change. The male takes on a generally lighter-tan body color, with little or no evidence of transverse bands, while his lips, chin, and throat become strongly darkened. Simultaneously with the color change, the male begins a series of fanning movements of the body and tail. These fanning movements may be continuous over a period of 15–30 seconds or intermittent, i.e., occurring in short pulses. Irregularly, during the courtship, the male returns to his shelter, where he will often show a burst of nest-cleaning activity, and then returns to the female. When the female follows the male into his nest, the courting movements cease, and the male's coloration usually darkens. At this point another phase of the prespawning behavior begins, more fully described in an earlier report (1954). During the course of the early phase of courtship, which takes place outside the shelter, the male produces the grunting sounds. These are synchronous with quick, sharp, downward thrusts of the head, but they are not synchronized with the fanning movements.

As described earlier (Tavolga, 1956a), courtship could be stimulated in an isolated male by the introduction of small amounts of ovarian fluids from a gravid female into his aquarium. This courtship pattern was complete in every way with respect to color change and fanning movements. Sound production during this courtship was normal and identical to that produced in the actual presence of a female.

DESCRIPTION OF THE COURTSHIP SOUNDS OF *Bathygobius* MALES

The original description of these sounds was made from recordings using inadequate and makeshift equipment. The present equipment was assembled with the view to introducing as little distortion as possible into the recordings. The analysis by means of the sound spectrograph was, therefore, more precise and reliable. In addition, numerous records were made from several animals and in different containers.

The quality of the sounds was not affected detectably by differences in the aquaria used. Recordings made in 2-gallon or 5-gallon glass tanks were not significantly different from those made in a large 96 × 48 × 6-inch wooden aquarium. The position of the male with respect to the hydrophone also made no difference in the quality or loudness of the recording. When a male produced these sounds from the inside of a shelter, the sound became directed by the aperture and shape of the shell. These differences were detectable through earphones but could not be confirmed from analysis of the tape recordings.

Sona-grams (Fig. 1) of the goby sounds showed that they were non-

During the initial 25–30 milliseconds, the component frequencies ranged almost continuously from 100 to 500 cps. The top frequency dropped in the latter portion of the sound to 150 cps. Rarely, frequency peaks as high as 3,000 cps were encountered, but these were extremely weak.

The major portion of the sound

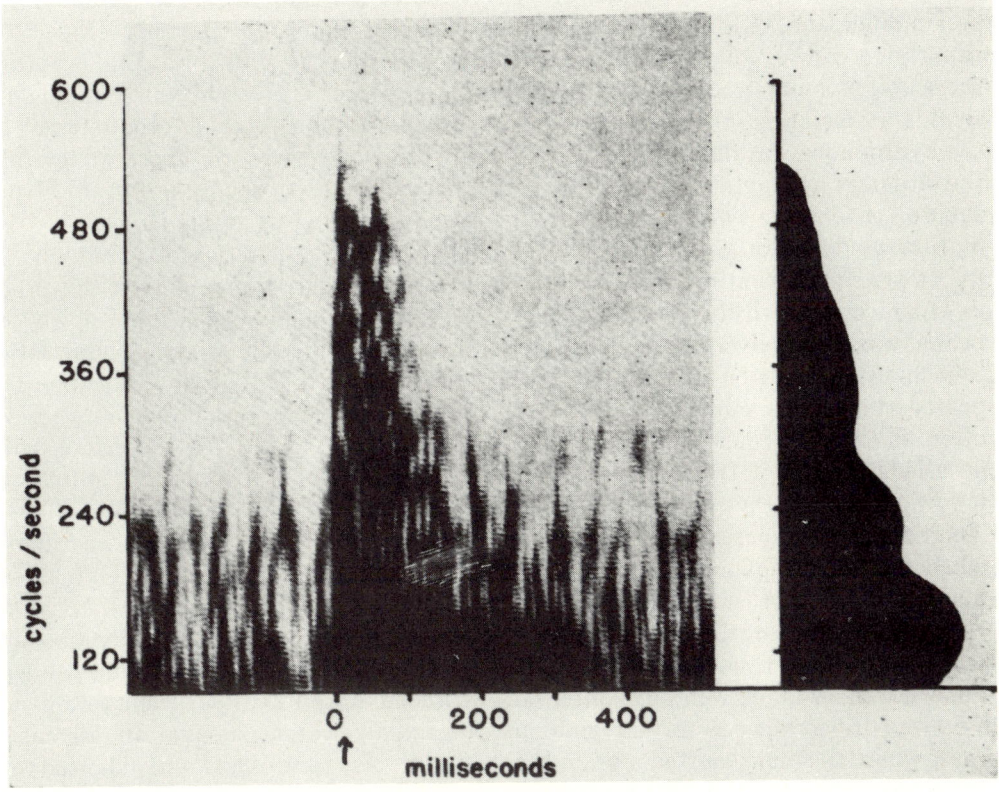

FIG. 1.—Sound spectrogram of a single sound produced by a male *B. sporator*. On the right is a frequency-amplitude section taken at the point indicated by an arrow on the time scale. Zero on the abscissa is set at the beginning of the sound, which is about 325 milliseconds in duration. Background noise is shown preceding and following the sound.

harmonic. The one illustrated can be considered typical of most. Their duration varied from 150 to 350 milliseconds, with an average duration of 225 milliseconds (average of 20 sona-grams). The duration of the sound is indicated in Figure 1, and the remainder of the tracing is background noise.

energy was concentrated between 110 and 150 cps, with a secondary energy peak around 240 cps, as shown in the frequency-amplitude section in Figure 1. The sound energy was highest at the beginning of the sound, dropping to about 30 per cent of its initial amplitude after 50 milliseconds.

The strength of the sound field generated could only be approximated. The order of magnitude of the sounds was extremely low—near the limit of sensitivity of the hydrophone. Therefore, both the ambient noise and the noise inherent in the amplifying equipment, although low, were sufficient to make amplitude measurements inaccurate. From data on the voltage gains of the preamplifier and tape recorder and the calibration curves of the hydrophone, the sound pressure produced by the goby sounds at distances of 6–8 inches from the hydrophone was determined to be in the order of a fraction of a dyne per square centimeter. The closest approximation would be 0.1–0.2 microbar (i.e., dynes per square centimeter).

The rate at which the sounds were repeated was extremely variable. When approaching a gravid female, the male repeated the sounds more frequently—as often as three to four per second, with occasional bursts of two to four sounds following one another without a break. The sounds also increased in frequency when the female followed the male into his nest. With a non-gravid female, courtship, if any, was short, and only a few grunts would be emitted by the male during a 2- or 3-minute interval. The repetition rate was more regular in cases where the male courted a gravid female who was confined in a wire cage. In one such situation, the male produced sounds for a full 6 minutes at a rate of one per second.

MECHANISM OF SOUND PRODUCTION

At the emission of a grunt, the male sharply thrusts his head downward. As he does so, he snaps his mouth closed and opens his opercula. By means of a long pipette, small amounts of carmine suspension were introduced into the water near the male's head while he was courting. By watching the path of small clouds of carmine, it was found that just prior to a sound burst, the male takes in a quick breath. At the sound burst, the head is snapped down, and carmine can be seen ejected from the dorsal sections of the opercular openings in short, strong jets.

The adult *Bathygobius*, as is true of most of the Gobiidae, possesses no swim bladder or any air reservoir. Dissection of the buccal and branchial cavities showed no structures which might be specialized for sound production. In short, there were no discernible analogues of the various pharyngeal structures described by Fish (1954) used by fishes in the production of sound.

A few surgical approaches to this problem were made. Stitching down or otherwise mutilating the tongue of a male *Bathygobius* had no effect upon the sound production and did not inhibit courtship. Attempts to interfere with passage of water through the gill chamber impaired respiration. When any blockage of the gill chamber was made, it was done on one side only. Thin sheets of plastic (Pliofilm) were used to block water passage, either partially or completely. The operculum on one side was stitched down in various ways to prevent or impair water passage. In all cases, the gobies survived and showed no respiratory difficulties. No courtship behavior was observed, however, and the operated males either showed no response to females or, at most, nudged them occasionally.

PLAYBACK OF *Bathygobius* COURTSHIP SOUNDS

Tape recordings of the courtship sounds were played from the tape recorder into a 20-watt power amplifier. The output (from the 500-ohm tap of the amplifier) went into the LF-400 transducer, which was placed in the water within a few inches of the fish. At first, the tape used was a 5-minute recording of a single male's output, but over this period there was a considerable variation in the quality and intensity of the sounds. A length of tape of about 3 seconds in duration was selected containing three grunts. This tape was spliced into a loop so that a continuous series of sounds would be produced over an indefinite period with little variation in pitch, intensity, quality, duration, and repetition rate.

The amplitude of the output was adjusted by monitoring the output of the LF-400 with another hydrophone, constructed from a Brush ceramic tube (see above for description) connected to an oscilloscope. The output of the LF-400 was adjusted so that an oscilloscope deflection from the monitor at a distance of 8 inches was approximately equivalent to that produced by a *Bathygobius* male at 8 inches from the monitor. As will be shown later, however, considerably higher sound intensities did not affect the responses of the animals.

RESPONSES OF FEMALES TO PLAYBACK

Mature females in a 5-gallon aquarium (one to five per tank) responded to the playback within 1 minute or less. The fishes showed a distinct increase in general activity. Their respiratory rate increased from a normal 1.4 to 2.2 per second (average of fifty observations, 1.7 per second) to a range of 2.0 to 3.7 per second (average of fifty observations, 3.2 per second). Their activity was not directed toward the hydrophone but rather to each other. There was an increase in the number of contacts between individuals, i.e., nipping, butting, approaching.

In a larger aquarium, a trough 96 × 48 × 6 inches deep, where no shelters were provided, the gobies normally remained along the sides or in corners. Twelve mature females were placed in this tank. At sound intensities close to normal (as described above), only those females within a range of about 12 inches showed any activity increase. When the sound level was increased about tenfold, i.e., producing a 10× normal oscilloscope deflection from the monitor, then all the animals showed a definite activity increase. Again there were no direct approaches to the hydrophone, although on occasion a single animal would swim by, stop, and then go on. The fishes darted frequently along the edges of the tank, stopping to nip or nudge one another.

In both these situations, distinct responses could be seen within less than 1 minute after the playback was begun. If the playback was continued, the activity of the animals would drop off after a few minutes, and after 5 or 6 minutes all the animals were quiescent.

RESPONSES OF FEMALES TO PLAYBACK WITH ADDITION OF VISUAL STIMULUS

Both laboratory and field observations showed that during normal prespawning behavior females will approach and follow courting males. The addition of a visual stimulus of another goby to accompany the playback of the sounds was accomplished by confining a mature male in a 100-ml. Erlenmeyer flask. This flask was then placed next to the hydrophone. The observations were made in the large trough, and the sound intensities were set at 10× normal levels. The animal in the flask was moderately active, bumping the sides and moving most of the time.

The visual stimulus without accompanying sounds had no effect other than an occasional approach by one of the females over a period of an hour. When the sound playback was begun, however, the females responded within a minute or less by approaching the confined goby. They bumped against the flask repeatedly and often turned to the hydrophone. Using their fused pelvic fins, they scrambled all over both the flask and the hydrophone. There was also a considerable amount of nipping and butting between the several females that approached the vicinity of the flask and hydrophone. Occasionally, one of the females would move away, only to return in a few minutes. This behavior could be maintained for as long as half an hour of continuous playback, after which the activity of the females would decrease, but they would remain within a 12-inch radius for as long as the sound kept up. When the sound was turned off, the females retreated to the edges and corners of the tank within 5–10 minutes. Although the groups of females observed included animals in both gravid and non-gravid conditions, no differences in their responses could be detected.

When the flask containing the visual-stimulus goby was placed at some distance (2 feet) from the hydrophone, the responses of the females were quite variable. In most cases, they approached the flask first, then turned toward the hydrophone, darted part of the distance toward it, and then returned to the flask. Some animals approached the hydrophone first and then, after some minutes, moved toward the flask. In all cases, however, the orientation of the females appeared to shift from the flask to the hydrophone and back again. Often a number of animals would remain moving about in the area between the two. When the sound was turned off, most of the fishes moved toward the flask and remained there for a few minutes before dispersing again toward the periphery of the tank.

RESPONSES OF MALES TO PLAYBACK

Single males in 2-gallon aquaria were provided with shelters (shells) and isolated for periods of 24 hours or more. The hydrophone was placed in the tank and left there for 15–30 minutes, to minimize the effect of the disturbance created within such a small tank. When the sound was turned on, the goby came out of his shelter and turned facing the hydrophone. His respiratory rate increased, as in the case of the females. In all cases, the response was rapid, and the animal approached the hydrophone, nudged it a number of times, climbed over it, and remained near it for from 5 minutes up to 1 hour, as long as the sound continued. During this period, the male often returned to his nest for brief intervals to sweep out some sand vigorously and then returned to the hydrophone. When the sound was turned off, the goby soon returned to his shelter and became quiescent.

When several males were placed in the large trough, as in the case of the females, they normally remained around the edges or corners, although there was considerably more nipping and fighting among males than among females under the same conditions. When the hydrophone was placed in the tank and the sound turned on, the response of the males became apparent within 1–5 minutes. The hydrophone was approached, and in a few minutes all the males (up to twelve in some observations) were either next to or on top of the hydrophone or within a radius of some 6 inches. Activity was high during the first few minutes, with a good deal of nipping, butting, and fighting. After 10–15 minutes of continuous sound output, the fishes moved away from the hydrophone, but individuals would return for short periods sporadically.

VARIATIONS IN THE RESPONSES OF MALES AND THE IMPORTANCE OF "SET"

Previous investigations of the pre-spawning behavior in *Bathygobius* males showed that prior recent experiences would alter the types of responses and levels of excitability (Tavolga, 1956b). It was found that the set of the males distinctly affected their responses toward the sound playback.

Isolated males were either exposed to a vigorous combat with another male or permitted to court a gravid female for several minutes. As shown earlier, this type of experience would produce a set either toward combat or toward courtship. A combat set could also be produced by placing up to six males together in a 5-gallon tank for 24 hours or longer.

Isolated males in small 2-gallon tanks responded to the sound playback in the same way, regardless of their set. Differences in the intensities of the responses could not be detected.

Groups of twelve males were tested in the large aquarium. Males that had been isolated or given prior recent courtship experience responded quickly to the sound playback as described above. Males in combat set, however, showed no response whatsoever to the sound playback. If a gravid female were placed with this group of males for several hours, then the residents would subsequently respond positively to the sound stimulus.

RESPONSES OF GROUPS AND PAIRS OF BOTH MALES AND FEMALES

Six males and six females were placed in the large box-tank with no shelters. When the sound was turned on, the responses were variable but distinct enough to be recorded as positive. There was an almost immediate increase in activity on the part of all the animals. The males approached the hydrophone first and moved about it actively, nudging it and frequently. There was some nipping and butting among the males. Within a few minutes, females began to move toward the hydrophone in short, quick darts. As long as the sound kept up, there were always some gobies of both sexes near the sound source. Although there was much activity among the animals and some of the females were gravid, no courtship was observed.

If shelters were provided in the same situation as above, the results were essentially the same except that males returned to their shelters for short periods of time, during which they fanned actively. Occasional short periods of courtship took place, but this was also true when the sound was off. During such courtship bouts, sounds emitted by the males could be detected. It was only in this situation that the males produced sounds. During approach responses to playbacks, no sounds produced by the fishes could be detected by the monitoring equipment.

Sound playbacks were attempted in small tanks where a male and a female were placed together. If the female was not gravid, there was little or no courtship by the male, and in this situation the male approached the sound source readily, followed occasionally by the female. When the sound playback was begun while courtship was in progress, neither animal showed any response to the sound.

RESPONSES TO ARTIFICIAL SOUNDS

METHODS

Twelve animals, either all males or all females, were tested in a large aquarium (a wooden box, 96 × 48 × 6 inches deep). The males had all been brought previously into a courtship set. Females were tested with the addition of a visual stimulus. No shelters were provided, and the gobies normally remained around the periphery of the tank. The hydrophone was placed near the center of the tank, and the confined goby in a flask, used when testing females, was placed next to the sound source. A response was judged to be positive if the majority of the animals approached the sound source within 10 minutes.

A sine-wave audio generator was the signal source. The signal was fed through a gating circuit. Either a continuous signal or signal bursts were fed into a power amplifier and from there into the LF-400 hydrophone. In some cases, tape recordings were made of groups of signal bursts, and these were played on a continuous loop through the amplifier into the transducer. The output of the transducer was monitored, as before, by another hydrophone and oscilloscope.

As a starting point, the artificially generated audio signal consisted of pulses of a 200-cps sine wave. The amplitude was adjusted to a level comparable to the normal output level for a *Bathygobius* male. The duration of each pulse, as controlled by the gating circuit, was 75 milliseconds. The pulses were repeated at a constant rate of one per second. This type of signal was chosen because of its resemblance to the sound pattern produced by a male goby.

Sound intensities were varied by the gain controls, and the frequency was controlled by the audio generator. By increasing the output of the power amplifier, an overloading effect was produced which clipped the sine-wave output to a square wave and produced some harmonic distortion. The duration of the sound pulses was controlled by the transistor gating circuit for short intervals and manually with a key for pulse durations over $1\frac{1}{2}$ seconds. Pulse repetition rate was controlled manually by means of the trigger switch on the gating circuit. Similar control of pulse duration and repetition rate was obtained by using continuous loops of tape and erasing or recording appropriate portions.

EFFECT OF INTENSITY VARIATION

Amplitude levels within the order of magnitude of the normal output of a *Bathygobius* male produced positive results only on those animals within an 8-inch radius of the sound source. A tenfold increase in amplitude affected all gobies in the large tank. Increasing the amplitude further to levels that were clearly audible to the observer elicited the same positive effect from the gobies, both males and females. Further increase in amplitude, to the maximum the instruments would permit, began to distort the sound output. The sine waves were clipped, and the oscilloscope showed presence of harmonics. Nevertheless, the fishes responded to these abnormal intensity levels exactly as they did to normal ones.

EFFECT OF FREQUENCY VARIATION

Frequencies below about 100 cps could not be tested because of the lack of response of the equipment. From 100 to 300 cps, the responses of the fish were positive and unmistakable. From 300 to 500 cps, the responses were difficult to judge and may have been slightly positive in some cases. Above 500 cps, there were no positive responses. Amplitudes used here ranged from normal levels, at which there was no detectable distortion, to intensities up to ten times normal, where some clipping of the sine waves occurred.

EFFECT OF CHANGING PULSE DURATION

Pulse durations of 75–150 milliseconds were most effective and elicited clear, positive responses. The shorter durations were effective only at high repetition rates of five or more pulses per second. At pulse durations of less than 25 milliseconds, the effective stimulus was basically only a click, and, if repeated rapidly enough, the component frequency within the pulse became meaningless.

Pulse durations of 0.5 second or longer elicited no responses from either males or females.

REPETITION RATE OF SOUND PULSES

Since the duration of each pulse was varied, the repetition of each sound in terms of pulses per unit time became meaningless. Instead, a measurement of the time interval between pulses was used. Repetition of pulses with intervals of from 0.1 up to 10 seconds produced positive responses on the part of the fishes. Shorter intervals of from 0.05 to 0.01 second had variable effects, depending upon the pulse duration. With durations of 50 or more milliseconds, there were few or no positive responses at these short intervals. The short intervals, however, were effective at pulse durations of less than 25 milliseconds.

Groups of pulses could be adjusted and spaced so as to produce a positive response. For example, pulses of 10 milliseconds each in groups of ten with 10-millisecond intervals between each pulse would elicit a positive response. Spike-form signals could be used as well as sine and square waves. The component and harmonic frequencies became immaterial in these conditions.

OTHER SOUND PATTERNS TESTED

Recordings were made of courtship sounds produced by a species of blenny, *Chasmodes bosquianus*. Males of this form produce low-pitched thumping sounds which are similar to those of *Bathygobius* in frequency and duration and conditions under which they can be detected. A more detailed description of the *Chasmodes* sounds will be reported separately (Tavolga, 1958, in press). Such recordings played back to *Bathygobius* males or females elicited positive responses indistinguishable from responses to normal *Bathygobius* sounds.

As a test of the low order of discrimination in the responses of the gobies, a tape recording of the author's voice saying "ugh-ugh" at 1-second intervals was played into the tank containing male gobies. The fishes responded positively.

FIELD OBSERVATIONS

The boat dock at the Cape Haze Marine Laboratory had a population of gobies living among the oyster shells attached to the pilings and littering the general area. At low tides, when the water was calm and clear, the activities of the gobies could be observed from the edge of the dock at several points.

All the major aspects of territorial and reproductive behavior were seen, i.e., nest-cleaning, combat, courtship, and spawning. It was noted that, as in the case of observations on groups of both sexes in the large aquarium, when a single male began courting, other males soon appeared in the vicinity. The largest of these males quickly became dominant, and the courted female followed him into his nest, even though, in some cases, his shelter was as much as 10 or 20 feet away from the spot where this activity first began. A number of smaller gobies always approached such scenes of activity, and it is quite probable that these included both males and females, but from the edge of the dock one could not be certain of the sex.

Attempts were made to pick up courtship sounds from these situations. Unfortunately, the lowering of the hydrophone dispersed the animals, and, although courtship was resumed within a few minutes, it took place several feet away from the hydrophone. As shown before, the sounds could be picked up only if the hydrophone were within a few inches of the courting male. Furthermore, the large number of snapping shrimp, crabs, pinfish, grunts, and other underwater noisemakers made the detection of the goby sounds difficult. Adequate detection of these sounds was finally achieved by lowering a gravid female *Bathygobius*, confined in a wire-screen cage, next to the hydrophone. Several males approached the confined female and courted her for several minutes at a time. Under these conditions the courtship sounds could be detected through earphones, but the other interfering noises made good recordings impossible, even after filtering out much of the high-frequency component of the snapping shrimp sounds.

The hydrophone was placed at several points where there were shelters inhabited by gobies. Recordings made in the laboratory and a series of pulses on a continuous loop of tape were played back through the transducer. The intensities used here were probably considerably above normal levels, since the monitor could not pick up the emitted sounds over the background noise. The responses of the gobies were rapid, as in the laboratory observations. Individuals that were most probably mature males frequently approached the hydrophone. There was a considerable amount of nest-cleaning activity within a radius of about 5 feet. Smaller individuals, probably females and small males, also approached the hydrophone. Contacts between individuals were frequent, and activities such as nipping, chasing, combat, and courtship were observed. When the sound was turned off, within a few minutes only sporadic activity was observed. In general, as many as fifteen or twenty gobies would be visible within a 5-foot radius of the hydrophone while the sound was on, whereas only three or four could be seen at any one time while the sound was off.

DISCUSSION AND CONCLUSIONS

Despite the low order of sound pressures produced by the males of *Bathygobius*, the range of effectiveness of these sounds may be greater than indicated by the present equipment. Although the background noise of snapping shrimp and other common marine noisemakers is considerably louder than the goby grunts, the frequency and general character of the grunts may be sufficiently different from the noise to stand out. The sensitivity of the sound-reception system system of *Bathygobius* is not known, but the frequency range of the grunts probably involves both the auditory and the lateral-line systems and possibly cutaneous touch receptors as well. It is not likely, however, that the normal effective range of the grunts is greater than 2 or 3 feet.

The selectivity and discrimination of the responses of both males and females are low. This is indicated by their positive responses to a wide range of variations in the characteristics of the sounds played back. In general, any sounds which show a slight resemblance to the normal grunts of a male goby will produce a positive approach reaction in males and also in females if coupled with a visual stimulus.

The effect of the sounds upon males raises their activity and probably their level of responsiveness. Orientation toward the sound source tends to increase the probability of several animals competing for a female. The females are also stimulated to more activity and responsiveness, and, therefore, encounters with males become more likely. Although females are not specifically oriented by the sounds, they become responsive to a visual stimulus.

The frequency levels of the sounds produced by males most closely resemble the types described by Fish (1954) that are produced by air bladders. Stridulatory mechanisms produce considerably higher frequencies. The only clue thus far to the mechanism of the *Bathygobius* sounds is the squirting of water through the gill openings.

The present report added to the author's earlier studies shows that there are three basic sensory modalities involved in the initial phase of pair formation in *Bathygobius*. All three—visual, olfactory, and auditory—operate to reinforce each other. An olfactory stimulus from a female induces a male into courtship, with or without a visual stimulus. The courtship movements and sounds stimulate a female to approach the male, thereby increasing the effectiveness of her stimuli upon him. Males are attracted by the sight, sound, and smell involved in the courtship processes. Their level of activity and responsiveness is increased, and their behavior subsequently may be channeled into nest preparation, combat, or competitive courtship.

SUMMARY

The males of the gobiid fish, *B. soporator*, produce low-pitched grunting sounds while courting females.

Sound spectrograms and other analyses of these sounds show durations of 150–350 milliseconds. Component frequencies range from 100 to 500 cps, with occasional low-energy peaks as high as 3,000 cps. Frequencies down to 80 cps occur toward the end of each sound. The sounds are non-harmonic, with most of the energy between 110 and 150 cps and during the first 10 milliseconds. The amplitude drops rapidly to about 30 per cent of its original value in 50 milliseconds. The strength of the sound field generated at 6–8 inches from the hydrophone is estimated at 0.1–0.2 microbar. The repetition rate is variable and is dependent upon the level of excitement of the male.

The mechanism of sound production is unknown, but it appears to involve the forcible ejection of water through the gill openings.

Playback of recorded sounds to the fish results in positive approach responses from males. Females respond with a general activity increase. Females approach the sound source only if a visual stimulus of a confined goby is also present. Males show a readiness to respond if they have recently had encounters with females.

Artificially generated sounds (pure sine waves) produce positive approach responses. The intensity of these sounds can be raised at least a hundred times over normal levels, component frequencies varied from 100 to 300 cps, and pulse durations varied from 75 to 150 milliseconds. Similar sounds from other species and even human voice imitations also elicit positive responses.

Observations and tests with gobies in their natural habitat confirm the results of laboratory experiments in aquaria.

LITERATURE CITED

Fish, Marie Poland. 1954. The character and significance of sound production among fishes of the western North Atlantic. Bull. Bingham Oceanog. Coll., **14**:1–109.

Fish, Marie Poland, Kelsey, A. S., Jr., and Mowbray, W. H. 1952. Studies on the production of underwater sound by North Atlantic coastal fishes. Jour. Marine Res., **11**:180–93.

Griffin, Donald R. 1950. Underwater sounds and the orientation of marine animals: a preliminary survey. Project NR 162-429, O.N.R. and Cornell University, Tech. Rept. No. 3. ATI No. 90329.

Moulton, James M. 1956. Influencing the calling of sea robins (*Prionotus* spp.) with sound. Biol. Bull., **111**:393–98.

Tavolga, William N. 1954. Reproductive behavior in the gobiid fish, *Bathygobius soporator*. Bull. Amer. Mus. Nat. Hist., **104**:427–60.

———. 1956a. Visual, chemical, and sound stimuli as cues in the sex discriminatory behavior of the gobiid fish, *Bathygobius soporator*. Zoologica, **41**:49–64.

———. 1956b. Pre-spawning behavior in the gobiid fish, *Bathygobius soporator*. Behavior (Leiden), **9**:53–74.

———. 1958. Underwater sounds produced by males of the blenniid fish, *Chasmodes bosquianus*. Ecology (in press).

Copyright © 1961 by The Ecological Society of America
Reprinted from *Ecology*, **42**, 274–282 (1961), with permission of Duke University Press

REPRODUCTIVE ECOLOGY AND SOUND PRODUCTION OF THE TOADFISH, *OPSANUS TAU*[1]

Grace-Ann Gray and Howard E. Winn

Zoology Department, University of Maryland, College Park

Introduction

The fact that fishes produce a variety of sounds has been clearly demonstrated in recent years (e.g. Fish 1954). These sounds are made under a variety of behavioral circumstances, yet few studies have been made of their functional significance. The toadfish, *Opsanus tau* (Linnaeus), was selected for studies on sound and its relation to breeding biology because it is known to produce 2 distinctly different sounds. Also, these fish nest in shallow water where they are accessible to observation and experimentation. The objective of this study is to describe the reproductive biology of the toadfish and at the same time to attempt to analyze the relationship of sound production to the behavior of the fish.

General descriptions of the life history of *Opsanus tau* have been published by Ryder (1886), Gill (1907), and Gudger (1910). They have pointed out that the toadfish utilizes old cans, jars, and the undersides of various other solid objects as nests which the males defend in the late spring and early summer in northern latitudes. The female enters the nest to deposit eggs on the underside of the object and then leaves. It has usually been assumed that in Chesapeake Bay the eggs hatch in 10 to 15 days and that the larvae remain hanging by a pedicel for another 3 to 4 weeks. Gudger (1910) stated that the diet of *O. tau* varies with the habitat, and noted that the blue crab is eaten most frequently, whereas Hildebrand and Schroeder (1928) claimed the fish to be omnivorous.

It has been known for many years that *O. tau* makes a harsh grunt and a more musical sound referred to as the boatwhistle call (Gill 1907). Sorensen (1884), Tower (1908), and Fish (1954) identified the air bladder as the organ of sound production and they demonstrated that contraction of the muscles attached to each side of the air bladder results in emission of the grunt. More intensive studies have been made in recent years where the sounds have been analyzed for their frequency, time, harmonic, and intensity characteristics (Fish *et al.* 1952, Fish 1954, Fish *et al.* 1959, Tavolga 1958b). Fish (1954) associated the boat-whistle call with prespawning and early egg-guarding activity, and because of this the sound was referred to vaguely as a mating call. On morphological grounds it was thought that both

[1] This is contribution No. 13 of the Chesapeake Bay Fund, Zoology Department, University of Maryland.

sexes could produce the boat-whistle and it was claimed that the grunt was produced throughout the year by both sexes (Fish 1954). Tavolga (1958b) compared the physical characteristics of sounds produced by *Opsanus tau* and *O. beta* (Goode and Bean), whereas Fish and Mowbray (1959) analyzed sounds emitted by a West Indian form, *Opsanus sp.* According to Knudsen *et al.* (1948) the boat-whistle call was not subject to any diurnal variation.

Materials and Methods

Most observations and all experiments were made in the shallow water beside the long pier at the Chesapeake Biological Laboratory, Solomons, Maryland, during the late spring and early summer of 1958 and 1959. The pier which was 6 feet wide served as a permanent location of testing areas and facilitated the placement of the equipment. From the dock a hydrophone could be placed directly over or beside a nest. The water depth along the dock varied from 0 to 12 feet but the area utilized in this study was 2 to 5 feet deep.

Tin cans and glass jars, open at one end (17.5 cm in length by 10.5 cm in diameter and 21.7 cm by 15.6 cm), were utilized as nests by the toadfish. These were placed at intervals of 10 feet along each side of the pier. Seventy-three nest containers were studied during each year. In 1958, 5 cans were placed in the water on June 17, 12 on June 18, 26 on June 29, 24 on July 1, and 6 glass containers on July 16. In 1959, all 73 containers were placed in the water on or before May 22.

At the onset of the 1958 testing program, some males were removed from the cans, sexed by means of the external urogenital papillae, measured, marked with numbered strap-tags (small mammal ear tags) on the left operculum and returned to the water. All of the tagged fish left the area so that in 1959 individuals were recognized by the distinctive characteristics of the fleshy projections around the mouth and eyes. On occasion, these were clipped underwater while the fish were still in the cans. Sex was determined on a different occasion.

Data were collected on the length of time males stayed at a nest, the time it took the eggs to hatch and the time the young remained in the cling stage. If the nests were checked every 24 hours this resulted in a range of 48 hours in any of the data, since the eggs were laid at an unknown time within a 24-hour period and hatched at an unknown time within a 24-hour period. For example, if no eggs were found in a particular nest on July 8 but were present on July 9, and the eggs were still there on July 15 but were cling young on July 16, then the hatching time was recorded as between 6 and 8 days. If the time between checks was greater than one day, the range would be greater.

One hundred and thirty-five specimens, collected in crab pots, were sexed by examination of the gonads in order to verify the reliability of the use of the urogenital papillae for the determination of sex. This method was 100% reliable with reproductively mature fish. The male papilla was smooth, round, and somewhat elongated, whereas the female papilla was short, swollen, irregular, and red in appearance.

In 1958, the recording equipment consisted of a Tandberg Tape Recorder (Model 3BF) and a Tandberg crystal microphone (flat response between 30 and 12,000 cps in air) waterproofed with a rubber membrane. In 1959, a Magnemite Portable Tape Recorder (Model 610DV) was used in conjunction with a hydrophone (Model LF-310A-A) equipped with an internal preamplifier.

The first underwater recordings of sound were made on June 24, 1958, after spawning had started and were continued until August 20 at which time all spawning and sound production in the study area had ceased. In 1959, intensive listening and recording began on May 22 when spawning had already begun and observations were continued until September 9. Data on sound production by various males in their nests were kept throughout these periods.

In order further to understand the behavior of a male at its nest in the various phases of reproduction (with and without eggs or young) a series of presentation tests were made in 1958 and 1959. The items presented were: male toadfish (either normal, anaesthetized, or confined in a jar); female toadfish with ovulated eggs; blue crabs, *Callinectes sapidus* Rathbun, 4 to 5 inches in carapace width; adult bloodworms, not from the Chesapeake Bay, *Polycirrus eximius* (Leidy); menhaden, *Brevoortia tyrannus* (Latrobe), 4 to 6 inches in total length; and oyster shells. These objects were tied to a dark-colored fish line and placed before a male in its nest. Sound recordings and visual observations of 50 seconds duration were made for each presentation experiment.

In 1959 four wire pens, 15 by 15 feet, were constructed of one-inch square heavy welded wire and placed along the pier. These pens permitted the study of known individual fish.

Results

Data on the reproductive seasons of 1958 and 1959 are presented in Table I, where information is given on the number of nests that were empty or occupied, as well as whether or not there were eggs, young free from the egg but still attached

TABLE I. A resumé of the condition of the toadfish nest containers and their occupants during the spawning season beside the Solomons Island Laboratory dock during 1958 and 1959

Nest Condition	May 29	June 8	15	16	18	22	23	24	25	26	29	30	July 1	2	3	7	8	10	15	16	17	23	24	27	28	30	August 4	5	6	11	13	17	18	Sept. 4
1958[a]																																		
Total empty						4		8	8	9	29		25	49	44	48			57	58		64	67		64	61		62						
Total occupied						1		9	9	8	14		18	18	23	19			10	9		9	6		9	12		11						
Single fish[b]						1		6	7	5	2		7	5	16	16			7	4		4	2		3	6		8						
Pair of fish						0		2	2	1	12		7	7	1	0			0	0		1	0		0	0		0						
Pair & eggs						0		0	0	2	0		2	0	1	0			0	0		0	0		0	0		0						
Male & eggs						0		1	0	0	0		2	6	5	3			2	4		1	1		2	2		1						
Male & cling young						0		0	0	0	0		0	0	0	0			1	1		3	3		4	2		1						
Male & free young						0		0	0	0	0		0	0	0	0			0	0		0	0		0	2		1						
1959[a]																																		
Total empty	46	45	41	40	45	40	42				39	43	43	43			43	47	55		57	60	61		63	63	65	68	70	67	63	68	64	64	66
Total occupied	27	28	32	33	28	33	31				34	30	30	30			30	26	18		16	13	12		10	10	8	5	3	6	10	5	9	9	7
Single fish[b]	7	4	5	3	2	11	8				7	2	2	2			1	3	2		4	0	0		0	2	3	2	1	6	10	5	9	9	7
Pair of fish	0	3	5	1	0	1	3				0	0	0	1			1	1	0		2	1	0		1	0	0	0	0	0	0	0	0	0	0
Pair & eggs	1	1	0	2	0	0	0				0	2	0	0			0	0	0		0	0	0		0	0	0	0	0	0	0	0	0	0	0
Male & eggs	19	18	12	12	11	5	5				12	10	12	10			13	9	6		0	2	2		1	0	0	0	0	0	0	0	0	0	0
Male & cling young	0	2	10	15	15	16	14				15	14	8	9			8	9	7		8	8	6		5	5	3	1	0	0	0	0	0	0	0
Male & free young	0	0	0	0	0	0	1				1	2	8	8			7	4	3		2	2	4		3	3	2	2	2	0	0	0	0	0	0

[a] See materials and methods for the date of placement of nest cans.
[b] These single fish were usually males.

to the nest by a pedicel (cling young), or detached free-swimming young. Additionally, eggs were not located on May 8, 1959, but on May 22 there were many cans which contained either a pair or a male with freshly laid eggs. The data (Table I) demonstrate that maximum reproductive activity occurred throughout the month of June and the first part of July. Actual spawning began in late May and usually was completed by early August. Based on the presence of eggs, the 1959 spawning season was several weeks longer than in 1958. The reduced number of cling young and attached young during 1958 were caused by manipulation of the nest cans.

In Table II data are presented on the sex ratios of some toadfish caught in crab pots in 1958. Although the data are extremely limited, the male catch from June 29 to July 20 suggested that most of the males were on nests and that the low catch of females represented a time when many were spawning. Early in July, the catch of females increased and this was correlated with the fact that most of the females had already laid their eggs. As the males began to leave the nests in large numbers at the end of July, the catch of males increased and was about equal to the number of females caught from July 27 to August 3. A progressive increase in the number of spent females was evident (Table II).

In the nests studied (Table I) a female spawned with a male in a can, and as soon as a clutch of eggs was deposited she left the immediate area and went into either beds of vegetation or deeper water. A male in a can usually received eggs from a single female. The males maintained residence in the cans until the young had attained the free-swimming stage. Twenty-nine fish, found in cans, proved to be males when sexed either by examination of the gonads or the urogenital papillae.

The males without eggs moved freely throughout the area and from can to can until they received eggs. Males with eggs or young also left, but less readily. Thirty-five fish that were strap-tagged in 1958 left their shelters or nests (24 were individual males, 10 were members of a pair and one was a male with eggs). Eight single males, seven with eggs, one with cling young, and four with detached young left their shelter or nest after they were handled during the 1958 season. Two males, one with eggs and one with detached young, had returned to their cans by the next day. When objects were presented to males in nests (experiments to be discussed later), 20 males in cans

TABLE II. The sex ratio and sexual condition of females of *Opsanus tau* captured in crab pots placed at Little Cove Point, 1½ miles southeast of Drum Point Light during the breeding season of 1958

	NUMBERS OF INDIVIDUALS			
Date	Male	Female		
		Ripe	Spent	Total
June 15	1	5	2	7
June 22	11	5	3	8
June 29	3	1	0	1
July 6	0	1	0	1
July 13	8	2	17	19
July 20	3	1	8	9
July 27	10	0	10	10
Aug. 3	23	0	21	21

without eggs, 3 in cans with eggs and one with cling young left the area and only one returned to its container. Nine males in 1959, identified individually by the fleshy projections around the head, were noticed in separate cans without eggs or young on one day, but were not there when the cans were checked one to 2 days later. One returned 4 to 7 days later to his original can and remained to receive eggs.

From periodic inspection of the nests (Table I) it was possible to obtain information on the length of time a male stayed in a nest. Marked males that were followed from the time a pair entered the nest until free young were present, stayed for at least 23, 25, 25, 36, and 37 days, whereas 4 unmarked fish stayed for at least 24, 28, 28, and 37 days. In all instances it was not known how long these males remained with the free young except where 2 unmarked individuals stayed a total of 31 to 36 and 32 to 37 days with the eggs and young. Other marked and unmarked males, in attendance at the nest from the egg to the free young stage, were studied, but the exact date of egg-laying and the length of stay with the young were unknown. The data were as follows (in minimum number of days): 20, 21, 22, 26, 27, 33, 33, 34, 35, 35, 37, and 46. It appeared as though the males usually stayed with the eggs, cling young and free young from 23 to 46 days, although, under other conditions of temperature, this time may be shorter or longer. The length of stay also appeared to be affected by disturbances.

The time that it takes the eggs to hatch was based on the following data: 6-8, 6-9, 6-9, 7-9, 6-8, 5-7, 7-12, 9-11, 9-11, 8-10, 7-9, and 8-10 days. The first 7 hatching times occurred between June 23 and July 3, 1959, whereas the next 3 times occurred between June 15 and 26. The shorter hatching times of the first group were correlated with high temperatures (Figure 1). The egg-hatching periods varied from 5 to 12 days. The data for the length of the cling stage in days were as follows: 4-6, 6-9, 9-11, 9-11, 9-16, 10-14, 10-17, 11-15, 11-15, 12-14, 14-16, 14-17, 15-23, 15-24, and 19-27. The first 2 times, which seemed significantly shorter than the others, occurred during the temperature peak of late June, 1959 (Figure 1). A salinity plot exhibited no obvious correlation. It appears as though the cling stage usually varied from 6 to 19 days with a potential range of 4 to 27 days. A figure of 9 to 15 days seemed to be more normal during the time of this study. The male stayed with the young in 4 instances from 5-6, 6-7, 7-18, and 10-15 days. This time was quite variable because the male readily left the container when disturbed.

The toadfish make grunt-like sounds when they are taken out of the water and handled. These also may be heard when a swimmer or another male toadfish approaches a male in a nest cavity. In order to test which kinds of stimuli evoked a grunt response from a male, a variety of objects was presented to males in various phases of the reproductive cycle. The responses of the fish are summarized in Table III. A male on his nest nearly always grunted at active male toadfish and he grunted somewhat less than 50% of the time at females and inactive males. However, inactive

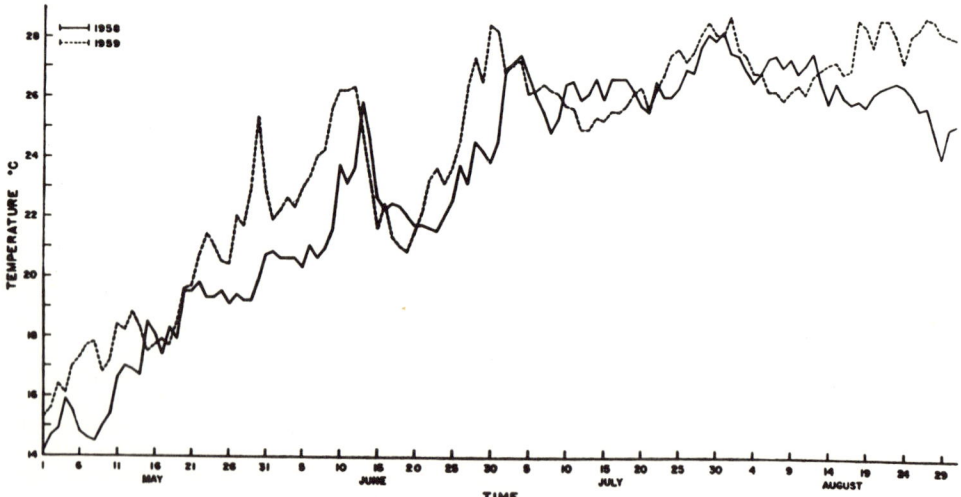

FIG. 1. Surface temperatures at the end of the pier, Chesapeake Biological Laboratory, Solomons, Maryland.

TABLE III. Grunt and attack responses of nest-guarding male toadfishes toward a variety of objects

Introduced material	Nest-guarding male toadfish with:	Number of tests	Responses					
			Grunt			Attack		
			Yes	No	% that grunted	Yes	No	% that attacked
Oyster shells.........	Eggs & young	22	2	20	9	2	20	9
	None present	22	0	22	0	6	16	27
Bloodworms..........	Eggs & young	3	0	3	0	3	0	100
	None present	2	0	2	0	1	1	50
Blue crabs...........	Eggs & young	37	15	22	41	28	9	76
	None present	23	0	23	0	16	7	70
Menhaden............	Eggs & young	10	0	10	0	4	6	40
	None present	1	0	1	0	0	1	0
Ripe female toadfish...	Eggs & young	33	13	20	39	5	28	15
	None present	13[1]	5	8	36	1	13	7
Active male toadfish...	Eggs & young	50	43	7	86	16	34	32
	None present	36	27	9	75	5	31	14
Inactive male toadfish[2].	Eggs & young	27	12	15	44	0	27	0
	None present	21	2	19	9	0	21	0

[1] Three individuals (9 tests) not included, responded with a boat-whistle call—see text.
[2] Includes fish, anesthetized or confined in a jar.

males appeared to cause fewer grunts from males with no eggs or young. Attacks were infrequent or absent with toadfish as the stimulus. When such objects as menhaden, bloodworms, and oyster shells were present, a grunt response was usually not evoked but it was frequently produced when crabs were placed in front of males guarding eggs or young. Attacks (or feeding responses) were frequent toward menhaden, blue crabs, and bloodworms. The oyster shells elicited a small percentage of attacks. Thus toadfish, especially active males, were the stimuli that elicited the maximum production of grunt responses. Males with eggs and young either grunted or attacked more frequently than males without eggs or young. This was further substantiated by the fact that in 20 tests males without eggs and young left the can at the start of an experiment (thus no response, and therefore not included in Table III, whereas only 3 males with eggs deserted their nests).

In an unusual circumstance many grunts were produced by several fish. On July 31, 1959, an underwater explosion occurred about 5 miles from the pier while underwater sound recordings were being made beyond the pier in water 30 feet deep. As soon as the sound of the explosion was heard, several toadfish emitted a series of grunts for several seconds.

Another sound produced by the toadfish is referred to as the boat-whistle call. Twenty individuals that emitted boat-whistles in cans or other cavities were all males. Many other toadfish (48 plus) that produced this sound were believed to be males on the basis that only males have been found to stay in the nest and guard the eggs. Not once was a female found that emitted the boat-whistle in the large area surrounding the pier. A particular effort was made to listen over females in the area. Several wire pens were placed in the water beside the dock and no sound was heard from 4 females in one pen, whereas in another pen the one male present emitted boat-whistles. In a 3rd pen one pair and a single male with eggs were present, and boat-whistle calls were heard only prior to the removal of the males.

During 1958, boat-whistles were heard from June 18 when the work began until July 15 when the call became infrequent. Finally, none of these sounds was heard from July 28 to August 12. No sounds were heard from toadfish known to be around the dock when checked for 45 minutes on November 7, 1958, 70 minutes on March 27, 1959, and over 15 minutes on April 25, 1959. On May 8, 1959, a few calls were heard that did not seem to represent the regular boat-whistle or the usual grunt but appeared to be some precursor to the usual boat-whistle call. On May 22 many odd grunts or short boat-whistles and one regular boat-whistle were heard and a few cans contained freshly laid eggs. From May 24 through early June many of these sounds were produced. They then decreased in frequency until none was heard from July 30 to September 4, 1959. Although no boat-whistles were heard at these times in the experimental area, they were prevalent from July 28, in water 30 feet deep located straight out from the laboratory pier, until August 18, after which date they were no longer produced.

Boat-whistles were heard from 39 males on nests without eggs, from 8 with eggs, and from one with cling young, and in no instance from a male with free young. In 22 of the 39 instances, when a lone male in the can produced boat-whistles, the male stayed in the same can and received eggs shortly thereafter.

In 46 experiments (Table III) where ripe female toadfish were presented to males with and without eggs or young, the male in over 50% of the cases made no observable reaction, whereas in somewhat less than 40% a grunt response was given. However, a different response was obtained in 9 experiments not included in Table III, as follows: On July 15, 1958, in 4 experiments when a live ripe female on the end of a string was moved past a silent male in a jar without eggs, he responded with a series of boat-whistles. This was repeated on July 24 with the same results and eggs were laid in the jar between July 24 and July 28. In a similar experiment on June 24, 1959 the male, which was boat-whistling earlier in the day, responded with a rapid series of boat-whistles after which the female entered the can. The same experiment was repeated later in the day and the male again responded with boat-whistles. Similarly in an experiment on the same day with another male, which was boat-whistling, the response was the production of a more rapid series of the same sound. The female then entered the can and turned upside-down. This was repeated later in the day and the male again responded with an increased rate of sound production.

DISCUSSION

The previously published information on the reproductive season of *Opsanus tau* is sketchy and there is a lack of evidence that anyone has determined the first egg-laying date for any population or that anyone has systematically checked nests throughout an entire reproductive season. Furthermore, local differences in shallow water and deep water populations and yearly climatic differences might be expected, particularly if temperature is causative. Based on several years records, Gudger (1910) stated that at Woods Hole, Massachusetts, eggs and young were found in June, July and August. The earliest recorded dates were June 3 and June 12. Other data from Woods Hole which are in disagreement are given as follows: spawning occurred in the month of June only, and freshly laid eggs were rarely found in July; eggs or young were found in June, July and August, and oviposition occurred about the middle of July (various references in Gudger, 1910). Gill (1907) found eggs on July 14 and young on July 21 at Noank, Connecticut, whereas Fish (1954) concluded that eggs were not laid until at least after the middle of June in Wickford Harbor, Rhode Island. Nichols and Breder (1926) gave June and July as the spawning months in the vicinity of New York City. Hildebrand and Schroeder (1928) stated that in the Chesapeake Bay, spawning apparently took place throughout the summer, as females with large eggs were taken from April 13 to October 25, 1922 (note that this did not state ripe eggs). Gudger (1910) found advanced embryos the first week in June and eggs in segmentation late in August in the Beaufort, North Carolina area and maintained that the higher temperatures at Beaufort allowed the breeding season to be drawn out. Gill (1907) said that in the southern states, the approach to shore and the reproductive season commenced in April or May in the Gulf of Mexico (a different species as the group is now understood). Breder (1941) stated that reproduction took place when the water temperature reached its approximate maximum, whereas, in the southern form, reproduction took place when it reached its approximate minimum. He further stated that these fish spawned for the most part in water in the middle and high sixties (°F) extending, where the temperature rise was rapid, into the seventies or even the low eighties. Breder gave the spawning season for *O. beta* at Palmetto Key on the west coast of Florida as February and March.

Our data for one local shallow water population of toadfish demonstrated that eggs were laid in water temperatures from 63.5°F (17.5°C) or slightly below to 80°F (27°C) or higher (compare egg-laying dates with Figure 1), and that this period extended from between May 8 and 22 to between July 10 and 15, 1959. Males with cling and free young were found up to August 6, 1959. The fact that boat-whistles were heard in deeper water until August 18, 1959, is interpreted to mean that individual males still may have been ripe and may have been guarding eggs in early stages of development. It seems obvious that there are inadequate data for a comparison among Atlantic Coast populations, but it is reasonable to assume that spawning occurs earlier in the southern part of the range. Certainly, Breder (1941) demonstrated that *O. beta* spawned in February and March, but whether or not they also layed eggs earlier was not clearly documented. On the basis of sound production and observation of our population we presume that the boat-whistling of *O. tau* in late August near St. Augustine and the first hooting of *O. beta* heard on August 20, 1957 on the west coast of Florida (Tavolga 1958b) represented the onset of reproductive activity. There was some evidence that 1958 reproduction of the Solomons Island population was slightly more protracted because eggs were found at least 15 days later than in 1959 (Table I).

In essence, the results are in agreement with previous authors on the following points. The male guards the nest and stays with the young for a short time after they are free in the nest (Gudger 1910, Breder 1941). The nest is a can, jar, under a shell, board or other object, and the eggs are attached to the roof, usually in a single layer (*e.g.* Gudger 1910, Ryder 1887). The fish are polygamous (*op. cit.*), although in most of our test cans only one clutch of eggs was deposited. This, in general substantiates Breder's (1941) statements that when nest materials are abundant there will be less chance that several females will deposit eggs in one nest. It is possible that a male without eggs is a more effective stimulus for the attraction of a female than a male guarding eggs and this would tend to reduce the number of clutches deposited in any one nest. This is supported by the fact that after eggs are deposited in a nest boat-whistles produced by the male become less frequent and soon disappear. Furthermore the small size of the cans may have limited the number of clutches. It was noticed that a few nests under large boards contained several clutches of eggs.

Various estimates have been made on the time it takes the eggs to hatch and the time it takes the cling young to free themselves as follows: At Woods Hole, Massachusetts, the fixed condition of the egg and embryo lasted for at least 3 to 4 weeks, and the egg-membrane was ruptured in about half that time (Ryder 1887), the cling stage lasted for 3 to 4 weeks (Clapp 1899), the eggs hatched in about 26 days while the young became free on the 42nd day, whereas at Beaufort, North Carolina, the egg burst the shell (now cling stage) 11 days after fertilization and the young stayed attached for 13 to 15 days (Gudger 1910). Again, as with the data on the reproductive season, it is impossible to tell how accurate these estimates are and there are certain conflicts in the information. It is difficult to rationalize a 26-day hatching period with information that eggs were not laid until at least the first or 2nd week of June in the Woods Hole area and were seldom or never found after the first of July. In summary, in the current study, eggs hatched in 5 to 12 days, the young usually remained clinging 6 to 19 days, and the male stayed with the free young 5 to 18 days. Most of the data were obtained from the middle of June to the end of July, and it is reasonable to assume that the incubation period is longer for eggs laid in late May when temperatures are lower. Although males move from can to can before the reception of eggs, once these are deposited the males stay for 25 to 46 days, as was established with marked individuals.

Many hypotheses have been formed in regard to the functions of sounds produced by fishes, but only Tavolga (1956, 1958a) has experimentally demonstrated the function of sound produced by a fish. He demonstrated that in combination with visual stimuli the short grunt made by male gobies during reproduction attracted the females. Potential functions of the 2 types of sounds emitted by the toadfish were postulated by Marie P. Fish (1954). She said that the grunt was no doubt a reaction to annoyance, fear and aggravation and might express both warning and alarm. She also presumed that the boat-whistle was some sort of mating call used only by individuals with mature sex products.

A primary stimulus to the production of the grunt is the approach of a male or unripe female toadfish. Except for the situation where blue crabs were introduced to a male with eggs, the grunt was usually not produced when food items were presented. The difference in the response to active and inactive males appears to be based on the difference in movement. The grunt is made less frequently by the male when he does not have young in the nest. Also it appears that toadfish outside of the breeding season and females ready to spawn are less prone to grunt than ripe males when handled. Thus a male guarding eggs has a greater internal stimulation to produce this call. It is reasonable to suppose that the grunt of a male on his nest is part of a stimulus situation that causes an approaching toadfish to leave the immediate vicinity of the nest.

The reaction to an introduced crab is less easily analyzed. Here is the only case where grunts were produced commonly when the presentation tests did not involve a male toadfish. Possibly in some manner the crab on approach to a nest is a stimulus situation that causes the guarding male to grunt and then bite the crab. There is no evidence that a crab can hear water-borne sound vibration. Whether this is primarily a food response or also an aggressive response is not amenable to analysis at this time. A food response, as well as supplying food, certainly prevents the crab from entering the nest. Fish (1954) referred to the crab as a natural enemy of the toadfish, but this is difficult to understand because the blue crab is an extremely common food item (Gudger 1910 and unpublished data).

In order to understand the potential functions of the boat-whistle sound, it was necessary to find out whether or not both sexes made this call. Fish (1954) said that since the air bladder was alike in both sexes and there was no apparent difference in their grunting, it seemed reasonable to believe that both male and female had the ability to produce this sound. But in the few cases where she located a sound producer, it was always a male. The data from the current study strongly suggest that only the male produces the boat-whistle call. There is no evidence to indicate that the female produced boat-whistles in the area studied. This is the first instance in fishes where, although only one sex makes the call, the sound producing apparatus is not structurally different.

Boat whistles were produced quite commonly throughout the shallow water area studied by us in constrast to their apparent rarity in the shallow waters of Wickford Harbor studied by Fish (1954). There is some indication that she did not listen early enough. This evidence that the boat-whistle is produced shortly before spawning and for a short time after the male has eggs, verifies Fish's statement that this call is associated with prespawning and early spawning activities.

A problem remains regarding the presence of boat-whistles in deeper water. Fish (1954) stated that throughout the prespawning period, before eggs were found in the shallow inlets of Wickford Harbor and nearby waters, a medley of boat-whistles arose from deeper water. Neither Fish (1954) nor Tavolga (1958b) have done more than listen to these sounds in deeper water. Possibly the fish produce these sounds just before migration to shallow water to nest as Fish suggested. It has been thought by most investigators that the toadfish move into deeper water in the winter. Although this is true for many individuals, we and personnel of the Solomons Island laboratory saw some toadfish during the winter in less than 10 feet of water off the dock in the study area. There is, of course, considerable movement into shallow water just before the spawning season. The underwater recordings demonstrated that when the fish first entered shallow water in May, 1959, the boat-whistle sound was uncommon but then became more frequent. Possibly the earlier calls in deeper water represent fish in nesting cavities ready to spawn with females. There is considerable movement of males from nest site to nest site until eggs are deposited and many males move into shallower water setting up various sites from which they call. There could be distinct deep water and shallow water populations wherein males of the deep water population mature several weeks earlier. In most instances in the animal kingdom, the male's sexual products are mature before eggs are ready for fertilization. This call is probably coincident with this maturation of the male's sperm although Tavolga (1958b) seems to think the call is also given when it has nothing to do with reproduction.

When ripe females were used as a stimulus, the male in his nest did not produce sound in over 50% of the experiments. Approximately 40% of the time grunts were emitted but in about 10% of the cases the males responded with a boat-whistle response. This was either by the production of the call (not previously emitted) or by an increase in the rate of sound production. It is possible to discount all the experiments (33 out of 46) where a ripe female was introduced to a male with eggs or young because most of the tests were conducted well after the middle of the reproductive season when most boat-whistling had discontinued. All of the 13 experiments were done with single males near the end of the spawning season (between June 29 and July 8, 1959, when most males had stopped calling; eggs were deposited in only one nest after June 29). Furthermore, a roughly handled female attached to a string and presented to a male on his nest might act abnormally. Also, the females that were utilized in the presentation tests might not have had ovulated eggs and thus might not represent the stimulus that elicits boat-whistling. On the basis of all the evidence, it would seem that boat-whistle emission in 9 instances is much more significant than the small number of responses would indicate.

The following is proposed as a working hypothesis: The boat-whistle call is given off by males at a nest when they are ready to spawn and this call is an attractive stimulus to females that are ready to lay eggs. The evidence to support this is as follows: only ripe males make the boat-whistle; this is emitted only from cavities; a male either starts to call or increases its rate when a ripe female is placed in front of the nest. In order to fully substantiate the above hypothesis, it will be necessary to find out more about the boat-whistle in deeper water and to carry out the critical playback experiments.

Summary

The reproductive habits and sound production of a shallow water population of the toadfish, *Opsanus tau* (Linnaeus), were studied at Solomons Island, Maryland during June, July and August of 1958 and 1959.

Maximum reproductive activity occurred throughout the month of June and the first part of July. Crab-pot catches of the male and female toadfish were correlated with reproductive activities. The males guarded the nests from egg deposition until after the young were free swimming (23 to 46 days). Until eggs were deposited in a nest with the males, they apparently moved from object to object. In the population studied the eggs hatched in 5 to 12 days and the young remained in the cling stage 6 to 19 days.

Experiments with the grunt-like sound demonstrated that it was most frequently produced by a male in his nest when other males or spent female toadfish approached and especially so when eggs and young were present. Blue crabs occasionally would evoke the grunt response, but such objects as other species of fish, worms, and oyster shells were not adequate stimuli. It is suggested that ripe females and both sexes outside of the spawning season produced the grunt much less frequently.

Only the male produces the boat-whistle on the nest even though the structure of the sound producing apparatus is similar in both sexes. It is hypothesized that this call is a stimulus for the attraction of females to the nest. Data are presented which support this supposition, but it is emphasized that several critical experiments must still be performed.

Acknowledgments

We wish to gratefully acknowledge the assistance of the following persons: Dr. L. Eugene Cronin, Director of the Chesapeake Biological Laboratory, for providing laboratory facilities; Dr. Frank Schwartz, Chesapeake Biological Laboratory, for his enthusiastic assistance; and Jane Goodrich, William Goodrich, John Stout, Joseph Marshall, Thomas Savage, and Anthony Picciolo for aid in conducting the presentation experiments. Mrs. Carolyn Winn kindly criticized the manuscript. Mr. Harten of Solomons Island generously supplied us with the toadfish used in dissections.

This study was supported by the Chesapeake Bay Fund, the U. S. Public Health Service, Institute of Neurological Diseases and Blindness (B-1668), and the Office of Naval Research (N.R. 104-489).

References

Breder, C. M., Jr. 1941. On the reproduction of *Opsanus beta* Goode and Bean. Zoologica 26: 229-232.

Clapp, C. M. 1899. The lateral line system of *Batrachus tau*. J. Morph. 15: 223-264.

Fish, M. P. 1954. The character and significance of sound production among fishes of the western North Atlantic. Bull. Bingham Oceanogr. Coll. 14: 1-109.

———, A. S. Kelsey, Jr., and W. H. Mowbray. 1952. Studies on the production of underwater sound by North Atlantic coastal fishes. J. Mar. Res. 11: 180-193.

——— and W. H. Mowbray. 1959. The production of underwater sound by *Opsanus sp.*, a new toadfish from Bimini, Bahamas. Zoologica 44: 71-76.

Gill, T. N. 1907. Life histories of toadfishes (Batrachoidids), compared with those of weevers (Trachinids) and stargazers (Uranoscopids). Smithsonian Misc. Coll. 48: 388-427.

Gudger, E. W. 1910. Habits and life history of the toadfish (*Opsanus tau*). Bull. U. S. Bur. Fish. (1908) 28: 1071-1109.

Hildebrand, S. F. and W. C. Schroeder. 1928. Fishes of Chesapeake Bay. Bull. U. S. Bur. Fish. (1927) 43: 337-338.

Knudsen, V. O., R. S. Alford and J. W. Emling. 1948. Underwater ambient noise. J. Mar. Res. 7: 410-429.

Nichols, J. T. and C. M. Breder, Jr. 1926. The marine fishes of New York and southern New England. Zoologica 9: 1-192.

Ryder, J. A. 1887. Preliminary notice of the development of the toadfish, *Batrachus tau*. Bull. U. S. Fish. Comm. (1886) 6: 4-8.

Sorensen, W. 1884. Om Lydorganer hos Fiske. En physiologisk og comparative-anatomisk Undersgelse. Kjbenhavn: 245.

Tavolga, W. N. 1956. Visual, chemical, and sound stimuli as cues in the sex discriminatory behavior of the gobiid fish, *Bathygobius soporator*. Zoologica 41: 49-64.

———. 1958a. The significance of underwater sounds produced by males of the gobiid fish, *Bathygobius soporator*. Phys. Zool. 31: 259-271.

———. 1958b Underwater sounds produced by two species of toadfish, *Opsanus tau* and *Opsanus beta*. Bull. Mar. Sci. of the Gulf and Caribbean 8: 278-283.

Tower, R. W. 1908. The production of sound in the Drumfishes, the Sea-Robin and the Toadfish. Annals N. Y. Acad. Sci. 18: 149-180.

Behavior, Diel Activities, and Stimuli that Elicit Sound Production and Reactions to Sounds in the Longspine Squirrelfish

HOWARD E. WINN, JOSEPH A. MARSHALL, AND BRIAN HAZLETT

The normal nonreproductive social organization of *H. rufus* is territorial. Both immature and adult fish defend territories with visual and acoustic displays. Peaks of sound production occur at dusk and dawn. Staccatos are produced less frequently at night when the fish are most active in feeding than during the day when they hover over their rock crevices. Various fishes introduced to a population of squirrelfish demonstrated that constantly moving fish caused many staccatos to be produced. Under certain circumstances more staccatos were produced toward introduced alien species than toward squirrelfish. A type of "mobbing" behavior with the production of staccatos occurred. When 3 types of sounds were played out underwater, the following modifications in behavior were recorded: sometimes the fish "jumped," they retreated into their crevices, they turned their heads toward the sound, and either during sound or after, the fish investigated the source. This did not occur with lobster sounds. The characteristic daily cycles of sound production of various fishes are discussed. The significance of the playback experiments is discussed particularly in relation to orientation to the source of sound. It is considered that fishes do orient to the source and the possibility of binaural localization should be considered more seriously as a possible mechanism. The grunt sounds of variable time intervals seem related primarily to aggressive behavior in territorial defense. The staccato is first accompanied by escape behavior, then a shift occurs to investigative behavior, and finally it is involved with a special "mobbing" behavior which is probably aggressive. The sounds in relation to the fish's behavior are quite complex and need further study. A large number of parallelisms occur between the way squirrelfishes utilize their sounds in a complex community and the way birds have organized acoustical signal systems.

INTRODUCTION

FOUR aspects of acoustical communication in squirrelfishes are considered. Two types of sounds, the grunt and staccato, in relation to territorial behavior, and diel or 24-hr cycles of sound and locomotory activity were studied. In a preliminary attempt to define some of the external stimuli that cause sound production, various fishes were introduced into populations of the longspine squirrelfish, *Holocentrus rufus* (Walbaum), and the resultant sound production was recorded. Finally, certain sounds were played through underwater speakers to the squirrelfish populations and any changes in behavior noted. The fact that populations could be established in large laboratory tanks wherein the individuals exhibited at least qualitatively, the same behavior as under natural conditions, was important to this analysis.

Previous studies on squirrelfish behavior and their sonic activity are scanty. A first paper by us (Winn and Marshall 1963) described in detail certain properties of the sound-producing organ. Moulton (1958) mentioned that body muscles and the air bladder were involved in sound production. Also, he stated that *H. ascensionis* was one of the most frequent sound producers in the Bimini area and he briefly described a thump-like single sound (grunt) and a rapid volley of sounds (staccato).

MATERIALS AND METHODS

The sounds of the squirrelfish and its activity over 24-hr periods were recorded at specific times. Field recordings were made with an LF-310 hydrophone connected to a Magnemite, DV 610, portable tape recorder. At Bermuda in 1959, 1960, and 1961, the samples were obtained over a small reef near the center of Whalebone Bay which contained over 50 individuals of *H. rufus* (150–233 mm total length) and only 1 or 2 individuals of *H. ascensionis*. In Puerto Rico (January and February 1961) the samples were taken from the laboratory dock, Institute of Marine Biology at La Parguera. Times other than those listed were monitored, but they did not deviate from the described patterns. The locomotory activity cycles were measured with an infrared photoelectric relay system placed across a small wooden tank (43 × 23 × 18 cm deep) with 2, 10-cm Plexiglas portholes opposite each other. Each crossing of the sensitive area was recorded on an Esterline-Angus Operations Recorder. Mechanical paddle-wheel recorders also were used. Sound recordings were taken over groups of 9 individuals of *H. rufus* in the concrete tank (see below) at Bermuda. The equipment consisted of Atlantic Research Corp. BC-50 and Chesapeake Instrument Corp. LF-310 hydrophones connected to Tandberg and Ampex 601 tape recorders, respectively.

At Bermuda, a series of introduction experiments were performed where one individual of various species of fishes was placed into a 46 × 279 × 46 cm concrete trough (Fig. 1) which had 9 individuals of *H. rufus* (14–25 cm total length) established in it. The trough contained 9 fruit juice cans (17.5 × 10.5 cm) which served as hiding places for squirrelfish. The species introduced, with their total lengths in cm, were as follows: the longspine squirrelfish, *H. rufus* (24); a mullet, *Mugil* sp. (27.5); the palometa, *Trachinotus glaucus* (20); the bluestriped grunt, *Haemulon sciurus* (22); and the spotted moray, *Gymnothorax moringa* (55). They were released into one end of the tank and the general behavior and the number of grunts and staccatos produced by the resident squirrelfish were recorded for 3-min periods.

At Puerto Rico, 6 individuals of *H. rufus* (7–11 cm total length) were established in a Durotex trough, 240 × 44 × 20 cm deep, containing 6 cans, 6 × 10 cm in length. Sounds were recorded under 3 sets of conditions: from this group by itself; when 6 new squirrelfish were introduced; and when 6 fish of other species were released into the tank. The species, with some variation in their combination, were: *Sparisoma* sp., *Holacanthus tricolor*, *Holacanthus ciliaris*,

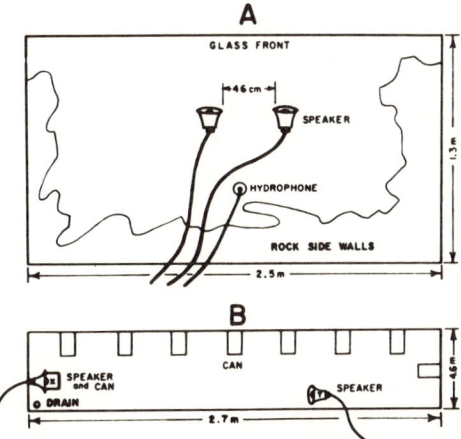

Fig. 1. Top view of the physical configuration of the tanks used in the playback experiments. (A) tank at Puerto Rico and (B) tank at Bermuda.

Haemulon flavolineatum, *Haemulon sciurus*, *Pomacentrus leucostictus*, *Balistes vetula*, and *Myripristis jacobus*. All were 7 to 15 cm in total length. In several other instances, 3 squirrelfish and 3 individuals of *Balistes vetula*, all of about the same size, were introduced into a group of 3 squirrelfish (*H. ascensionis*, in this instance) established in the small display tank (1.38 × 1.21 × 0.86 m deep). Crevices in the coral rock along the sides and on the bottom were available to the fish as hiding places.

Sounds were played to the same group utilized earlier in the introduction experiments at Bermuda. The sounds were first recorded on continuous message cartridges. University submergence-proof speakers (MM 2L) were connected to a Mohawk Message Repeater, Senior Model. The following sounds, equal in amplitude to that emitted by the fish, were played out of the speakers at X and Y (Fig. 1): staccato call of *H. rufus*; boatwhistle call of the toadfish (*Opsanus tau*); sounds of hand-held bigeye scad (*Selar crumenophthalmus*); and spiny lobster (*Panulirus argus*) sounds.

Some playbacks of squirrelfish staccato sounds were made at the Institute of Marine Biology, Puerto Rico. The tank was of the public aquarium type (1.38 × 2.5 × 0.86 m deep) containing 7 adult-sized *H. rufus* (Fig. 1) in nonbreeding condition. The same equipment and sound amplitudes were used as in the Bermuda playbacks.

The speaker was set on the bottom in the center of the tank. In many of the experiments, a silent second speaker was placed 46 cm from the speaker in use (Fig. 1). Sounds were continuously monitored. All experiments were performed after normal darkness but with illumination supplied by low-intensity laboratory lights. The observer–recorder sat opposite the large viewing glass in relative darkness. A second person was in back of the tanks, out of sight of the fish and observer. He signaled electrically with a dim red light near the observer–recorder the start of the following periods: the first and second minutes, a period of 3 min, the sixth and seventh minutes. The sound was played out randomly either the second or sixth minute without the knowledge of the observer–recorder. This allowed for a control not present in the Bermuda playbacks. It was soon obvious that, based on the behavior of the fish, the observer could guess with 90% accuracy, when the staccato sound was played out. These experiments were repeated in a small public aquarium tank (1.38 × 1.21 × 0.86 m deep) and in a Durotex trough (240 × 44 × 20 cm deep).

GENERAL BEHAVIOR AND SOUND PRODUCTION

This generalized description of the behavior of immature squirrelfish is based on observations of natural populations (over 5 hr) and laboratory populations (over 20 hr). No major qualitative differences between laboratory and field populations were discerned. In a later paper, a quantitative analysis will be presented particularly to show how the elements of the territorial displays are related to each other.

Holocentrus rufus is an abundant fish on the reefs at Bermuda. During the day they can be seen hovering over any kind of crevice. At night with underwater lights, they can be seen swimming around the reef areas but not far from their daytime haunts. Immature individuals of both *H. rufus* and *H. ascensionis* are common in the shallow waters around Bermuda. In Whalebone Bay over 90% of the individuals in the size class studied were *H. rufus*, whereas in the reefs just west of Longbird Bridge, *H. ascensionis* was more abundant. Breeding adults of both species were found during June, July, and August in water deeper than 4 to 8 fathoms.

Three basic types of sounds are produced by immature individuals from 7 cm in total length, up to mature adults: (1) a single grunt or groups of grunts with long and variable time intervals between; (2) a staccato call which consists of a variable number of grunts repeated very rapidly but with a uniform time interval; and (3) a series of grunts produced rapidly or slowly in series with variable time intervals, but the grunt element characteristically longer than in the first 2 types. This latter sound was obtained only when the fish was hand-held. Presumably, it is produced when a fish is injured or attacked by a predator. The physical analysis and time measurements of these calls will be presented elsewhere. For our present purposes we are interested only in the single grunts and staccato calls.

Each individual maintains a territory and, where crevices are common, many territories are maintained adjacently. If one squirrelfish enters the territory of another, one of several things may occur. The resident squirrelfish erects its fins and dashes at the intruding fish and frequently makes a grunt call. Upon meeting, they both may shudder (the body vibrates but it has a greater amplitude than quivering) or only the resident shudders. Head-shaking may also occur. Rarely the resident squirrelfish may make a staccato call when abruptly confronted by another squirrelfish. If the intruder and resident are near the border of their territories, they may shudder, erect fins, nip, and present themselves parallel to each other. Caudal nipping seems to be more frequent than nipping at other areas. Frequently from the parallel position they move forward in a tail display and form a narrow V with their caudal fins touching or almost so. At the end of this forward movement of 30 to 90 cm, they usually separate and return to their territories. A good picture of this position by a Pacific species of squirrelfish was published by Herald and Dempster (1957) except that for *H. rufus* the tails do not cross. Large numbers of bluestriped grunts frequently are found near the squirrelfishes.

Reactions also were given toward intruders of other species. When a bluestriped grunt moves into or within a territory, a squirrelfish may shudder, grunt, or "staccato." A squirrelfish not uncommonly attempts a tail display with a bluestriped grunt, and if the grunt moves forward the squirrelfish follows in tail display with no equivalent response by the other fish. When large fishes of different species approach or enter a territory, especially if they suddenly appear, staccatos, accompanied by fin erection, are produced. This latter behavior frequently is exhibited towards a human observer. If the large fish enters the territory, the resident squirrelfish may either retreat or come up to the animal and staccato at it in a sort of mobbing behavior. Several squirrelfish may do this toward a moray eel. Chasing is by far the most common activity. Nips, shudders, head shakes, chases, lateral display, and fin erection all may occur as single elements with or without grunts. Sometimes, the last 3 elements are accompanied by staccatos, although these sounds are produced most frequently in combination with fin erection when the squirrelfish is confronted suddenly by any strange object. All 6 behaviors were directed primarily toward individuals of their own species.

Behavioral Patterns

Diel cycles. — In general, the squirrelfish produces few staccato sounds at night, more during the day, and a large number at dawn and dusk. The grunt sounds follow a somewhat similar pattern. However, in field recordings the grunt sound of the squirrelfish is much more difficult to distinguish from other fish sounds than is the staccato. These cycles were somewhat similar but more variable in laboratory populations. Locomotory activity was low during the day and then built up at dusk to a high level that was sustained until dawn (Table 1 and Fig. 2), and 1 Esterline-Angus recording out of the 17 made on 3 fish is shown in Fig. 2. There was no suggestion of crepuscular peaks in locomotory activity. Fishing, underwater observations at night on laboratory populations, and analyses of stomach contents all suggest a high nocturnal feeding activity.

The number of staccato calls for 10-min periods in field recordings was analyzed for the dawn and dusk periods. The crepuscular peaks were very sharp for 10 to 20 min at 0420–39, 0430–49, 1730, and 1740 hr, at Bermuda for 11 and 12 July 1960. These were somewhat longer and occurred at different times in Puerto Rico. Official sunrise and sunset at Bermuda for 11 July 1960 were 0520 and 2029 hr.

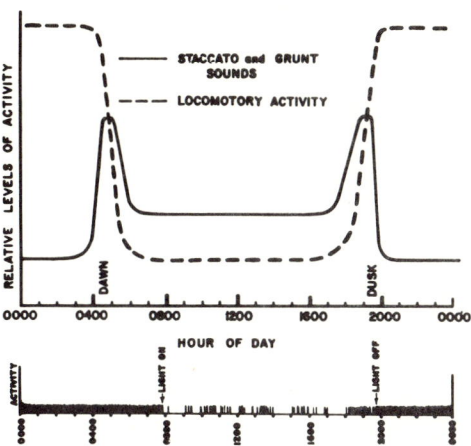

Fig. 2. The upper graph is a summary of the 24-hr activity cycle of locomotion and staccato sounds. The lower graph is a typical recording of activity where the vertical strokes indicate movements and the light was controlled by a clock.

Introduction experiments. — Grunt sounds were produced *on the average* of 1 per introduction when squirrelfish, mullets, grunts, and morays were introduced (none were produced in 12, 9, 7, and 10 experiments, respectively, out of 20 each). The results of 1 series of introduction experiments are given in Table 2. However, the squirrelfish emitted 4.5 grunts per test toward palometa (only 1 test with no grunts). The moray and palometa produced a higher stimulus for staccato production than the squirrelfish and mullet, and the bluestriped grunt was intermediate. Staccato sounds typically were much longer in duration (made up of more grunts) in the moray introductions and very short toward palometa. Most frequently, the fish whose territory was invaded made the staccato sounds although there were exceptions, as when several individuals displayed themselves in front of a moray's head. When a new individual of any suitable species was placed in a tank, the squirrelfish produced sound either as it swam towards the newly introduced fish or when the new fish passed its crevice. The range of squirrelfish behavior in this situation included shuddering, placing of mouth against introduced fish, nipping, occasionally tail displaying, fin erect-

TABLE 1. DIEL ACTIVITY OF *H. rufus*; THE NUMBER OF GRUNT AND STACCATO SOUNDS PRODUCED IN THE FIELD (F) AND IN THE LABORATORY (L) FOR 40-MIN PERIODS AT BERMUDA (B) AND AT PUERTO RICO (PR); THE NUMBER OF CROSSINGS OF AN INFRARED PHOTOCELL. All lights were turned out at 1900 hr.

Activity Recorded	Where Recorded	Date	Number of Actions Recorded for 40-min Periods; Beginning of Time Periods													
			0410	0500	0600	0700	0900	1300	1600	1720	1810	1900	2000	2200	0200
Staccato call	F, B	11 August 1959[1]		106			29	23	29		54[2]	59		11	16
	F, B	11 July 1960	37	40	22		10	14	29		32	75	14	13	2
	F, B	12 July 1960	47	14	14		21	12	10			36	13	20	2
	F, PR	5 February 1961	8	4	31		88	42			181	27		5	
	F, PR	14 February 1961			220	104[2]		48		58	54	29		11	
	F, PR	16 February 1961		25	120	33									
	L, B	10 August 1959		8			2	3	5			7	3	1	3
	L, B	13 August 1959		19			3	0	9			13		6	4
	L, B	26 July 1960	1	4	14		2	2	3			0	1	1	1
	L, B	27 July 1960	4	1	5			2	2		1	0	2	7	5
	L, B	28 July 1960	5	7	8		3				3	8	3	4	4
Grunt call	F, B	11 July 1960	91	39	26		20	15	78		86[2]	82	41	25	20
	F, B	12 July 1960	106	31	16		44	21	52		41	158	41	32	18
	L, B	26 July 1960	5	7			7	3	3			11	10	8	8
	L, B	27 July 1960	10	4	9		7	6	3		0	1	3	7	11
	L, B	28 July 1960	9	15	11						8	25	15	9	3
Locomotory activity	L—one squirrelfish at a time	July 1960	116	101	12		7	18	19		22	37	110	112	132
		August 1960									6	68	78		
		August 1960									20	101	299		
		August 1960									76	125	385		
		August 1960									165	311	435		

[1] 30-min periods for this date only.
[2] Actually recorded for only 20 min but the numbers were doubled.

TABLE 2. THE NUMBER OF GRUNT AND STACCATO SOUNDS PRODUCED BY TWO GROUPS OF SQUIRRELFISH (*H. rufus*) WHEN ONE INDIVIDUAL OF VARIOUS SPECIES OF FISHES IS INTRODUCED INTO THE GROUP FOR 20 PERIODS (10 PER POPULATION) OF 3 MIN EACH, BERMUDA, 1960 (See Materials and Methods for the physical arrangement).

Fish Introduced	Average (Range) of Number of Sounds	
	Grunts	Staccatos
Squirrelfish	0.6 (0–4)	2.7 (0–14)
Mullet	1.0 (0–4)	2.2 (0–9)
Grunt	0.9 (0–3)	4.0 (0–13)
Palometa	4.5 (0–12)	7.7 (0–23)
Moray	0.9 (0–4)	8.5 (0–22)

ing, and dashing at the fish. None of these reactions were particularly correlated with sound production except the grunt and chasing. Quick, brief dashes were the most frequent response toward the palometa. The above displays were particularly conspicuous toward a moray in a type of mobbing behavior. Up to 6 squirrelfish approached and displayed within 5 to 7 cm of the moray's head and then followed it around the tank.

During the experiments, only 3 grunts and no staccatos were produced when 2 resident squirrelfish were aggressive towards each other. We have the general impression (not documented) that more grunts were produced in this way after the introductions than either before, or during the tests.

At Puerto Rico, recordings were made of resident squirrelfish populations, in various tanks, and then from the same population when a new group of squirrelfish or a mixed group of other species was added (Table 3). Once a group of squirrelfish had become adapted to its new habitat (1 to 3 days), sounds were produced infrequently. Although more data are needed, it appeared that newly introduced squirrelfish caused the residents to produce some staccatos but more grunt calls. Other species placed in the tank caused the production of some grunts and many staccatos. More staccatos were produced toward strange species than toward strange squirrelfish, but the production of grunts was more variable in this respect. This was further verified with a few introductions to *H. ascensionis* in which case staccatos were directed in large numbers against other species, and grunts were directed more commonly against other squirrelfish (Table 4). At this time we know of no differences in the relation of sound production to behavior between *H. rufus* and *H. ascensionis*.

Sound playbacks.—When sounds were played out underwater to the squirrelfish at Bermuda, various behavioral changes occurred. The staccato, toadfish boatwhistle, and scad sounds caused a significant number of fish to enter cans and stay there for the first minute during sound emission (Table 5), so that there was less free swimming activity at this time. A significant difference between behavior during test and

TABLE 3. THE NUMBER OF GRUNT AND STACCATO SOUNDS PRODUCED BY SQUIRRELFISH (*H. rufus*): (1) OF A RESIDENT POPULATION ONLY; (2) OF THE RESIDENTS PLUS AN EQUAL NUMBER OF NEWLY ADDED SQUIRRELFISHES; AND, (3) OF THE RESIDENT SQUIRRELFISHES PLUS AN EQUAL NUMBER OF NEWLY ADDED DIFFERENT SPECIES OF FISHES. The first 4 introductions were in trough and the last 2 were in a small display tank. The number of individuals refers to the number in the resident population and the number of new squirrelfish or other species introduced to the resident population, not the total of both (See Materials and Methods). All periods were 20 min duration, Puerto Rico, 1961.

Introductions	Number of Individuals	Number of Sounds Produced—Average and Range					
		Residents Only		Plus Squirrelfish		Plus Other Species	
		Grunts	Staccatos	Grunts	Staccatos	Grunts	Staccatos
6	5	0	0	0	1.0 (0–4)	1.2 (0–3)	2.5 (0–7)
4	6	0.8 (0–2)	0.3 (0–1)				
6	6			9.7 (0–20)	2.7 (0–10)		
5	6					15.0 (0–27)	56.2 (1–127)
1	3	1	0				
2	3			4.0 (2–6)	2.0 (0–4)	2 (1–3)	80.5 (41–120)

TABLE 4. SOUNDS PRODUCED UPON INTRODUCTION OF 3 INDIVIDUALS OF *H. ascensionis* AND 3 INDIVIDUALS OF OTHER SPECIES TO A POPULATION OF 3 *H. ascensionis* IN A SMALL DISPLAY TANK ON THE FIFTH AND SIXTH DAYS AFTER THE POPULATION HAD BECOME ESTABLISHED (20-MIN PERIODS).

Day	Fish Introduced	Grunts	Staccatos
5	(Residents only)	1	0
	3 *H. ascensionis*	2	0
	3 Other fishes[1]	1	41
6	3 *H. ascensionis*	18	12
	3 *Balistes vetula*	9	360

[1] 1 *Sparisoma viride*, 1 *Epinephalus guttatus*, and 1 *Balistes vetula*.

control periods was not obtained for lobster sounds or control background noise. After the first minute in cans, some fish emerged slowly. Those fish whose territories were near to (13 out of 18 experiments, the information was not obtained in tests beyond 18) or included the speaker (15 out of 18 experiments), after a short period of less activity with the sound on, went over to the speaker with their head facing toward it and then moved quickly away. From 1 to 5 fish, regardless of whether or not they were close to the speaker, did this in 19 of 30 experiments (2 for lobster sound, 6 for staccato, 6 for boatwhistle, 5 for scad sound). In 12 of 26 experiments, fish went to the sound source within 1 min after the sound was turned off. Other types of reactions were noted for fish up to about 100 cm from the source. In almost every experiment fish in cans did the following things when the sound was turned on: the fish's head was turned towards the speaker or, if its head was already directed toward it, the head was turned away and back again, and a fish near the speakers backed several centimeters into its can and then slowly stuck its head out several centimeters while the sound was still on. Fish near the speakers frequently "jumped" slightly when the sound was turned on. Sometimes, fish whose territory included the speaker stayed inside the can or behind the speaker, and other times they swam around the general area with no apparent change in behavior. More rarely, fish that were inactive became active with the onset of sound. At times, some individuals exhibited no overt change in behavior. One fish that regularly traveled from the end of the tank where the sound-emitting speaker was located to the other end of the tank, stayed at that end until the sound was turned off, and then returned to the can at the end of the tank near the speaker.

Eleven playback experiments of staccato calls were performed in Puerto Rico (8 in large display tank, 2 in small display tank, and 1 in trough). In 7 of 10 experiments (no notation for 1 experiment) the fish moved actively until the sound was turned on, then with the introduction of sound the fish stayed in or returned to the wall crevices. They appeared less active for most of the minute with sound. This was especially true for those fish whose territory was near but did not include the speaker. The 1 fish in the large tank, whose territory included the speaker, stayed in his territory, but usually jumped when the sound came on. In 6 of the 7 experiments where less activity was noted, all the fish became active again within a minute after the sound was stopped. In 8 of the 11 experiments with the sound on, one or more fish quickly went over and faced the opening of the speaker and then returned to its crevice. This behavior was exhibited only 3 times before sounds were played, although fish frequently passed by or stopped near the speaker.

Fish "looked into" the speaker 5 times just after the sound was turned off. In 4 experiments the fish were quite inactive before the sound was emitted, but in 3 of these, several fish went to the speaker that was emitting sound and some of the fish were more active. After the sound was turned off they became inactive again. Three times, fish near the speaker "jumped" when the sound was turned on. Several times, some fish backed into crevices and then moved their heads out, but this behavior was not as fully documented as for the Bermuda playbacks.

DISCUSSION

Patterns of sound production throughout the 24-hr cycle are common among fishes as among birds. The squirrelfish produces most sounds during the crepuscular periods and fewer at night than in daylight. The period of low production corresponds to the time when the fish actively move about the reef in search of food. Unpublished evidence suggests that at night territorial

TABLE 5. VARIOUS SOUNDS WERE PLAYED OUT TO A GROUP OF 9 SQUIRRELFISH. The number of squirrelfish outside of their cans actively dashing from place to place was recorded for consecutive times as follows: The last minute of a 3-min period without sound, the first minute of a 3-min period with sound played out a speaker at 1 of 2 positions, X and Y (see Fig. 1), and the first minute of a 3-min period without sound. The sounds used were a squirrelfish staccato, a toadfish boatwhistle, a bigeye scad tooth-scraping sound, and a spiny lobster stridulatory sound. Similar observations were noted below on 3 controls when only background noise was played into the tank.

Playback Position	Number of Fish Out of Cans for Designated Periods											
	Staccato			Boatwhistle			Scad Sound			Lobster Sound		
	No	Yes	No	No	Yes	No	No	Yes	No	No	Yes	No
x	4–6	5	5–9	6	2	4+	7	4	6	7	5	7
x	7	4	6	6	5	5	7	5	7	7	6	6
x	6	1	5	7	5	7	8	5	8	9	9	8
y	7	2	5	8	5	6	8	3	7	7	7	7
y	6	1	6	7	3	8	8	3	9	6	6	6
y	6	1	5	6	4	7	–	–	–	8	2	9
x	–	–	–	5	2–3	4	–	–	–	5	5	5
y	5	4	5	7	3	7	–	–	–	6	6	8
Chi-square significance	0.05	0.05		0.05	0.05		0.05	0.05		–	–	
Average difference	3.4	3.0		2.8	2.3		3.6	3.4		1.1	1.1	

Control (no sound, ambient sound, no sounds) : 7, 7, 7; 8, 7–9, 7–9, and 5, 5, –.

boundaries break down to some extent but that the fish do not stray extensively from their normal crevice and that one fish remains at the same crevice for periods of at least several weeks. We did not obtain any evidence of extensive nightly migrations or dispersal. Other documented diel cycles of sound production by fishes are: few knock sounds in the morning with increasing frequency up to dusk and then practically stopping at dark, correlated with increased chasing activity (*Corvina nigra*, Dijkgraaf 1947); chorus began at sunset, increased for 2 to 3 hr and tapered off to rare outbursts (one or more Sciaenidae, Johnson 1948); drumming started at noon and increased until midafternoon, then decreased, until at sundown only a few fish could be heard, probably associated with spawning activity (*Aplodinotus grunniens*, Schneider and Hasler 1960); increase in sounds before and after dusk in one case correlated with spawning (*Gadus callarias*, Brawn 1961); peak at 2030 with curve symmetrical, possibly coincided with active feeding (possibly *Micropogon undulatus*, Dobrin 1947); a slight peak of sound production at dusk and a major peak in the early morning after sunrise (*Cynoscion regalis*, Fish 1954); and nocturnal sound production (*Galeichthys felis*, Tavolga 1960). It is clear that many fish species exhibit daily cycles of sound production. At least for marine fishes, crepuscular cycles are common but each one is slightly different, in its relation to dawn and dusk. Some have unequal peaks with either a high morning or high evening peak, although the squirrelfish usually seems to have equally high dawn and dusk choruses. In some instances there is preliminary evidence which relates the cycles of sonic activity to either spawning or feeding activity. Tavolga (1960) suggested that the nocturnal sounds of the sea catfish are for maintenance of aggregations. In 2 instances, ecological factors other than the light cycles have been related to sound production. Peaks of sound activity by croakers were correlated with tidal changes and possibly with feeding activity (Fish 1954), and the freshwater drum stopped sonic and probably spawning activity on cloudy days with decreased temperature (Schneider and Hasler 1960). In general, comparisons of the daily cycles of activity such as feeding, resting, spawning, etc. have not been related to each other in detail. The fact that sound was lowest when feeding activity of the squirrelfish was highest demonstrates a relationship between various activities. The patterns of sound activity are clearly related to territorial behavior and occur only at certain times of day. In this way fishes exhibit patterns similar to those of mammals, birds, and insects, wherein it is known that sounds have function. In fact, the more known about fish sounds, the more they show most of the features associated with sonic activity of, for instance, birds.

In the series of introduction experiments, fish that moved constantly in front of the squirrelfish territories (moray and palometa) elicited the most staccatos. The mullet, that moved constantly but released few staccatos, swam at the surface. Introduced squirrelfish and mullets elicited the fewest staccatos. The high number of grunts produced toward the palometa is interpreted as an unusual aspect of sound production in that the fish went by the squirrelfish so fast at times that only a grunt, one element of a staccato, was produced. This, however, is not the usual situation in which grunts are emitted. When the palometa swam more slowly, the most frequent result was a staccato call. Typically, the staccato was much longer, in that it contained more vertical pulses, when produced toward a moray than toward a palometa. This suggests a differentiation of this sound, dependent on the type of stimulus.

Resident squirrelfish only rarely produce staccatos at each other, as shown in the population introduction experiments (Table 2), except possibly at dawn and dusk. Grunts are produced normally between adjacent territorial squirrelfish but not at the high rate as when new squirrelfish or other species are placed in the tank. Species other than squirrelfish seem to elicit many more staccatos than squirrelfish.

When sound was played out of an underwater speaker, there were several obvious changes in behavior. The fishes "jumped," slowly backed away, turned their heads towards the speaker, became inactive, and then at times went to the sound source. If, however, the group was inactive when the sound began, they seemed to become more active, and then after the sound stopped they frequently went to the sound source. Finally, discrimination was quite poor in that they reacted to three sounds (sounds of the squirrelfish, scad, and toadfish) but not to the lobster call. The behavioral reactions to the 3 sounds were the same, but the important difference was that when sudden visual objects appeared, the squirrelfish responded with a call that resulted in and was accompanied by escape, approach, and sonic reactions. The warning signal is under control of the fish whereas external sources of sounds which can produce the same effect are not under their control.

It has been stated that fish are unable to localize the source of the sound (Dijkgraaf 1960). However, most of the evidence considered by Dijkgraaf is concerned with *Phoxinus laevis*, an ostariophysan with the single chain of bones (the Weberian apparatus) connecting the air bladder to the inner ear. Kleerekoper and Chagnon (1954) gave evidence that other minnows could go to a source of sound possibly by following complicated lines of increasing or high amplitude. One factor in these experiments, and in almost all works on fish sounds in small tanks, is the complexities of reverberation and standing waves. Data, possibly in support of Kleerekoper and Chagnon, are given by Delco (1960) where he stated that one species of minnow eventually spent more time at one end of a tank with a speaker emitting its own sound as opposed to a speaker emitting the sound of another species at the opposite end of the tank. We have presented only suggestive evidence that squirrelfishes can localize the source of the sound. Head movements away from and toward the source may indicate that the animal is picking up a differential signal on some receptor and finally equalizing it by directing its head towards the source of sound. Some squirrelfish also went to the source of sound after it was turned off, and this included only the speaker that had emitted sound when a dummy speaker was also present. Admittedly, we desire to verify these results with more experiments. Unpublished data (Winn and Marshall) suggest that toadfish, *Opsanus tau*, under certain circumstances go straight and rapidly towards the source of sound 7 meters away. Even though the reactions may be visual in part, the object that emits sound is chosen over the same object not emitting sound. Sharks can react by turning toward the sound and swimming directly to it (Kritzler and Wood 1961) even when the speaker's position was randomly changed. It has been said that binaural localization is impossible underwater on theoretical grounds. The 3 most important objections are that *most* of the sound passes through the fish's body due to similar densities of fish and water, the ears are very close together, and that mostly low frequencies with little shadowing are involved. Kuroki (1957) with a fish and Dijkgraaf (1958) with the underwater frog, *Xenopus*, have shown the lateral-line system to be capable of giving directional information to underwater *vibrations*. Porpoises are able to carry out very refined sound localization but have ears more separated in space, have them imbedded free from the cranium proper in a possible sound absorbing material, and utilize high-frequency sounds. The argument may be looked at from the viewpoint of function. It is difficult to see how the endogenously driven boatwhistle call of the males of *Opsanus tau* can serve a useful function unless it can be localized. This call of the toadfish (and the staccato sound of squirrelfish) satisfies the criteria necessary to locate a sound by land vertebrates (Marler 1959). Briefly, this is a short, repetitive sound made up of low and high pitches. A large part of the stimulus value of sounds, namely its localization utilized by higher vertebrates, would be absent in fishes, yet sound production by fishes is frequent, especially during reproduction. We suggest that fishes can localize the source of a sound and that it need not involve a path of increasing amplitude. This means, then, that free nerve tactile or vibration receptors, the lateral line, and the ears are the most likely receptors involved. Evidence has been presented by others that there could be *bilateral line localization*, and, similarly, the touch receptors are capable of this in close proximity to the sound. This may be true also for the hair cells in the labyrinth which are similar to those in the lateral line. *Holocentrus rufus* and some other squirrelfishes have a double-lobed connection to the inner ears and might in some unknown way give differential information to the 2 ears which is less likely to occur in the single-chain system of ostariophysans. Possibly the complex of tissues sufficiently dampen the sound so that direct differential information is received or it is possible that phase differences might give information for localization. Dijkgraaf (1960) suggested that if the otoliths responded to the vibratory motion of the sound wave, then this would potentially allow for sound location. Both conditioning and neurophysiological experiments are in progress to further elucidate this problem.

Some fishes, such as serranids and sciaenids (e.g., Fish 1954, Moulton 1958, Tavolga 1960), produce sounds when startled. Others, like *Corvina nigra* (Dijkgraaf 1947), *Pomacentrus leucostictus* (Moulton 1958), *Gadus callarias* (Brawn 1961), *Notropis analostanus* (Winn and Stout 1960), etc., produce sounds in territorial defense. Primarily, the sounds are involved in reproductive territorial defense but for some this is not clear. Several species appear to utilize defense sounds outside the breeding season (Brawn 1961). The startle and aggressive sounds of some species are similar. The longspine squirrelfish has elaborated a more complicated system of startle and defense sounds than the above fishes as far as present knowledge extends. At least a double signal system has evolved, the grunt, for aggressive behavior, and the staccato, primarily for attentive and escape behavior. A third sound is produced when the fish is held in the hand, and is analogous to one of the "fear" calls of birds. The staccato also appears to be involved in a special aggressive or mobbing reaction. All are used in the setting of a nonreproductive territorial society by mature and immature individuals of both sexes.

The squirrelfishes, both *H. ascensionis* and *H. rufus*, are very important sound producers in the shallow reefs around Bermuda, which is in agreement with Moulton's (1958) findings at Bimini. Although the species are mixed in many shallow areas (immature forms), there are large areas where one predominates. Presumably, they also contribute considerably to the sound level in approximately 4 to 10 fathoms where reproductively mature individuals are caught in traps. The 2 types of sounds considered here, the grunt and staccato, are produced while the fish, regardless of sex or maturity, are guarding their territories centered around crevices. Nothing is known at this time about their reproductive habits and its relation to sound production.

Moulton (1958) was the first to recognize that *H. ascensionis* (presumably) produced single thump-like sounds at irregular intervals (the grunt) and rapid volleys (the staccato). We have not been able to recognize any real difference in the types of sounds produced by the 2 species (*H. rufus* and *H. ascensionis*). Moulton's comments can in most respects apply to both species. He further noted that startled fish produced staccatos associated with fin erection and movement into its territorial crevice.

Preliminary hypotheses concerning the significance of the grunt and staccato sounds are as follows. The grunt is produced toward more familiar objects (fish) that enter a squirrelfish's territory. This is particularly true of the relationship among squirrelfish with established adjacent territories, and usually is associated with aggressive reactions, such as chasing an adjacent territorial fish. The staccato, on the other hand, is the result of the sudden presentation of any strange object (many unusual visual changes). In the normal environment this usually means the sudden appearance of other large fishes such as those used in the introduction tests (e.g., moray, palometa, etc.). Small fishes and the commonly seen bluestriped grunts usually do not elicit the staccato call. Sudden movements by larger fishes habitually nearby can cause staccatos, but sometimes if these fishes move slowly into a squirrelfish's territory they will be displayed at or chased with the associated grunt frequently being produced. Even the sudden appearance of a squirrelfish can cause staccatos to be produced. Grunts seem to belong primarily to the aggressive system and do not habituate readily whereas staccatos are associated with escape tendencies and habituation readily occurs. That is, if the strange animal stays around for a time, staccato production is stopped. When a squirrelfish produces staccatos it seems to have a more attentive posture (presumably, threshold of reaction is lowered), and frequently retreats into a crevice. But under certain circumstances the fish then comes out to perform aggressive displays and staccatos against some of the animals as in the introduction experiments (e.g., moray). This "mobbing" reaction, in clear-cut form, was seen only in the laboratory. There, under special circumstances the staccato became part of aggressive behavior against potential predators (alarm system). The reaction of a squirrelfish to the staccato of an adjacent squirrelfish seems to bring out another potential element. The reactions to playbacks of staccatos were a slow escape reaction but soon afterward the fish came out to investigate the sound source with no aggressive or escape behavior displayed unless a sudden visual stimulus or intruding squirrelfish appeared. If the fish are inactive and deep in their holes, one sees only the investigative part. It would seem that the investigative (attentive) behavior is another important element.

Under normal circumstances, 1 of 4 things occurs. The fish sees or hears nothing new (that is, the stimulus that caused an adjacent squirrelfish to produce a staccato has left the area) and returns to its normal territorial activities, or else aggressive mobbing with staccatos, territorial defense with grunts, or escape behavior with staccatos takes place. The adaptive significance of staccatos upon a nearby squirrelfish where it involves escape behavior might be reduction in effective predation. The staccato of the producer might tend to cause intruding animals to escape or be startled, in turn protecting the squirrelfish and maintaining the normal territorial organization. However, this is only surmise. The grunt, on the other hand, appears to be part of a complex aggressive stimulus situation which is followed by escape reactions of intruders, especially squirrelfish in adjacent territories.

The dawn–dusk peak of sounds, both of grunts and staccatos, is poorly understood. The data for the grunts are less reliable. A preliminary hypothesis is that a relationship exists between decreased light, with decreased clarity of surrounding objects, and the squirrelfishes' increased activity, which causes them to encounter more frequently unfamiliar or moving objects (especially other squirrelfish). Also, at this time the diurnal species outside the reef are moving in for the night and the nocturnal species are becoming active, and they must pass through the squirrelfish territories. The crepuscular sound peaks resemble the large din of sounds from flocks of birds as they roost at dusk and reawaken at dawn. Snow (1958) described a similar situation for blackbirds where individuals move into the first part of the group attempting to maintain a perch and the evening chinking seems to serve as an assertion of roost-position rights that are habitually invaded by strange birds. To our knowledge these types of bird calls are not clearly understood. Perhaps the aggressive sounds aid in the positioning of individual birds in each place of the roosting area. This could well be true of the sounds produced in crepuscular peaks by the squirrelfish. As they change over from territorial defense to the presumed nonterritorial period of feeding activity, there is a brief period of increased territorial defense and is associated with a period of shifting internal drives. This would also mean there is a relationship of the staccato to the maintenance of a territory and thus is indirectly linked with aggressive behavior. As they return in the morning, the resettling into their territories could also involve much greater-than-normal contact with adjacent squirrelfish and other fish in the period of "shift-over" of internal drives coupled with exogenous factors (light and discreteness of objects).

The acoustical behavior of *H. rufus* seemingly resembles certain avian soundmakers in land communities. Upon the approach of certain mammalian or avian predators (or merely large animals before habituation) into the community, many species of birds produce alarm calls. Most other birds remain quiet. Certain aspects of attention, escape, and investigative behavior of the birds presumably are modified just as in the squirrelfish. So far we have not attempted to discover whether other fishes, which remain quiet under this situation, have their behavior modified by the squirrelfish's staccato call. The "mobbing" of the moray by the squirrelfish in the laboratory certainly resembles the mobbing of predators (e.g., owl, cat) by birds, with associated sound production.

It is significant that there are a large number of parallelisms between the development of certain alarm calls by birds and the alarm call (staccato) of squirrelfish (see summary for birds in Thorpe 1961). Briefly these are: (1) both an attractive and dispersive effect on members of the same species; (2) mobbing accompanied by sound; (3) attention and alarm components; (4) some differentiation of the alarm call; (5) similar calls at dusk and dawn of certain species; (6) at a minimum, a different distress (sound of squirrelfish when held in hand), alarm, and territorial aggressive call are developed; and (7) a potential differentiation in fishes as in birds of responses to a predator on the bottom and one swimming off the bottom, the short and long staccato (bird predator on ground or in flight).

Thus it seems that on the community level certain relationships of the utilization of information at the interspecific level (here, sounds for warning) exhibit parallel evolutionary development. Fish in the coral reef in many ways ecologically replace the birds of land communities. One might say that in a complex community like the coral reef or forest, the use of sound in the way the squirrelfish or certain birds utilize it is adaptive. Undoubtedly we need to know more about information transfer here at both the intraspecific and interspecific level. It is suggested that one of the efficient signal systems is the one discussed because of its development in such remotely related communities. It appears that sonic behavior and its relation to environmental factors and other behavior are quite complex in the squirrelfishes. Obviously much more information must be gathered before we will understand satisfactorily the functional relationship of sound production to behavior.

Acknowledgments

Dr. Anthony Picciolo, Mr. Michael Salmon, and Mr. Thomas Savage kindly helped make the diel recordings. This paper is Contribution No. 340 of the Bermuda Biological Station and is a contribution of the Institute of Marine Biology, Mayaguez, Puerto Rico. The directors of these two stations, Drs. William Sutcliffe and Juan Rivero, kindly facilitated our research in every possible way. This research was supported by O.N.R. Contract N.R. 104-489, U.S.P.H.S. Grant NB-03241, and the General Research Board of the University of Maryland. The manuscript was prepared while the first author was a Guggenheim Fellow. Dr. Robert Ficken kindly criticized the manuscript.

Literature Cited

Brawn, V. M. 1961. Sound production by the cod (*Gadus callarias* L.). *Behaviour* 18:239–255.

Delco, E. A., Jr. 1960. Sound discrimination by males of two cyprinid fishes. *Texas J. Sci.* 12:48–54.

Dijkgraaf, S. 1947. Ein Töne erzeugender Fisch im Neapler Aquarium. *Experientia* 3: 1–4.

———. 1958. Elektrophysiologische Untersuchungen an der Seitenlinie von *Xenopus laevis*. *Ibid.* 12:276–278.

———. 1960. Hearing in bony fishes. *Proc. Royal Soc., B*, 152:51–54.

Dobrin, M. B. 1947. Elements of underwater noise produced by marine life. *Science* 105: 19–23.
Fish, Marie P. 1954. The character and significance of sound production among fishes of the western North Atlantic. *Bull. Bingham Oceanogr. Coll.* 14:1–109.
Herald, E. S., and R. P. Dempster. 1957. Courting activity in the whitelined squirrelfish. *Aquarium J.* 28:43–44.
Johnson, M. W. 1948. Sound as a tool in marine ecology, from data on biological noises and the deep scattering layer. *J. Marine Res.* 7:443–458.
Kleerekoper, H., and E. C. Chagnon. 1954. Hearing in fish, with special reference to *Semotilus atromaculatus atromaculatus* (Mitchill). *J. Fish. Res. Bd. Can.* 11:130–152.
Kritzler, H., and L. Wood. 1961. Provisional audiogram for the shark, *Carcharhinus leucas*. *Science* 133:1480–1482.
Kuroki, T. 1957. Biophysical studies on the auditory characteristic of fish. I. Directional audibility through one lateral line. *Bull. Jap. Soc. Sci. Fish.* 23:400–404.
Marler, P. 1959. Developments in the study of animal communication. *In* P. R. Bell (Ed.), Darwin's biological work: some aspects reconsidered. Cambridge University Press, Cambridge. Pp. 150–206.
Moulton, J. M. 1958. The acoustical behavior of some fishes in the Bimini area. *Biol. Bull.* 114:357–374.
Schneider, H., and A. D. Hasler. 1960. Laute und Lauterzeugung beim Süsswassertrommler *Aplodinotus grunniens* Rafinesque (Sciaenidae, Pisces). *Zeitschr. vergl. physiol.* 43:499–517.
Snow, D. W. 1958. A study of blackbirds. George Allen and Unwin Ltd., London. 192 pp.
Tavolga, W. N. 1960. Sound production and underwater communication in fishes. *In* W. E. Lanyon and W. N. Tavolga (Eds.), Animal sounds and communication. Am. Inst. of Biol. Sciences, Washington, D. C. Pp. 93–136.
Thorpe, W. H. 1961. Bird-song. The biology of vocal communication and expression in birds. Cambridge University Press, Cambridge. 143 pp.
Winn, H. E., and J. A. Marshall. 1963. Sound producing organ of the squirrelfish *Holocentrus rufus*. *Physiol. Zoöl.* 36:34–44.
———, and J. F. Stout. 1960. Sound production by the satinfin shiner, *Notropis analostanus*, and related fishes. *Science* 132:222–223.
Woods, L. P. 1955. Western Atlantic species of the genus *Holocentrus*. *Fieldiana-Zool.* 37:91–119.

Department of Zoology, University of Maryland, College Park, Maryland.

19

Copyright © 1964 by Pergamon Press, Inc.

Reprinted from *Marine Bio-Acoustics*, W. N. Tavolga, ed., Pergamon Press, Inc., Oxford, 1964, pp. 213–230

THE BIOLOGICAL SIGNIFICANCE OF FISH SOUNDS*

H. E. WINN

Zoology Department,
University of Maryland

FOR SEVERAL centuries it has been known that fish produce sounds but it was not generally appreciated how prevalent and significant these were until M. P. Fish published two papers only 10 years ago (Fish *et al.*, 1952; Fish, 1954). She established and reviewed the following facts about the underwater sounds of fishes: (1) that the sounds can occur in seasonal and daily cycles; (2) that sound production can be restricted to one sex, usually the male; (3) that sounds can be produced in particular behavioral contexts; and, finally, (4) that the sounds possess a qualitative species specificity. From this baseline others have investigated how the sounds act as effective stimuli in communication. Recent studies have demonstrated that sounds stimulate courtship activities, possibly attract females to males, stimulate aggressive and escape behavior and serve as warning signals to produce escape and investigatory activities. Other functions and variations on the above themes undoubtedly will be demonstrated in the near future.

It is the purpose of this paper to examine the variety of behavioral conditions under which sounds are produced by fishes and to present hypotheses about the possible functions of the sounds. The nature of the acoustic signals of a few fishes will be discussed to illustrate how they are organized in amplitude, frequency and time so that they can transmit information both intra and interspecifically. Finally, experimental analyses of the functional significance of sounds produced by the frillfin goby (Tavolga, 1956, 1958a), the satinfin shiner (Stout, 1963a, b), the longspine squirrelfish (Winn *et al.*, 1963) and the oyster toadfish (Gray and Winn, 1961; Winn and Marshall, work in progress) will be reviewed. The latter two species produce sounds with muscle-air bladder organs but in the first two the sound producing mechanism is unknown.

---

* This project was supported in part by U.S. Public Health Research Grant (NB-03241) from the Institute of Neurological Diseases and Blindness, the Office of Naval Research (N.R. 104-489), and the General Research Board of the University of Maryland. Mr. Dean Holt helped in preparing the spectrograms. Dr. Robert Ficken kindly criticized the manuscript.

It should be stated at the outset that almost all information about sounds produced by any one species is far from complete and that the generalizations and hypotheses presented may be restricted, modified, or discarded in the future. Another limitation is that data are available only for fishes that can be conveniently studied under controlled conditions. A further assumption that may or may not be justified is that at least some of these animals are near the apex of complexity with respect to the development of their acoustical signalling systems (e.g. squirrelfishes, toadfishes and marine catfishes).

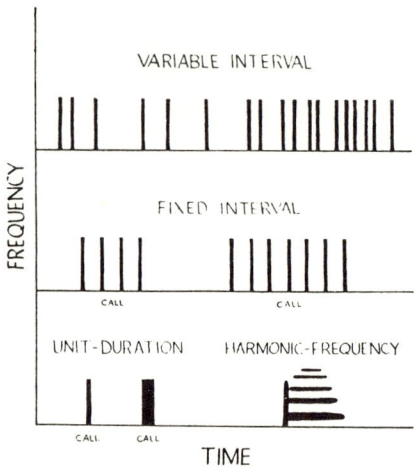

Fig. 1. Diagrammatic representation of fish sounds. See text for explanation.

SIGNAL CHARACTERISTICS

The sounds produced by any species of fish are organized in a simple fashion and give only gross information about environmental and internal physiological states (major moods). Furthermore, the responses, dependent on acoustic signals alone, are relatively simple. There seems to be a lack of a finely graded series of signals, as among birds, and of a subdivision into many kinds and degrees of responses. Perhaps experiments to date have not been refined enough to detect these subtleties, but this seems unlikely because of the rarity of more than transient pairing in fishes.

The properties of acoustical signals in some fishes may be categorized in five basic ways (Fig. 1). The first and second are *variable-time-interval* and *fixed-time-interval* signals. The time between any unit which is heard and viewed as one sound on a spectrogram (grunt, knock, etc.) may be variable or fixed. This does not include pulses which are smaller parts of a sound unit produced by a muscle-swim bladder mechanism. These pulses can be resolved by slower playback, and are the result of single muscle contractions (Winn and Marshall, 1963). If the sound is produced by an air bladder mechanism, the variable and fixed-interval calls are usually non-harmonic sounds containing frequencies from 2000 or 3000 cps down to frequencies below 100 cps with the greatest amplitude in the lower frequencies. Pharyngeal teeth and other stridulatory sounds may have the greatest amplitudes spread more uniformly throughout the various frequencies and can range to well over 6000 cps. In the third and fourth types of signals, the duration of any unit of sound is lengthened (*unit-duration* signal) or the amount of time during which units are produced is varied (*time-length* signal). The first is uncommon as a single feature and the second is apparently frequently done with fixed interval sounds. *Harmonic-frequency* signals comprise the fifth category. In some fishes at least the variable-time-interval type can be graded by varying the interval while the fixed-interval and harmonic-frequency types can be graded by varying the length of time they are emitted. It seems unlikely at this time that fish make much use of variations in amplitudes to communicate information even though the distance of detection changes. We can expect intermediates among the above basic types. The various kinds of signals are schematically shown in Fig. 1.

Another type of classification is based on the proximate cause for production of a sound. Most seem to require general or specific changes in the environment (stimuli). These would be exogenously driven sounds. The other type occurs when the animal is in the required environmental and physiological state and "spontaneously" produces sounds i.e. endogenously driven sounds. The only well-known call of this type is the "boatwhistle" call of *Opsanus tau* and other toadfishes. There is a suggestion that serranids, catfishes, sciaenids, sea robins and some other fishes have endogenously driven signals but none has been adequately documented.

Different species utilize the above types to variable degrees. A large number of species known to produce sounds will not be considered here, but rather I shall concentrate on those that are known in more detail.

One of the better understood is the oyster toadfish (*Opsanus tau*) which gives off grunt sounds as variable-interval signals and these may grade into a fixed-interval type signal when produced at the maximum rate (growls, Fig. 2). The other call of the oyster toadfish is endogenous (i.e. given by males in the nest), and is a harmonic-frequency signal, of long but relatively constant duration. The other species of toadfishes seem to have a similar repertory.

From what little is known, the northern midshipman (*Porichthys notatus*) emits variable-interval, fixed-interval, and possibly a harmonic-frequency sound that is not as well developed as in the toadfishes (Cohen and Winn, manuscript in preparation). It seems that the midshipman does not have as

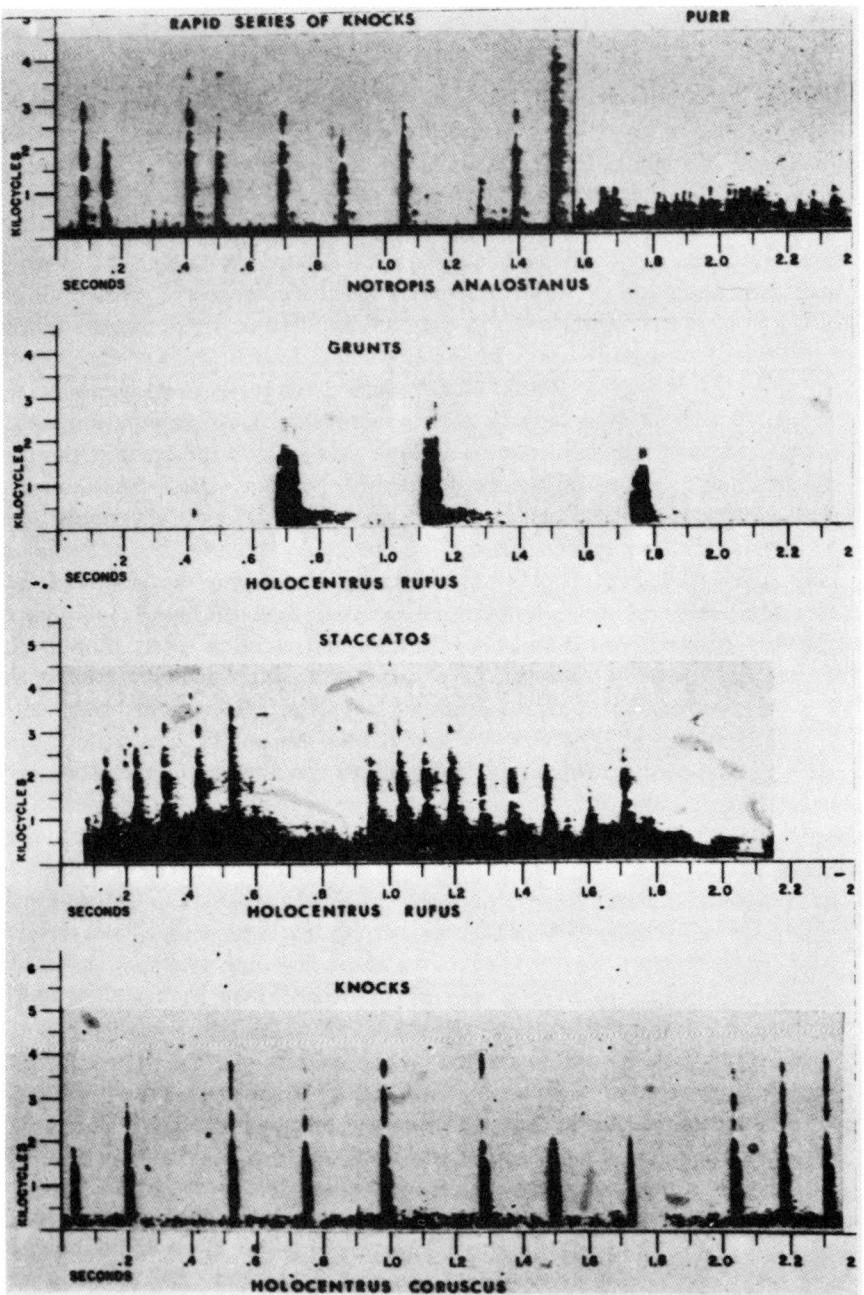

Fig. 2. Sonograms of representative samples of the various types of fish calls. The rapid series of knocks produced during display fighting of *N. analostanus* are a variable-interval type which grades into what is almost a fixed-interval call, the purrs produced during courtship (from Stout, 1963a). The organization is similar for the aggressive grunts and growls of *Opsanus tau*. The grunts and knocks of *H. rufus* and *H. coruscus* are variable-interval calls whereas the two staccatos and the thumps of *H. rufus* and *M. bonaci* are fixed-interval type

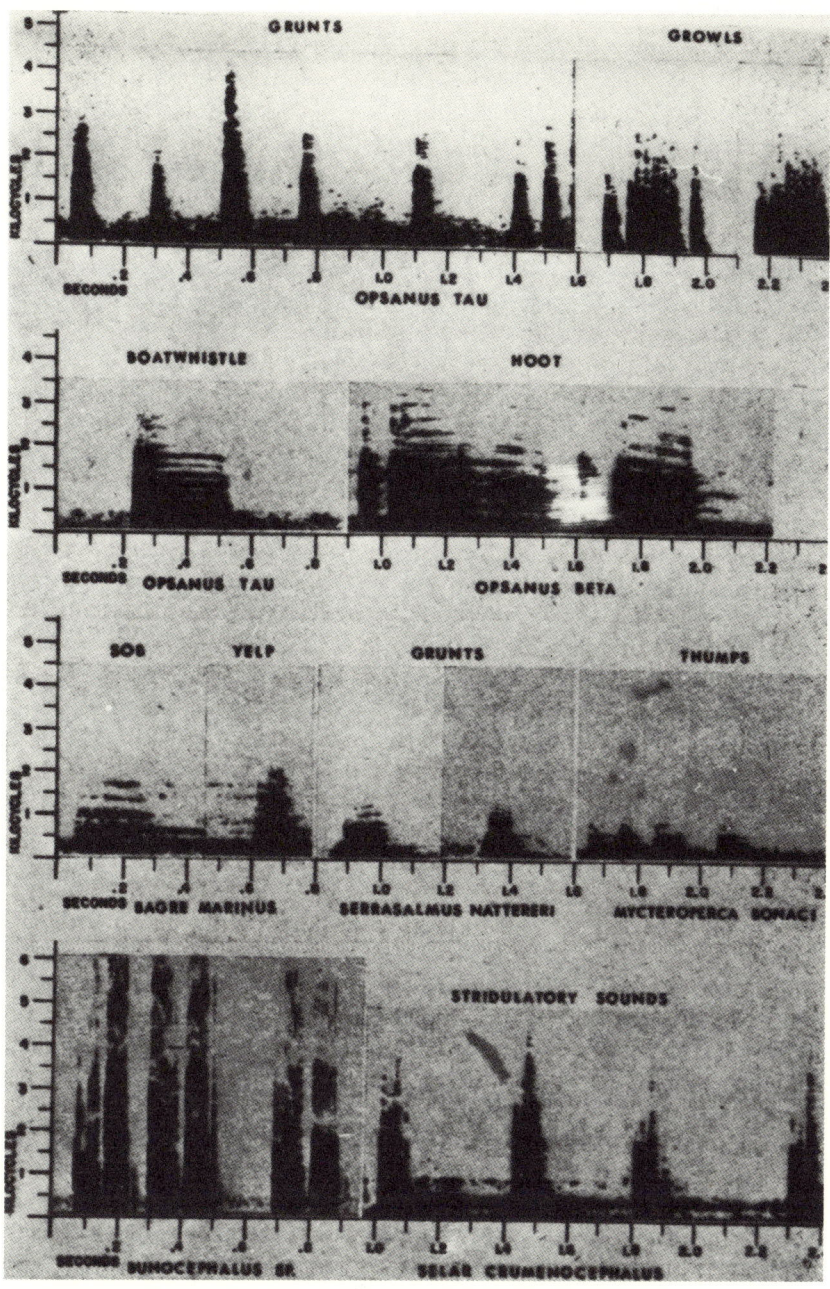

signals. The harmonic-frequency type is represented by the boatwhistle and hoot of *O. tau* and *O. beta* as is the sob of *B. marinus*. The yelp and grunts of *B. marinus* and *S. nattereni* exhibit weaker harmonics.

Stridulatory sounds are represented by those of a banjo catfish, *Bunocephalus* sp. (pectoral fin), and a bigeye scad, *S. crumenocephalus* (pharyngeal teeth). The sounds used for sonagraphic analysis of *O. beta*, *B. marinus*, and *M. bonaci* are from Tavolga (1960).

highly developed a system as *Opsanus*. There is doubt as to whether or not some sounds of catfishes (Tavolga, 1960), *Myripristis jacobus*, a piranha (Winn, unpublished observations; see Fig. 2) and other fishes (Fish, 1954) which have weak harmonics close together should be classified as biologically significant harmonic sounds. Many of them are intermediate in development and as a first thought it appears that the harmonic content of the sound is unimportant in the communication of information. *Bagre marinus* has two variable-interval sounds, a low pitched grunt and a high pitched yelp, plus a harmonic-frequency sound, the "sob" (Tavolga, 1960; see Fig. 2), but the harmonics do not originate in the air bladder (Tavolga, 1962). *Galeichthys felis* produces long (100 msec or over) grunts under duress and short grunts when in the company of other individuals of the same species. Here is the first example of two sounds differing primarily in duration of the unit (a grunt). The satinfin shiner has variable-interval and fixed-interval types of sound (Fig. 2; Stout, 1963a, b). These are produced in different contexts (see later discussion). *Holocentrus rufus* has a variable-interval grunt-like signal and an entirely distinct fixed-interval staccato call. Here, there is no gradation of one into the other (Fig. 2). The length of staccato call depends on the strength of the stimulus. The captive black grouper (*Mycteroperca bonaci*) emits fixed-interval sounds with about five beats in each group (Tavolga, 1960; Fig. 2). The reef squirrelfish, *Holocentrus coruscus*, has only a variable-interval signal which is graded according to the temporal and strength aspects of the stimulus (Winn, manuscript in preparation; Fig. 2). These sounds were recorded under several conditions, particularly during territorial behavior, but never during reproductive behavior, when it is possible that other types of sounds are emitted. The croaking gourami (*Trichopsis vittatus*) seems to produce only variable-interval sounds (Marshall, work in progress). Tavolga (1956) stated that *Bathygobius soporator* gives variable-interval signals and that the interval decreased as the courting male and female increased the vigor and closeness of courtship. This may grade into an occasional fixed-interval-type signal because he mentioned that under maximum stimulation sounds are sometimes given in fours or fives in quick succession.

When sounds are categorized into variable-interval, fixed-interval, unit-duration, time-length, and harmonic-frequency signals several important principles become apparent. Firstly, most sounds if not all are obviously different in their temporal patterning and this is the feature that may communicate intraspecific information in courtship, territorial defense, and reactions to predators. Information can be graded by varying the intervals and length of time units are emitted. Amplitude and frequency characteristics of most acoustical signals do not seem to contain enough information. This is seemingly true of *Bathygobius soporator*, *Holocentrus coruscus*, the grunts of *Opsanus tau*, and *Trichopsis vittatus*. It must be cautioned that considerable experimental evidence is necessary. Tavolga (1958a) demonstrated that positive approach by males of *B. soporator* could be induced by artificial sounds such as pure sine waves and other sounds that were different than the natural sounds (positive responses to increased sound levels of one hundred fold, frequencies from 100–300 cps and pulse durations from 75–150 msec). Moulton (1956) obtained sound mimetic behavior with artificial sea robin calls of low frequencies but suppressed calling with sounds of 200–600 cps. Stout (1963a, b) obtained differential responses of males and females of *N. analostanus* (see last section).

It has been shown that *Phoxinus laevis*, a cypriniform, can discriminate between close frequencies (cf. review by Dijkgraaf, 1960). It may tentatively be presumed that *Notropis analostanus* has this ability but its acoustic signals are variable and fixed-interval type sounds which would imply that only temporal patterning is used for discrimination. Frequency and amplitude characteristics might well be used to differentiate its own sounds from other sounds of the environment. Schneider and Hasler (1960) suggested that repetition rate might be the important factor for species recognition in the Sciaenidae.

The harmonic type call is uncommon. The best known one is the boatwhistle call of the male oyster toadfish. Here the primary features of the call are its frequency characteristics and its long but fixed duration compared to grunts. A first hypothesis is that the most important characteristic of the signal is its frequency pattern although further information could be obtained from the intermittent sequence with which it is given and its long duration.

On first inspection, it appears that fixed-interval signals are more commonly associated with warning behavior and with the exceptionally close approach of a rival to a territorial male, whereas harmonic-frequency signals most frequently occur in lone ripe males.

In general, we have ignored the patterning of the high frequency stridulatory sounds about which little is known. These tend to grade from variable-interval into fixed-interval types (Fig. 2). Their general value may be in eliciting slight escape reactions by nearby fish when one is attacked by a predator or another alarming stimulus. Specific variations in patterning may be less significant here because there may be no advantage for discrimination. In fact, it could be of survival value to lack discrimination.

CORRELATIONS OF SOUND PRODUCTION WITH ENVIRONMENTAL AND BEHAVIORAL VARIABLES

As soon as it was known that fish sounds were common underwater, it became evident that they were correlated with certain environmental and behavioral circumstances. These facts resulted in the hypothesis that the sounds have varied stimulus values.

The literature on cyclic activity of sound production correlated with environmental variables has been reviewed by Winn, Marshall, and Hazlett (1963) and the subject will be briefly summarized at this time. There are diel rhythms of sound production. Some are diurnal, some crepuscular, some nocturnal, and others involve various subpatterns and combinations of the first three types.

The satinfin shiner (*N. analostanus*) apparently produces sound only during the day, the "singing fish" of Ceylon (apparently a catfish; Smith, 1927; Lange, 1953), and *Galeichthys felis* (Tavolga, 1960) only at night, whereas the boatwhistle sounds of *Opsanus tau* are more arhythmic (Knudsen *et al.*, 1948; Winn, unpublished observations). *Holocentrus rufus* produces most staccato sounds at dawn and dusk and fewer at night than in daylight (Winn, Marshall, and Hazlett, 1963). Other documented diel cycles of sound production are as follows: A few "knock" sounds occur in the morning, increase in frequency up to dusk, and then practically stop when it becomes dark (*Corvina nigra* by Dijkgraaf, 1947); chorus began at sunset, increased for 2 to 3 hr and tapered off to rare outburst (one or more Sciaenidae by Johnson, 1948); drumming started at noon and increased until mid-afternoon, then decreased, until at sundown only a few fish could be heard (*Aplodinotus grunniens* by Schneider and Hasler, 1960); increase in sounds before and after dusk (*Gadus callarias* by Brawn, 1961); peak at 8 p.m. (possibly *Micropogon undulatus* by Dobrin, 1947); height of sound making at dawn and dusk (*Micropogon undulatus* by Fish, 1954); and, a slight peak at dusk and a major peak of sound production in the early morning after sunrise (*Cynoscion regalis* by Fish, 1954). Also, peaks of sound activity by croakers were correlated with tidal changes and possibly feeding activity (Fish, 1954). The freshwater drum stopped sonic and probably spawning activity on cloudy days with decreased temperature (Schneider and Hasler, 1960). When the water temperature fell below 4°C codfish stopped producing sounds (Brawn, 1961). Many of the documented diel cycles of sound production are crepuscular or nocturnal with each a different species-specific pattern. *Holocentrus rufus* emits most aggressive territorial sounds and warning staccatos when it is hovering over its crevice and not at night when it displays greatest locomotor activity.

Seasonal cycles of sound activity are less well documented but it is known that seasonal cycles are common especially in relation to reproductive activity. The knocks and purring sounds of *Notropis analostanus* are much more abundant during the spawning season even though some knocks can be obtained in the laboratory at other times. Purring has been heard only during spawning (Winn and Stout, 1960; Stout, 1963a, b). The toadfish boatwhistle is restricted to a few months from late May through early August at the dock of the Solomons Island Laboratory, Maryland. Fish (1954) found a similar cycle of activity in Rhode Island. Also, grunts are more readily evoked from individuals in the spawning season than at other times (Gray and Winn, 1961; and Winn, work in progress). It has been stated that characteristic choruses of various species of croakers are synchronized with spawning migrations (Fish, 1954). *Aplodinotus grunniens* only drum from April to August (Schneider and Hasler, 1960). A cod (*Gadus callarias*) has two peaks of sound production, one in the fall and one during the February-March spawning period (Brawn, 1961). Steinberg, Kronengold and Cummings (1962) have reported diel and seasonal cycles of sounds at Bimini that were most likely animal (fish) sounds. Once the fishes are identified, valuable information on cycles will be obtained. Cummings *et al.* in another article in this book give evidence for lunar and tidal cycles of sound production.

There is seemingly more sonic activity in fishes during reproductive seasons. This is certainly true for acoustic signals in higher vertebrates. The use of sounds as an aid to solve the problems of aggregation, synchronization and stimulation of reproductive activities is further corroborated by the restriction of sound production to one sex. The grunt is made more frequently by male oyster toadfish during the spawning season than by the female or by both sexes at other times of the year. Over 100 individuals were located and it was always a male that produced the boatwhistle call (Gray and Winn, 1961, and Winn, work in progress). Over fifty per cent of the males made sounds but a sound was never heard from a female, as the result of handling over 150 females and 300 males of *Porichthys notatus* during the spawning season. The muscle on the air bladder is large in the male and considerably smaller in the female (Cohen and Winn, manuscript in preparation). Only male individuals of most species in the family Sciaenidae, like *Cynoscion regalis*, possess the drumming muscles and produce sounds (Fish, 1954, and Schneider, 1961). The males of *Bathygobius soporator* and *Chasmodes bosquianus* make sounds during courtship (Tavolga, 1956, 1958a, b). Stout (1963a, b) has established that the males of *N. analostanus* produce sounds. Although no sounds were ever heard from females, he could not positively eliminate it as a possibility. On the other hand, Delco (1960) stated that the females of *Notropis lutrensis* and *N. venustus*, both in the same subgenus with *N. analostanus*, made sounds, although the method utilized may be questionable (Stout, 1963a, b). Recording in a stream suggested that female fallfish (*Semotilus corporalis*, family Cyprinidae) made sounds in a pre-spawning aggregation (Winn, unpublished observations). Brawn (1961) demonstrated that both sexes of a codfish (*Gadus callarias*) produce sounds outside the spawning season but that during prespawning and spawning only the males emit sounds. Marshall (work in progress) found that both sexes of the croaking gourami produce similar sounds during courtship but those of the male are much louder. Both sexes of *Opsanus tau* produce grunts but the spawning males produce them more readily and frequently from their nests (Fish, 1954; Gray and Winn, 1961). Both sexes of several species of squirrelfishes produce a variety of sounds outside of the reproductive season, and unrelated to sexual maturity. Most of the sounds of hand-held and startled marine fishes are produced by both sexes (Fish, 1954; Moulton, 1958; Burkenroad, 1931). The several sounds of marine catfishes studied by Tavolga (1960, 1962) are produced by both sexes.

It has become obvious in recent years that sounds are produced in a variety of other behavioral circumstances. In some, the signal is the same for several behaviors. In others, the sound is produced in only one specific aspect of behavior, whereas a few fish have evolved several different sounds for different situations. In fact, the marine catfishes (Tavolga, 1960) and the squirrelfishes (Winn, Marshall and Hazlett, 1963) which have two to four different sounds for non-reproductive activities, may have others for courtship and spawning. A brief summary of the correlations between sounds and various behaviors is presented in Table 1.

In non-reproductive behavior (Table 1) sounds have been reported during competitive feeding, territorial defense, aggregating behavior, new stimulus situations, escaping, and during migration. Not all of these types are well documented as for instance in the croakers during competitive feeding and migration. The sounds all could be simply aggressive reactions which result in proper spacing between individuals but this seems unlikely. The category of new stimulus situations includes items of heterogeneous origin like prodding, handling, approach by other fish, etc. It seems that at least the following natural conditions are involved: aggressive sounds, warning sounds, startle sounds, and sounds when caught by a predator ("fear" sounds).

During reproductive behavior of various species, sounds are produced in the following circumstances: stationary defense of nest; chasing; display fighting; various phases of courtship; at the nest site; and, finally, in a breeding aggregation. A correlation between a sound and a particular behavior needs careful analysis because the call may function in a different way. The knocks produced by a male of the satinfin shiner when first encountering ripe and spent females were interpreted by Stout (1963a, b) as aggressive rather than courtship sounds. A second sound was produced in courtship once the male had "determined" from the female's behavior that she was ready to spawn. From these correlations between sound and behavior there seems little question that sounds are an integral part of complex stimulus situations which are used to communicate information for the solution or problems that relate to individual and population survival. This requires at different times that individuals be kept out of contact with each other, maintained at a certain distance or brought together for aggregational maintenance or spawning. This is accomplished by way of increases and decreases in activity. Orientation to the source of sound is perplexing and controversial at this time.

TABLE 1. CORRELATIONS BETWEEN SOUNDS AND SPECIFIC PHASES OF BEHAVIOR IN SOME FISHES.
(For more detailed explanation see text.)

Behavior (*Non-reproductive*)	Sound and Species	References
Competitive Feeding	Suspected maximum number of sounds produced near midnight when feeding, croakers, Chesapeake Bay.	Dobrin, 1947
	Possibly *Epinephalus striatus*.	Moulton, 1958
	Aggressive behavior and sounds increased after feeding but not during the feeding of *Stenotomus chrysops*, *Prinonotus carolinus* and others.	Fish, 1954; Brawn, 1961
Territorial Defense	Grunts against own species. *Holocentrus rufus*.	Winn, Marshall and Hazlett, 1963
"Spontaneous" Sounds in Aggregation or School	Grunt sounds, of nocturnal schools, *Galeichthys felis*.	Tavolga, 1960
	Sob and yelp sounds at night when in small groups, *Bagre marinus*.	Tavolga, 1960
	Sounds of *Pygocentrus piraya*.	Meschkat, in Schneider, 1961.
	Sounds of *Therapon jarbua*.	Schneider, 1961
Confronted with New Stimulus Situations (e.g. Hand-held, Prodding, etc.)	Prodded grunt sounds of *Galeichthys felis*.	Tavolga, 1960
	Prodded grunt sounds of *Bagre marinus*.	Tavolga, 1960
	Prodded distress sounds and sounds when approached by another species of fish of *Mycteroperca bonaci*.	Tavolga, 1960
	Grunts and growls of *Opsanus tau* and *Opsanus beta*.	Fish, 1954; Gray and Winn, 1961; Tavolga, 1960
	Warning staccato of *Holocentrus rufus* and *Holocentrus ascensionis* and different hand-held grunts.	Moulton, 1958; Winn, Marshall and Hazlett, 1963
	Warning sounds of *Epinephalus striatus*.	Moulton, 1958
	Aggressive defense sounds of *Gadus callarias*.	Brawn, 1961
	Other fishes.	Fish, 1952, 1954
Escaping	Grunts when escaping from attacks and grunts of another individual, *Porichthys notatus*.	Greene, 1924
	Escape sounds after aggressive encounter of codfish.	Brawn, 1961
Migration	Croaker migrations after reproduction.	Johnson, 1948
Exploring New Environment	Clicks of *Hippocampus hudsonius*.	Fish, 1953, 1954

TABLE 1 *(continued).*

Behavior (*Reproductive*)	Sound and Species	References
Spawning Migration	Many Sciaenidae.	Fish, 1954
Stationary Defense of Nest	Grunts and growls by male on nest against many objects but more toward other males, *Opsanus tau*.	Gray and Winn, 1961
	Grunts of *Porichthys notatus*.	Greene, 1924; Cohen and Winn, (in MS)
	Snoring sound of *Gobius jezo*.	Kinzer, 1961
Chasing	Isolated knocks of *Notropis analostanus*, probably males only.	Winn and Stout, 1960; Stout, 1963a, b
	Grunts by males of *Corvina nigra*.	Dijkgraaf, 1947
	Grunts of *Pomocentrus leucostictus*, possibly also non-reproductive territory.	Moulton, 1958
	Sounds in male to male encounters of *Gadus callarias*.	Brawn, 1961
Display Fighting	Rapid series of knocks of *Notropis analostanus*, probably males only.	Winn and Stout, 1960; Stout, 1963a, b
Courtship	Male sounds of *Bathygobius soporator*.	Tavolga, 1956, 1958b
	Male sounds of *Chasmodes bosquianus*.	Tavolga, 1958c
	Purring sounds, and occasional knocks of *Notropis analostanus*, probably only males.	Winn and Stout, 1960; Stout, 1963a, b
	Many sounds only at beginning of courtship, *Trichopsis vittatus*.	Marshall (work in progress)
	Some grunts produced by *Gadus callarias*.	Brawn, 1961
	Clicks of a pair of *Hippocampus hudsonius*.	Fish, 1953, 1954
Spawning	United pair of *Hippocampus hudsonius*.	Fish, 1954
"Spontaneous" Sounds by Male at Nest Site	*Opsanus tau* boatwhistle (possibly also *Opsanus beta*).	Fish, 1954; Gray and Winn, 1961; Tavolga, 1958c
	Probably staccato call of *Prionotus* spp.	Moulton, 1956
	Snoring sound of *Gobius jozo*.	Kinzer, 1961
Spontaneous Sounds of a Breeding Aggregation, Not Territorial	Male drumming of *Aplodinotus grunniens* seems to fit this category, in spawning aggregation.	Schneider and Hasler, 1960
	Possibly scratching sounds of a prespawning aggregation of female *Semotilus corporalis*.	Winn, unpublished

EXPERIMENTAL EVIDENCE AND BEHAVIORAL VALUE OF THE STIMULUS

Several types of functions have been reported for sounds in non-reproductive activity. Escape reactions to sounds have been reported by several authors and there have been many reports of no response to playbacks (see reviews: Warner, 1932; Maliukina and Protasov, 1960; Moulton and Backus, 1955; Schneider, 1961). Stridulation and swimming sounds of *Caranx latus* were played back to schools of *Anchoviella choerostoma*, *Caranx ruber*, and *C. latus* (Moulton, 1960). Increased activity and in some instances movements away from the sound were noted. Conversely, the swimming sounds of *Anchoviella* when played to *C. latus* caused quickened movements. The staccato call of *Holocentrus rufus* produced startle effects followed by slow escape into cavities and then an approach to the sound source (Winn, Marshall and Hazlett, 1963). Some other sounds produced the same effect. There are a large number of parallelisms between the development of certain alarm calls by birds and the alarm call (staccato) system of the longspined squirrelfish (cf. summary for birds in Thorpe, 1961). Briefly, these are: (1) both an attractive and dispersal effect on members of own species; (2) mobbing accompanied by sound; (3) attention and alarm components; (4) some differentiation of the alarm call; (5) similar calls at dusk and dawn; (6) at a minimum, different distress (hand-held sound of squirrelfish), alarm (staccato) and territorial aggressive calls (single grunts) are developed; and (7) a potential differentiation of responses to a predator on the bottom and one swimming off the bottom (the short and long staccato).

It is presumed, but not adequately demonstrated, that sounds produced by territorial fish outside of the breeding season are part of the effective stimulus pattern causing dispersal of the intruder (see Table 1). Another kind of sound is produced when the fish is hand-held. Again, it is presumed, but not experimentally verified, that this would allow other individuals to escape more easily when one is attacked by a predator, much like the "alarm substance" given off by many fishes. The use of the signal in confusing or reducing the effectiveness of the predator seems inadequate as a hypothesis.

Several analyses of the use of acoustic signals in reproductive behavior are now available although much more needs to be known about each system. Tavolga (1956, 1958a), in the first published papers of this sort, experimentally modified the behavior of males and females of *Bathygobius soporator* by playback of the sounds produced by males during courtship. Males approached the speaker whereas the females only approached if there was the visual stimulus of another goby in the vicinity. Activity was clearly increased and in some manner males localized the source of sound, if only at a short distance. Thresholds of reactivity were probably lowered (more attentive). Acoustic, visual, and chemical signals in combination were utilized in the synchronization of spawning behavior. Stout (1963a, b) recently analyzed the functions of the sounds produced by the territorial males of *Notropis analostanus*. These include *single knocks* when chasing and during courtship, *rapid series of knocks* during display fighting and *purring* during courtship activities of approach, circling and male passing over the future egg site (solo spawning). Briefly, the results were as follows: The "rapid series of knocks" stimulated the dominant male to increase aggressive behavior toward submissive males, tended to hasten the exit of the submissive male from a dominant male's territory, inhibited the courtship activity of a male, and increased the movements of gravid females. Single knocks appeared to be of insufficient stimulus value to effect the male's aggressive behavior in the experiments but did increase the male's courtship activity. Stout suggested that the "rapid series of knocks" and "single knocks" are involved only with aggressive behavior. "Purrs" stimulated the aggressive activity of a dominant male, increased courtship activity of the male, and reduced movements of a gravid female thus keeping her near the male. These sounds are all used in conjunction with visual stimuli and when the fish are less than a foot from each other. Orientation to the sound source was not observed. Delco (1960) seemed to show that males of two species of minnows could recognize and eventually locate the calls of the females of their own species when the sounds were played out simultaneously.

The functional significance of another type of sound, the toadfish boatwhistle, is being analyzed (Fish, 1954; Gray and Winn, 1961; Winn and Marshall, work in progress). This call is produced by males when they first establish nest sites in the late spring and is emitted at regular intervals without a specific stimulus change. Once eggs are laid, the sound rapidly diminishes in occurrence even though the male remains guarding the eggs. The boatwhistle and similar sounds may be analogous to song in birds. Some of the characteristics of song (Thorpe, 1961) are as follows: they are related to reproduction and territorial behavior; long and made up of a particular pattern of notes; must be of a type to allow for localization; are given for long periods at regular intervals; are internally driven ("spontaneous"); and are more frequently produced by unmated males and early in the reproductive season. The most likely hypothesis is that the boatwhistle attracts females to the nest. Tavolga (1958c) suggests it is a territorial call, but this could be a secondary function. On the basis of our work on one small population at Solomons Island, Maryland, the sound seems to attract females. The approach of toadfishes and other objects elicits aggressive grunt sounds from the guarding male during all phases of reproduction. The introduction of males to nest-guarding males has the highest stimulus value for production of grunts. Experiments are underway in pens on Chesapeake Bay to test our hypothesis concerning the function of the boatwhistle and the hypothesis that the grunts cause other toadfish to leave the area of the nest. In order to test the function of the boatwhistles, underwater loudspeakers were placed in the four corners of the pen. Females with ovulated eggs and also males were released in the center of the pen about 10 ft from all speakers. From 0900 to 1730 hr 8 females and 17 males showed no response to the sound or control speakers. From 1730 to 1830 hr, seven females out of a total of 18 (39 per cent) went in almost a straight line to the speakers emitting sound and most went inside the tin can attached to the front of the speaker. Three of a total of 31 males (4 per cent) went up to the speaker (only one entered). Twelve spent females did not go to the speaker emitting sound. In all instances there was no reaction to the control speaker. The predominance of positive responses by females supports the proposed hypothesis but final judgment will have to wait for completion of the experiments. Moulton (1956) played back staccato calls to individuals of *Prionotus* spp. in the field and obtained staccato calls in return. The sound was suppressed with the playback of other sounds. It may be that the staccato is a "spontaneous" spawning call similar to the toadfish boatwhistle, but its actual relation to behavior remains obscure.

Many other functions of fish sounds may exist but remain to be explored. Suggestive of recognition signals was Moulton's (1958) observation that two black angelfish (*Pomacanthus arcuatus*), probably a pair, came together immediately after the acoustical signals of one changed from short calls to moan-like sounds. The potential attraction of a predator is indicated by Moulton's (1960) observation of the appearance of barracuda when he was playing out underwater sounds of *Caranx latus*. Sounds produced by sea-horses when introduced into a new environment (Fish, 1953, 1954) suggests another unexplored function of acoustic signals.

The simplest responses to stimuli involve an increase in activity, inhibition or decrease of activity, change in the attentive level, and orientation to the stimulus. These all can be involved with the various types of adaptive behavior such as feeding, escape, migration and phases of reproduction. The few experimentally demonstrated examples of functions of fish sounds have shown that the sounds can function in increasing and decreasing various activities such as escape responses, investigative behavior, in changing attentive levels and, finally, as possible orientational cues. This latter role has been discussed by Winn, Marshall and Hazlett (1963). In all the previously mentioned instances, acoustic signals have been used when the fish were close to each other so that several sensory modalities are potentially involved in localization. Theoretically, the fish ear cannot localize a source of sound underwater and Dijkgraaf (1962) seemed to rule out lateral-line localization of a sound source at any distance (at least using propagated pressure waves). Van Bergeijk in this volume discusses the problem in relation to the lateral line. If localization does not occur at any reasonable distances, fishes would be the only group of animals with highly developed acoustical signalling systems with no value in localization. This makes the sensory world of fishes limited as compared to that of insects and terrestrial vertebrates. However, it is still possible that in some unknown way sounds are localized in a more accurate manner than the following of high amplitude levels as demonstrated by Kleerekoper and Chagnon (1954) or to near-field effects. The direct response of toadfishes and squirrelfishes to the sound source suggests localization but the sounds may be loud enough to allow the use of sensory receptors other than the ear. A visual response is ruled out in the toadfish experiments where preliminary results suggest localization up to at least ten feet. Schneider and Hasler (1960) suggested that the drumming of the males of the freshwater drumfish may be used for localization. The use of spontaneous sounds in aggregations especially nocturnal ones (see Table 1) suggests that the signals may aid in the maintenance of the group.

Another question that cannot be as yet answered satisfactorily is whether or not sound differences can be of value in speciation. All of the toadfishes and squirrelfishes so far studied produce different sounds except that the two sympatric species *Holocentrus rufus* and *Holocentrus ascensionis* produce sounds only slightly different on a statistical basis. These latter ones are used in non-reproductive activities where interspecific responses may be selectively advantageous. The boatwhistle sounds of the toadfishes are the ones that may be used in sexual behavior and it is these calls that differ most (not the grunts and growls) (Tavolga, 1958c). Delco (1960) provided preliminary evidence to suggest that different sounds of two minnows can aid in species isolation. The lack of specific responses to the sounds of gobies studied by Tavolga (1958a) suggests that the sounds are not important in speciation at their present stage of evolution. The key to the problem is to study the acoustic signals of sympatric species of fishes as has been done with birds, (e.g. Marler, 1959).

SUMMARY AND CONCLUSIONS

There is a lack of adequate evidence in fishes for many of the functions of sounds found in birds and mammals. Some of these are: echolocation, confusion choruses, young-parental sound relation, all-clear signals, recognition of individual call differences and many others. It is clear that sounds by fishes are not as finely organized and only major moods and responses can be obtained. The all important question of localization needs to be clarified but it appears that the sensory world of fishes is spatially more limited than that of land-dwelling vertebrates.

The organization of sounds into variable-interval, fixed-interval, harmonic-frequency, unit-duration and time-length patterns was briefly examined. It seems as though temporal patterning is the most important factor for discrimination and for sending variable information except for the harmonic frequency calls. Diel and seasonal cycles of sound production are common and often only one sex emits the sounds during reproduction. Different sounds are correlated with a wide variety of non-reproductive and reproductive phases of behavior. Some are elicited only by specific stimuli whereas others are so-called "spontaneous" sounds.

Experimental analyses of the stimulus value of sounds have demonstrated that increases in general, aggressive and courtship activities can be obtained with sound as the only controlled stimulus. Localization may be more general than that obtained when the fish are close to each other. A system of warning signals was described that closely resembles the systems developed by birds and sounds like the boatwhistle of the oyster toadfish may be comparable to bird song. Discrimination of signals occurs. The use of acoustical signals in species isolation is yet to be studied adequately.

Future definitive knowledge about fish sounds will depend on detailed studies of ecology, behavior and sensory physiology. It will be necessary to find experimental subjects that can be studied under the proper controlled conditions.

REFERENCES

BRAWN, V. M. (1961) Sound production in the cod (*Gadus callarias* L.). *Behaviour*, **18**, 239–255.
BURKENROAD, M. D. (1931) Notes on the sound-producing marine fishes of Louisiana. *Copeia*, **1931**, 20–28.
DELCO, E. A. JR. (1960) Sound discrimination by males of two cyprinid fishes. *Texas J. Sci.* **12**, 48–54.
DIJKGRAAF, S. (1947) Ein Töne erzeugender Fisch im Neapler Aquarium. *Experientia*, **3**, 1–4.
DIJKGRAAF, S. (1960) Hearing in bony fishes. *Proc. Roy. Soc.*, Series B, **152**, 51–64.
DIJKGRAAF, S. (1962) The functioning and significance of the lateral-line organs. *Biol. Rev.* **38**, 51–105.
DOBRIN, M. B. (1947) Measurements of underwater noise produced by marine life. *Science*, **105**, 19–23.
FISH, M. P. (1953) The production of underwater sound by the northern seahorse, *Hippocampus hudsonius*. *Copeia*, **1953**, 98–99.
FISH, M. P. (1954) The character and significance of sound production among fishes of the Western North Atlantic. *Bull. Bingham Oceanogr. Coll.* **14**, 1–109.
FISH, M. P., A. S. KELSEY, JR. and W. H. MOWBRAY (1952) Studies on the production of underwater sound by North Atlantic coastal fishes. *J. Mar. Res.* **11**, 180–193.
GRAY, G.-A. and H. E. WINN (1961) Reproductive ecology and sound production of the toadfish, *Opsanus tau*. *Ecology*, **42**, 274–282.
GREENE, C. W. (1924) Physiological reactions and structure of the vocal apparatus of the California singing fish, *Porichthys notatus*. *Amer. J. Physiol.* **70**, 496–499.
JOHNSON, M. W. (1948) Sound as a tool in marine ecology, from data on biological noises and the deep scattering layer. *J. Marine Res.* **7**, 443–458.
KINZER, J. (1961) Über die Lautäusserungen der Schwarzgrundel *Gobius jozo* L. *Aquar. und Terrar.* **7**, 7–10.
KLEEREKOPER, H. and E. C. CHAGNON (1954) Hearing in fish, with special reference to *Semotilus atromaculatus* (Mitchell). *J. Fish. Res. Bd., Canada*, **11**, 130–152.
KNUDSEN, V. O., R. S. ALFORD and J. W. EMLING (1948) Underwater ambient noise. *J. Marine Res.* **7**, 410–429.
LANGE, J. W. (1953) The singing fish of the Batticoloa Lagoon. *Brit. Roy. Asiatic Soc. Ceylon*, new ser. **3**, 12–24.

MALIUKINA, G. A. and V. R. PROTASOV (1960) Hearing and "voice" in fish and their reactions to sounds. *Russian Rev. Biol.* **50**, 215–227.

MARLER, P. (1959) Developments in the study of animal communication. Pages 150–206. In P. R. Bell, Ed., *Darwin's Biological Work. Some Aspects Reconsidered.* Cambridge Univ. Press.

MOULTON, J. M. (1956) Influencing the calling of sea robins (*Prionotus* spp.) with sound. *Biol. Bull.* **111**, 393–398.

MOULTON, J. M. (1958) The acoustical behavior of some fishes in the Bimini area. *Biol. Bull.* **114**, 357–374.

MOULTON, J. M. (1960) Swimming sounds and the schooling of fishes. *Biol. Bull.* **119**, 210–223.

MOULTON, J. M. and R. H. BACKUS (1955) Annotated references concerning the effects of man-made sounds on the movements of fishes. Fisheries Circular No. 17, Department of Sea and Shore Fisheries, Maine.

SCHNEIDER, H. (1961) Neure Ergebnisse der Lautforschung bei Fischen. *Naturwiss.* **48**, 513–518.

SCHNEIDER, H. and A. D. HASLER (1960) Laute und Lauterzeugung beim Süsserwassertrommler *Aplodinotus grunniens* Rafinesque (Sciaenidae, Pisces). *Z. vergl. Physiol.* **43**, 499–517.

SMITH, H. M. (1927) The so-called musical sole of Siam. *J. Siamese Soc. Nat. Hist.* **7** (suppl.), 49–54.

STEINBERG, J. C., M. KRONENGOLD and W. C. CUMMINGS (1962) Hydrophone installation for the study of soniferous marine animals. *J. Acoust. Soc. Amer.* **34**, 1090–1095.

STOUT, J. F. (1963a) Sound communication during the reproductive behavior of *Notropis analostanus* (Pisces: Cyprinidae). Ph.D. thesis, University of Maryland.

STOUT, J. F. (1963b) The significance of sound production during the reproductive behavior of *Notropis analostanus* (family Cyprinidae). *Animal Behaviour* (in press).

TAVOLGA, W. N. (1956) Visual, chemical and sound stimuli as cues in the sex discriminatory behavior of the gobiid fish, *Bathygobius soporator. Zoologica*, **41**, 49–64.

TAVOLGA, W. N. (1958a) The significance of underwater sounds produced by males of the gobiid fish, *Bathygobius soporator. Physiol. Zool.* **31**, 259–271.

TAVOLGA, W. N. (1958b) Underwater sounds produced by males of the blenniid fish, *Chasmodes bosquianus. Ecology*, **39**, 759–760.

TAVOLGA, W. N. (1958c) Underwater sounds produced by two species of toadfish, *Opsanus tau* and *Opsanus beta. Bull. Mar. Sci. Gulf and Carribean*, **8**, 278–284.

TAVOLGA, W. N. (1960) Sound production and underwater communication in fishes. In Lanyon, W. E. and W. N. Tavolga, Eds. *Animal Sounds and Communication. Amer. Inst. Biol. Sci., publ. No.* 7, pp. 93–136.

TAVOLGA, W. N. (1962) Mechanisms of sound production in the ariid catfishes *Galeichthys* and *Bagre. Bull. Amer. Mus. Nat. Hist.* **124**, 1–30.

THORPE, W. H. (1961) *Birdsong. The Biology of Vocal Communication and Expression in Birds.* Cambridge University Press, Cambridge.

WARNER, L. H. (1932) The sensitivity of fishes to sound and to other mechanical stimulation. *Quart. Rev. Biol.* **7**, 326–339.

WINN, H. E. and J. A. MARSHALL (1963) Sound-producing organ of the squirrelfish, *Holocentrus rufus. Physiol. Zool.* **36**, 34–44.

WINN, H. E., J. A. MARSHALL and B. HAZLETT (1963) Behavior, diel activities, stimuli that elicit sound production, and reactions to sounds in the squirrelfish (*Holocentrus rufus*). *Copeia* (in press).

WINN, H. E. and J. STOUT (1960) Sound production by the satinfin shiner, *Notropis analostanus*, and related fishes. *Science*, **132**, 222–223.

AUTHOR CITATION INDEX

Addens, J. L., 276
Agassiz, J. L., 40, 57, 158, 276
Albers, V. M., 40
Alexander, R. M., 40, 294
Alford, R. S., 42, 216, 278, 319, 349
Allis, E. P., Jr., 276
Aronov, M. I., 43

Backus, R. H., 40, 43, 44, 350
Bainbridge, R., 216
Barber, S. B., 40
Beccari, N., 276
Berg, L. S., 276
Bergeijk, W. A. van, 294
Berkelbach van der Sprenkel, H., 276
Bhimacher, B. S., 277
Biot, 81
Black, D., 277
Boas, J. E. V., 86
Bohls, 79
Bohr, 81
Bondesen, P., 40
Borelli, 99
Brahy, B. D., 40, 45
Brawn, V. W., 40, 331, 349
Breder, C. M., Jr., 40, 319
Bridge, T. W., 40, 59, 60, 65, 142, 158, 216, 277
Bright, T. J., 52
Broughton, W. B., 40
Burke, T. E., 52
Burkenroad, M. D., 40, 58, 158, 216, 277, 349
Busnel, R.-G., 40

Caldwell, D. K., 40
Caldwell, M. C., 40
Chadwick, L. E., 277
Chagnon, E. C., 278, 332, 349
Chapman, C. J., 42, 43
Charbonnel-Salle, L., 103, 104
Chranilov, N. S., 277
Clapp, C. M., 319
Clarke, G. L., 216

Clarke, W. D., 40
Cohen, M. J., 40
Cooper, S., 277
Craddock, J. E., 40
Cummings, W. C., 40, 45, 52, 350
Cushing, D. H., 40
Cuvier, G., 56, 142

Dann, R., 42
Davis, L. I., 40
Day, F., 85
De Beer, G. R., 277
Delaroche, F., 100
Delco, E. A., Jr., 216, 331, 349
Dempster, R. P., 332
Demski, L. S., 52
Dietrich, G., 40
Dijkgraaf, S., 41, 216, 277, 331, 349
Dobrin, M. B., 41, 162, 277, 332, 349
Dorai Raj, B. S., 41
Dôtu, Y., 41
Dufossé, M., 41, 65, 71, 76, 142, 158, 216, 277
Duméril, A., 83, 86

Eccles, J. C., 277
Edgeworth, F. H., 277
Eisenberg, J. F., 43
Emling, J. W., 42, 216, 278, 319, 349
Erman, 87
Evans, H. M., 277
Everest, F. A., 177, 201
Evermann, B. W., 142

Fänge, R., 41
Fawcett, D. W., 41, 277
Fay, R. R., 52
Fish, C. J., 201
Fish, J. F., 52
Fish, M. P., 41, 163, 201, 216, 222, 277, 310, 319, 332, 349
Ffowcs-Williams, J. E., 41
Frings, H., 41

Author Citation Index

Frings, M., 41
Frisch, K. von, 277
Fürbringer, M., 277

Gainer, H., 41
Gegenbaur, 92
Geoffroy St. Hilaire, I., 41, 277
Gerald, J. W., 52
Gill, T. N., 158, 319
Gilmour, D., 277
Goodrich, E. S., 277
Gray, G.-A., 41, 349
Gray, J. E., 83
Green, W. C., 42
Greene, C. W., 41, 158, 216, 349
Gregory, W. K., 278
Griffin, D. R., 41, 216, 278, 310
Gudger, E. W., 150, 319
Günther, A., 83, 96, 112, 142

Haddle, G. P., 44
Haddon, A. C., 40, 59, 60, 65, 93, 216, 277
Haedrich, R. L., 40
Hagman, N., 216
Haller, B., 278
Hardenburg, J. D. F., 216
Harrington, R. W., Jr., 278
Harris, G. G., 42, 294
Hashimoto, T., 42
Hasler, A. D., 44, 332, 350
Hawkins, A. D., 42, 43
Hays, E. E., 45
Hazlett, B. A., 42, 46, 350
Heinecke, P., 43
Hemmings, C. C., 43
Herald, E. S., 332
Herrick, C. J., 278
Herrniknd, W. F., 40
Hersey, J. B., 44
Hester, F. J., 42
Hildebrand, S. F., 319
Hoff, I., 43
Holbrook, J. E., 57, 142
Horch, K., 52
Horton, J. W., 42, 278
Hubbard, G. J., 216
Hubbs, C. L., 158, 278
Humboldt, 81
Hyrtl, J., 88

Ihering, R. von, 46, 106

Jäger, A., 142
Jayaram, K. C., 278
Jobert, 88
Johnson, M. W., 42, 177, 201, 332, 349

Jones, F. R. H., 42, 216, 278
Jordan, D. S., 142, 278
Joseph, N. I., 278

Karandikar, K. R., 278
Kellogg, W. N., 42, 278
Kelsey, A. S., Jr., 41, 310, 319, 349
Kilarski, W., 42
Kindred, J. E., 278
Kinzer, J., 42, 349
Klancher, J. E., 41
Klausewitz, W., 42
Kleerekoper, H., 278, 332, 349
Kner, R., 88, 96
Knudsen, V. O., 42, 216, 278, 319, 349
Koczy, F. F., 45
Koschkaroff, D. N., 278
Kramer, E., 43
Kritzler, H., 332
Kronengold, M., 42, 45, 350
Krumholz, L. A., 278
Kuroki, T., 332
Kusano, K., 41

Landois, L., 76
Lange, J. W., 278, 349
Lanyon, W. E., 42
Lanzing, W. J. R., 52
Lichtenfelt, H., 142
Loewenstein, J. M., 42
Loye, D. P., 42, 177

MacBain (Spires), J. Y., 45
MacDonnell, R., 83
Mackenzie, K. V., 42
McMurrich, J. P., 278
Mahajan, C. L., 42
Maliukina, G. A., 42, 350
Maniwa, Y., 42
Mansueti, R. J., 45
Markl, H., 52
Marler, P., 42, 332, 350
Marshall, J. A., 42, 46, 332, 350
Marshall, N. B., 43, 216, 278, 279
Masurekar, V. B., 278
Mathewson, R. F., 41
Mead, G. W., 40
Meder, E., 43
Merriman, D., 279
Mettenheimer, C., 63
Meyer, E., 43, 294
Midttun, L., 43
Minnaert, F. M., 294
Mitchill, S. L., 106
Möbius, K., 89

Author Citation Index

Moreau, A., 43, 67, 81, 104, 113, 142
Morse, P. M., 294
Moulton, J. M., 43, 216, 310, 332, 350
Mowbray, W. H., 40, 41, 319, 349
Müller, J., 43, 69, 79, 100, 279
Myrberg, A. A., Jr., 43, 52

Nelson, K., 43
Nichols, J. T., 319
Nishimura, M., 42
Nursall, J. R., 43

Owen, R., 114

Packard, A., 43
Parrish, B. B., 43
Parvulescu, A., 52
Pearce, G., 42, 278
Peters, 82, 83
Pfeiffer, W., 43
Pfriem, H., 294
Plattner, W., 216
Poey, F., 86
Poggendorf, D., 279
Popper, A. N., 52
Prosser, C. L., 279
Protasov, V. R., 42, 43, 44, 350
Proudfoot, D. A., 42, 177

Rashcheperin, V. K., 43
Rauther, M., 43, 216
Regan, C. T., 279
Reickel, A., 44
Revel, J. P., 41, 277
Richardson, E. G., 44, 216, 279
Roggenkamp, P. A., 278
Romanenko, E. V., 44
Ryder, J. A., 319

Salmon, M., 44, 52
Sartori, J. D., 52
Schevill, W. E., 44
Schneider, H., 44, 332, 350
Schneirla, T. C., 44
Schroeder, W. C., 319
Schultze, F., 80
Schwarz, A., 53
Sebeok, T. A., 44
Segemehl, M., 60
Shishkova, E. V., 44, 216
Shores, D. L., 40
Skoglund, C. R., 44, 279
Skudrzyk, E. J., 44

Smith, H. M., 44, 142, 150, 158, 279, 350
Snow, D. W., 332
Sørensen, W., 44, 57, 59, 70, 115, 142, 158, 216, 279, 319
Sorgente, N., 44
Spallanzani, L., 80
Spires, J. Y., 40, 52
Stampehl, H., 45
Steinberg, J. C., 45, 350
Stetter, H., 277, 279
Stout, J. F., 45, 46, 217, 332, 350

Tavolga, W. N., 42, 45, 53, 216, 217, 279, 294, 310, 319, 332, 350
Taylor, M., 45
Teal, J. M., 40
Tennent, E., 66
Thilo, O., 159
Thorpe, W. H., 332, 350
Tokarev, A. K., 217
Tower, R. W., 45, 58, 159, 167, 217, 222, 319
Tschiegg, C. E., 45
Tsvietkov, V. I., 43

Uchida, K., 217

Verheijen, F. J., 217
Vigoureux, P., 46, 279
Vincent, F., 46

Walters, V., 46
Warner, L. H., 350
Watkins, W. A., 46
Weber, E. H., 279
Weiss, O., 159
Wenz, G. M., 46
Weston, D. E., 46
White, J., 142
Wienz, G. M., 163
Wilder, B. G., 86
Wing, A. S., 40
Winn, H. E., 40, 41, 42, 44, 46, 53, 217, 332, 349, 350
Wittenberg, J. B., 41
Wodinsky, J., 294
Wood, L., 332
Woods, L. P., 332
Wright, R. R., 279

Young, R. W., 177, 201

Zacharias, O., 89

SUBJECT INDEX

Acipenser, 78
Acoustics, underwater, 6–10, 220, 280–294
Aelurichthys (Bagre), 230
Ageniosus, 229
Aggressive behavior, sounds in, 33, 47, 210, 215, 305, 315, 322–323, 325, 329–331, 339–340, 342, 344–347 (see also Territorial behavior, sounds in)
Ailurichthys (Bagre), 230
Air bladder (see Swim bladder)
Alarm response, 33 (see also Schreckstoff)
Alarm sounds, 5, 35–36, 51, 58, 151–152, 163, 193, 200, 322, 329–331, 338, 342–343, 345
Albula vulpes, 195
Alectis ciliaris, 205
Alewife (see Pomolobus pseudoharengus)
Alpheus, 178
Alutera schoepfii, 194, 198
Amblycepidae, 229
Ameiuridae, 228
Ameiurus, 237, 266–267, 275
Amia, 63, 78, 86, 266–267
Ammodytes americanus, 195
Amphipnous, 88
Amphiprion, 12
Anabantidae, 207
Anacanthini, 202
Analysis, spectral
  of ambient noise, 9, 165–167, 172–175, 180, 183–185
  of fish sounds, 5, 22–30, 38, 191–192, 196–200, 203, 205, 206, 208, 210, 258, 262–263, 269–270, 272–273, 276, 301, 311, 336–337 (see also Spectrograms)
Anchoviella, 209
  A. choerostoma, 203–204, 345
Anchovy (see Anchoviella)
Ancylodon, 108
Angelfish, 178
  black (see Pomacanthus arcuatus; Angelichthys ciliaris)
Angelichthys ciliaris, 207

Anguilla rostrata, 194, 196
Apeltes quadracus, 12, 195
Aplodinotus grunniens, 14, 35, 327, 340–341, 344, 348
Apodes, 204 (see also Eel; Anguilla rostrata)
Arapaima, 78
Ariidae, 206, 229–230, 232, 235, 267
Arius, 229, 233, 234
  A. felis, 51 (see also Galeichthys, felis)
  A. pidada, 232
  A. platystomus, 229, 232
  A. thalassinus, 232
ASDIC, 37, 162 (see also Echolocation)
Attraction of fish by sounds, 37, 38
Auchenipteridae, 206
Auchenipterus, 229
  A. nodosus, 206
Auxis thazard, 195

Bagre, 230, 289
  B. marinus, 12, 16, 19, 205, 226, 229, 236–237, 258, 262–263, 266, 269–274, 276, 337–338, 343
  skull of, 233–234, 243, 250–257
Bagridae, 229–230
Bagroides, 62
Bagrus, 230
Bairdiella chrysura, 123, 145–146, 155–156, 205–206, 210–211
Balistes, 12–13
  B. aculeatus, 89
  B. carolinensis, 194, 197, 205, 207
  B. vetula, 89, 205, 207, 321, 326
Balistidae, 11, 209
Barnacle, 178
Barracuda (see Sphyraena barracuda)
Barrelfish (see Palinurichthys perciformis)
Bass, sea (see Centropristes striatus)
  striped (see Roccus lineatus)
Bathygobius, 215
  B. soporator, 203, 207, 212, 298–310, 338–339, 341, 344–345
Batrachia, 79

355

Subject Index

Batrachoididae, 26, 153, 209
Batrachus grunniens, 101
Behavioral significance of sounds, 5, 39, 47 (see also Aggressive behavior; Reproductive behavior; Schooling; Territory)
Beluga, sturgeon (see *Huso huso*)
Berycomorphi, 203
Bibliography, 5, 37
Blenny, 29, 35, 207, 212, 215
Blueback (see *Pomolobus aestivalis*)
Bodianus costatus, 107
Bonefish (see *Albula vulpes*)
Bones in sound production (see Stridulation)
Botia hymenophysa, 13
Brain, 260
  stimulation of, 50–51
Brevoortia tyrannus, 195, 312
Brosmius brosme, 206
Broutulidae, 16, 17, 215
Bubble, air, 20, 21
Bunocephalidae, 229
Bunocephalus, 337
Burrfish, 188, 200 (see also *Chilomycterus schoepfii*)
Butterfish (see *Poronotus triacanthus*)

Callichrous, 62
Callichthyidae, 229
Callichthys, 81, 88
Cancer, 178
Carangidae, 152, 157, 205, 209
Caranx, 24
  C. crysos, 194, 196, 205
  C. hippos, 152, 205
  C. latus, 203–205, 209, 345, 347
  C. ruber, 203–205, 345
Carassius vulgaris, 87
Cariburus zaniophorus, 214
Cartesian diver, 229
Cat, internal rectus muscle of, 271
Catfish, 4, 12, 14–15, 19, 20, 25–26, 34–36, 57–58, 166, 177–178, 202, 207, 215, 220, 288, 291, 338, 340, 342 (see also under generic and family names; Siluridea; Siluroidea)
  banjo (see *Bunocephalus*)
  gaff-topsail, 27–28 (see also *Bagre marinus*)
  sea (see *Galeichthys, felis*)
Centrarchidae, 47
Centropristes striatus, 194, 196, 199, 207
Ceratodus, 78, 85–86
Chacidae, 228
Chaetodipterus faber, 152, 157, 194, 197

Chaetodontidae, 209
Chanos, 88
Characidae, 35, 61, 202
Carp, 202
Chasmodes bosquianus, 207, 212, 341, 344
Chatoessus jacunda, 88
Chemical sense in water, 31
Chilomycterus schoepfii, 194, 198, 205
  C. spinosus, 153
Chloroscombrus chrysurus, 152, 205
Chorusing (see Schooling, sounds in)
Cichlasoma centrarchus, 47
Cichlidae, 35, 47
Clarias, 63–66, 88, 105
  C. macracanthus, 97
Clariidae, 229
Claustrum, 60, 61, 228 (see also Weberian ossicles)
Clownfish (see *Amphiprion*)
Clupanodon aureus, 88
Clupea harengus, 195
Clupeidae, 194, 204–205
Cobitidae, 61, 87, 207
Cobitis, 79
Codfish, 26, 202 (see also *Gadus*)
Coelorhynchus caribbaeus, 213
Coenotropus labyrinthicus, 88
Communication, defined, 31–32, 35, 49–50 (see also Playback experiments)
  levels of interactions, 32–33
  underwater sound in, 3, 5, 28, 31–37, 47–50, 210, 215, 296–298, 308–309, 311–331, 333–349
Conger conger, 195
Congiopodus, 19
  C. leucopaecilus, 223–224
Corvina, 35
  C. nigra, 327, 329, 340, 344
  C. ronchus, 106
  various species, 111–112
Cottidae, 209
Cottus scorpius, 76
Courtship (see Reproductive behavior, sounds in)
Crab, 178, 317
  blue, 311 (see also *Callinectes*; *Cancer*; *Portunus*)
Crangon, 178, 179
Cranium, catfish, 228–257
Croaker, 119, 165, 177–179, 181, 183, 187, 189, 289, 343 (see also *Micropogon undulatus*; Sciaenidae)
Cryptopterus, 62
Cunner, 199 (see also *Tautoglabrus adspersus*)
Curimatus vittatus, 88

Cycles, behavioral (see Rhythms)
Cynoscion, 210
    C. arenarius, 153, 155
    C. nebulosum, 123, 143–144, 155–156, 206, 211
    C. nothus, 155, 165
    C. regalis, 120–122, 194, 197, 205–206, 327, 340–341
    C. xanthulus, 48
Cyprinidae, 61, 65–66, 205, 207–208, 215
Cypriniformes, 228, 291
Cyprinus carpio, 91
Cyprinus cyprinoides, 88

Dab, sand (see Hippoglossoides platessoides)
Dactylopteridae, 18
Dactylopterus, 17–18
    D. rolitans, 76–77, 124
Damselfish, 48, 50 (see also Eupomacentrus; Pomacentrus)
Dasyatis centroura, 195
Decapterus macarellus, 195
Decibels, 7
Defense, sounds in (see Aggressive behavior; Territorial behavior)
Demoiselles (see Pomacentridae)
Denticles, pharyngeal (see Stridulation)
Depth, effect on sound velocity, 6, 7
Diodon hystrix, 205
Diodontidae, 209
Dipnoi, 86, 87, 108
Directional acoustic sense (see Orientation, acoustic)
Displacement, acoustic, 4, 8–10, 21–22, 29, 31–32, 38, 51, 280–294, 348
Dolphin, 162 (see also Porpoise)
Doradidae, 206, 229, 232, 235
Doras, 88, 116, 226, 229, 269
    D. maculatus, 72–75, 92–94, 115, 205
Drumfish, 35, 119–123, 165, 178, 181, 210, 298 (see also Aplodinotus; Pogonias; Sciaenidae)
Dynes/cm² (see Microbar)

Echeneis naucrates, 195
Echo-ranging (see Echolocation)
Echolocation, 37, 51, 213, 215, 275, 348
Ecology, sound in, 5
Eel, 14 (see also Apodes)
    common, 199 (see also Anguilla)
    conger (see Conger conger)
    moray, 323, 325 (see also Gymnothorax)
Egypt, 56–57
Elasmobranchii, 11, 30, 202

Elastic spring, 4, 16, 18, 20, 64–117, 153, 206, 215, 226, 229, 232–236, 269, 272–276, 289
Electrical sense, 31
Ephippidae, 152, 157
Epinephelus guttatus, 19
    E. striatus, 14–15, 207, 343
Equipment, recording, 190–191, 268–269, 299, 312
Erythrinus, 79, 105
Esocidae, 204
Esophagus, 14
Esox, 103
Euanemus, 229
Eupomacentrus dorsopunicans, 47
    E. fuscus, 166
    E. leucostictus, 48
    E. partitus, 47–48
    E. planifrons, 48

Fallfish (see Semotilus corporalis)
Far-field (see Pressure, acoustic)
Feeding sounds, 11, 202, 209, 215, 343 (see also Stridulation)
Felichthys felis (Arius felis), 167, 226, (Bagre marinus), 155, 226, 230 (see also Bagre marinus; Galeichthys, felis)
Filefish, 188, 203
    orange (see Alutera schoepfii)
Fin spines in sound production (see Stridulation)
Fishing, use of sounds in, 4, 5, 37–38, 56, 128
Flounder, 178
    sand (see Lophopsetta aquosa)
    winter (see Pseudopleuronectes americanus)
Flow, water, 10
Fluids, ovarian, 300
Fluke, northern (see Paralichthys dentatus)
Foolfish (see Monacanthus hispidus)
Frequency analysis (see Analysis, spectral)
    range of hearing, 8 (see also Hearing in fish)
    relation to wavelength, 6
Fright sounds (see Alarm sounds)
Frog, 298

Gadidae, 206, 209
Gadus, 237
    G. aeglefinus, 114, 206
    G. callarias, 206, 327, 329, 340–341, 343–344
    G. morhua, 35, 114
    G. pollachius, 206

Subject Index

G. *virens*, 206
Galeichthys, 27
  G. *feliceps*, 62
  G. *(Arius) felis*, 16, 36, 205, 226, 229, 234–237, 258–262, 264–266, 269–274, 276, 327, 338, 340, 343
  G. *milberti (Arius felis)*, 153–155, 157 (see also *Arius felis*)
  skull of, 232–233, 241–249
Ganoidi, 108, 237
Garibaldi, 178 (see also *Hypsipops rubicundus*)
Gas bubble sounds, 14, 204, 215
Gasterosteus aculeatus, 12, 195
Glandulocauda inequalis, 14
Gobiidae, 209, 302
Gobius jozo, 344
Goby, 29, 35, 296
  frillfin, 333 (see also *Bathygobius soporator*)
Gonostoma javanicum, 88
Gourami, croaking, 341 (see also *Trichopsis vittatus*)
Grand Canyon Suite, 48
Grouper, 20, 26, 35, 178
  black (see *Mycteroperca bonaci*)
  Nassau (see *Epinephelus striatus*; Serranidae)
Grunt, 178, 323, 325 (see also *Haemulon*; Haemulidae; Pomadasyidae)
  blue-striped, 322 (see also *Haemulon sciurus*)
  white (see *Haemulon plumieri*)
Gurnard, 113, 127, 206 (see also Dactylopteridae; *Dactylopterus*)
Gymnarchus, 78
Gymnodontidae, 153
Gymnothorax moringa, 321
Gymnura micrura, 195

Haddock, 26 (see also *Gadus (Melanogrammus) aeglefinus*)
Haemulidae, 58, 149–150, 156, 205, 209 (see also Pomadasyidae)
Haemulon, 11
  H. *album*, 10
  H. *plumieri*, 10, 149–150, 205
  H. *sciurus*, 149, 205, 321
Hagfish, 178
Hake, silver (see *Merluccius bilinearis*)
Haplodoci, 203
Hardtail, 200 (see also *Caranx crysos*)
Harmonic analysis (see Analysis, spectral)
Harmonics, 25, 28
Hearing, in fish, 38, 51, 207, 215, 228, 280, 328–329, 347–348

Herring, 199 (see also *Clupea harengus*; Clupeidae thread; *Opisthonema oglinum*)
Heterotis, 78, 88
Hind, red, 20, 21 (see also *Epinephelus guttatus*)
Hippocampus, 12, 206
  H. *brevirostris*, 89–90, 212
  H. *hudsonius*, 194, 196, 212, 343–344
Hippoglossoides platessoides, 195
Hogfish (see *Orthopristis chrysopterus*)
Holacanthus, 152
  H. *ciliaris*, 321
  H. *tricolor*, 321
Holocentrus, 26–28
  H. *ascensionis*, 19, 207, 320–322, 325–326, 329–330, 343, 348
  H. *coruscus*, 47, 336, 338
  H. *rufus*, 15, 19, 47, 320–331, 336, 338, 340, 343, 345, 348
  H. *sogho*, 89
Homalopteridae, 207
Homeostasis, 32
Hormones, effect on sound production, 208
Huso huso, 29
Hydrodynamic sounds, 10, 21–22, 28–29, 34, 203–204, 215
Hydrophone, 5, 25, 30, 38, 165, 190, 192, 268, 299–300, 320
Hymenocephalus cavernosus, 213, 214
Hyodon claudalus, 88
Hypleurochilus geminatus, 207, 212
Hyporhamphus unifasciatus, 151–152, 157
Hypostomus, 88
Hypsipops rubicundus, 166
Hypsoblennius hentz, 207, 212

Intercalarium, 60, 61, 62, 228 (see also Weberian ossicles)
Identification of sound source, 6, 23, 38, 163, 214
Incus (see Intercalarium; Weberian ossicles)
Insect, flight muscles, 271
Interaction, levels, 32–33

Jacks, 35 (see also Carangidae; *Caranx*)
John Dory, 203 (see also *Zeus faber*)

Kingfish (see *Menticirrhus saxatilis*)

Labrus grunniens, 106
  L. *squeteague*, 106, 128

Lagodon rhomboides, 12, 152
Larimus dentex, 107
   L. fasciatus, 153, 155–156
Lateral line, 10, 32, 329, 347 (*see also* Hearing in fish; Orientation, acoustic)
Launce, sand (*see Ammodytes americanus*)
Leiostomus chrysurus, 145
   L. humeralis, 111
   L. xanthurus, 123, 155–156, 166–167, 210–211
Lepidosiren, 78–79, 83–84
Lepidosteus, 78, 86
Lepomis, 47
Liocassis, 62
Loach (*see Botia hymenophysa*)
Lobster, spiny, 178, 328 (*see also Panulirus argus*)
Lonchurus depressus, 112
Lophopsetta aquosa, 194
Loricaria, 88
Loricariidae, 229
Lota lota, 206
Luminescence, in deep-sea fishes, 213
Lyre, 130–131

Mackerel, cero (*see Scomberomorus regalis*)
   common (*see Scomber scombrus*)
   frigate (*see Auxis thazard*)
   Spanish (*see Scomberomorus maculatus*)
Macrones, 62, 64
Macrouridae, 17, 202, 206, 213, 215
Macrourus fabricii, 206
Malacocephalus laevis, 213
   M. occidentalis, 213–214
Malapteruridae, 206, 235
Malapterurus, 229
   M. electricus, 94–95
Malleus (*see* Tripus; Weberian ossicles)
Margate fish (*see Haemulon album*)
Mechanical sense in water, 31
Mechanisms, sonic, 10–22, 39, 50, 59–117 (*see also* Swim bladder; Stridulation)
Medulla, 50
Melanogrammus aeglefinus, 35
Meletta thryssa, 88
Melichthys piceus, 205, 207
Menhaden (*see Brevoortia tyrannus*)
Menidia, 237
   M. beryllina, 195
   M. menidia notata, 195
Menticirrhus americanus, 155–156
   M. saxatilis, 194, 197, 205
Merluccius bilinearis, 194, 198, 207
Meterological scales, 171
Microbar, 7

Micropogon undulatus, 107, 122–123, 127, 143–144, 155–156, 166, 206, 209–210, 327, 340–341
Midbrain, 51
Midshipman, 178 (*see also Porichthys*)
Mines, acoustic, 162, 208–209
Minnow, 35, 328, 348 (*see also Phoxinus laevis*)
Mobbing response, 329, 330, 331
Mola mola, 194
Molva molva, 206
Monacanthidae, 209
Monacanthus hispidus, 155–157, 194, 197
   M. pardalis, 89
Monomitpus, 16
Moray, 328, 331
   spotted (*see Gymnothorax moringa*)
Moridae, 215
Mormyridae, 204, 207
Mugil, 321
   M. cephalus, 195
Mullerian ramus, 229, 231, 232, 233, 234–236, 266, 274, 276 (*see also* Elastic spring)
Mullet, 323, 325, 328 (*see also Mugil*)
Muscle, sonic, 10, 14–20, 39, 50, 58, 65–70, 72–117, 120–141, 143–148, 152–153, 155–158, 166, 177, 197–199, 206–207, 210–211, 213–214, 220–247, 258–260, 269–271, 274, 276, 335, 341
   contraction rates of, 19, 21, 50 (*see also* Tetanization)
   innervation of (*see* Nerves, to sonic muscles)
   sarcoplasmic reticulum of, 19, 20, 271
   swimming, 204
   tensor tripodes, 229
Muttonfish, 178
Mycteroperca bonaci, 28, 337–338, 343
Myliobatis freminvillii, 195
Myoglobin, 236
Myoxocephalus, 13
   M. octodecimspinosus, 194, 198, 206
Myripristis, 51
   M. jacobus, 47, 321, 338
   M. pralinius, 48
   M. violaceus, 48

Naucrates ductor, 195
Near-field (*see* Displacement, acoustic)
Nebris microps, 111
Negaprion brevirostris, 34
Nematognathi, 228
Nerves, cranial, 230, 232, 234, 237, 260, 266
   hypoglossal, 20

Subject Index

occipital, 20, 232, 237, 258, 260–263, 266–267, 269, 272–273
spinal, 20
spino-occipital (see Nerves, occipital)
to sonic muscles, 17, 19–20, 39, 220, 226, 237, 258, 260–263, 266–267, 269, 272–274
to swim bladder, 139
Nervous system, 50
Neurophysiological experiments, 220, 222–224, 268–275
Nezumia, 213–214
Noise, ambient, 5, 8–9, 23, 48, 162–166, 168–188, 202
flow, 22, 28, 202
levels, 185–187
man-made, 184–185
water motion, 170–177
Notropis, 215
N. analostanus, 208, 329, 336, 339–342, 344, 346
N. lutrensis, 208, 341
N. venustus, 208, 341

Occipital nerves (see Nerves, occipital)
Operculum, in sound production, 13
Ophidium, 206
Ophiocephalus, 88
Opisthonema oglinum, 194, 196
Opsanus, 17, 19, 26, 35, 48–50, 119, 124, 203, 338
O. beta, 36, 210, 312, 337, 343–344
O. tau, 19, 25, 36, 125–127, 151, 166, 192, 194, 198, 206, 210, 222, 271, 298, 311–319, 321, 328–329, 335, 337, 340–341, 344, 346
Orientation, acoustic, 31–32, 36, 51, 204, 275, 298, 328–329, 342, 347–349 (see also Echolocation)
Orthopristis chrysopterus, 150, 156, 166, 205
Osphromenus, 88
Ostariophysi, 20, 60–117, 202, 204, 207, 215, 228, 266
Osteoglossidae, 204
Otolithus, 106, 108, 111
Oyster shell, 312

Palinurichthys perciformis, 195
Palometa, 323, 325, 328 (see also Trachinotus glaucus)
Pangasiidae, 206, 235
Pangasius, 99, 229
Panulirus argus, 321
Paralichthys dentatus, 195
Parrotfish (see Scaridae)

Pectoral girdle, 63–64, 91–94
Pellona lechenaultii, 88
Pelteobagrus, 205
Perca, 103
Perch, 203
Percomorphi, 203
Peristedion cataphractum, 89–91
Pharyngeal teeth (see Stridulation)
Phoxinus laevis, 205, 328
Phycis mediterraneus, 206
Physostomi, 60, 204
Pigfish (see Orthopristis chrysopterus)
  New Zealand (see Congiopodus)
Pike (see Esocidae)
Pilotfish, 194 (see also Naucrates ductor)
Pimelodidae, 66–68, 115, 117, 206, 229
Pimelodina, 98
Pinfish, 35 (see also Lagodon rhomboides)
Pipefish, 302 (see also Syngnathus)
Piramutina piramuta, 99
Piranha, 338 (see also Serrasalmus)
Platystoma, 97, 229
P. orbignyanum, 74–75, 116
P. tigrinum, 97, 99
Playback experiments, 35, 37, 48–49, 208–209, 212, 302–308, 320–322, 325–326, 339, 345 (see also Communication)
Plecostomus villarsii, 97
Plectognathi, 203, 205
Plotosidae, 228
Pneumatic duct, 14, 65–66, 68, 117–118, 120, 153, 204, 226, 264, 273
Pogonias, 114, 158
P. cromis (chromis), 105, 106, 108, 119–120, 128, 131, 155–156, 210–211
P. fasciatus, 106
Polyacanthus, 88
Polycirrus eximius, 312
Polynemus, 76
Polypterus, 63, 78, 86
Pomacanthus arcuatus, 207, 347
Pomacentridae, 166
Pomacentrus (Eupomacentrus) leucostictus, 321, 329, 344
Pomadasyidae, 10, 34 (see also Haemulidae)
Pomolobus, 195
Pompano, 178
Porcus, 230
Porgies (see Sparidae)
Porichthys, 5, 17, 18
P. notatus, 18, 25, 36, 153, 206, 335, 341, 343–344
P. porossissimus, 151, 153
Poronotus triacanthus, 194
Porpoise, 177–178, 181, 184, 329 (see also Dolphin)

Subject Index

Portunus, 178
Predation, 204, 209, 215
Prespawning behavior (see Reproductive behavior)
Pressure, acoustic, 4, 7–9, 22, 29, 31, 38, 204, 280–294, 347
    noise levels, 172–175
Priacanthidae, 11
Prionotus, 17, 18, 77, 119, 147–148, 206, 298, 344, 347
    P. carolinus, 124–126, 167, 194, 198, 343
    P. evolans, 194, 198
    P. punctatus, 153
    P. scitulus, 19
    P. tribulus, 153
Pristipomatidae, 106
Protopterus, 78, 82, 86
Protractor muscle (see Muscle, sonic)
Pseudaroides, 97
    P. clarias, 73–75, 116, 205
Pseudobagrus, 62
Pseudopleuronectes americanus, 195
Pseudosciaena aquila, 131
Puffer, 57, 118, 203 (see also *Spheroides maculatus*)
Pygocentrus piraya, 343

Radiation, acoustic, 280–294
Rain noise, 177
Raniceps raninus, 206
Rat-tail fishes (see Macrouridae)
Ray, 178, 194, 202
    butterfly (see *Gymnura micrura*)
    eagle (see *Myliobatis freminvillii*)
    electric (see *Torpedo marmorata*)
    northern (see *Dasyatis centroura*)
Razorfish, 178
Recording equipment (see Equipment, recording)
Red body (red gland), 18, 119–120, 125, 273–274
Reference level, acoustic, 7, 9
Reflection, acoustic, 6, 23, 51, 272
Reproductive behavior, sounds in, 5, 11, 33–37, 47–49, 51, 128, 207–213, 215, 296, 298–309, 311–319, 326, 333, 338–341, 343–349
Resonance, 20–21
Respiration, 79, 80, 81–89
Rheotaxis, 10
Rhinecanthus rectagulus, 13
Rhyncobdella, 88
Rhythm, behavioral, 36, 47, 165–166, 182, 209, 320–331, 340–341, 345, 348–349
Roccus lineatus, 194, 196, 199
    R. saxatilis, 207

Rudderfish, 200 (see also *Seriola zonata*)

Saccobranchus, 81, 88
Salinity, effect on sound velocity, 6
Salmon, 37
Salmonidae, 204
Sarcoplasmic reticulum, 19–20, 271
Sargo, 178
Scad, 325, 328
    mackerel (see *Decapterus macarellus*)
    wall-eyed (see *Selar crumenophthalmus*)
Scaphium, 60, 61, 228 (see also Weberian ossicles)
Scaridae, 12
Schilbeidae, 228
Schooling, sounds in, 5, 11, 33, 36, 39, 203–204, 208, 327, 343, 348
Schreckstoff, 33
Sciaena adusta, 106
    S. aquila, 89–90, 106, 110–112
    S. fusca, 106
    S. obliqua, 111
    S. ocellata, 107, 111
    S. diacanthus, 111
    S. xanthurus, 111
Sciaenidae, 4–5, 14, 19–20, 26, 35, 48, 76, 101, 105, 119–123, 128–129, 131, 141, 153, 155–158, 205–206, 209, 298, 327, 339, 341, 344
Sciaenops ocellatus, 155–156
Scleroparei, 203
Scomber, 266
    S. brachyurus, 118
    S. scombrus, 195
Scomberomorus maculatus, 195
Scombridae, 194
Scorpaenidae, 15, 209
Sculpin, 19 (see also *Myoxocephalus*)
Scup, 199–200 (see also *Stenotomus chrysops*)
Sea-anemone, 178
Sea bass, 26 (see also Serranidae)
Sea horse, 199, 202 (see also *Hippocampus*)
Sea robin, 19–20, 36, 123–126, 139–141, 166, 178, 181, 184, 199, 339 (see also *Prionotus*; *Trigla*)
    common (see *Prionotus carolinus*)
    red-winged (see *Prionotus evolans*)
Sebasticus, 15, 17
Selachians, 237
Selar crumenophthalmus, 321, 337
Semotilus corporalis, 341, 344
Seriola zonata, 194, 196, 205
Serranidae, 13, 209, 329
Serrasalmus, 50
    S. nattereri, 337

Subject Index

Shad, hickory (see *Pomolobus mediocris*)
Shark, 34, 202, 328 (see also Elasmobranchii)
   lemon (see *Negaprion brevirostris*)
   sucker (see *Echeneis naucrates*)
Sheepshead, 178
Shiner, blacktail (see *Notropis venustus*)
   red (see *Notropis lutrensis*)
   satin fin, 333, 338 (see also *Notropis analostanus*)
Ship, noise of, 184–185
Shrimp, 162
   mantis, 178
   snapping, 177–181, 189, 203, (see also *Alpheus*; *Crangon*; *Synalpheus*)
Signals, acoustic, 334–339, 348–349
Siluridae, 59–117, 229–230
Siluroidea, 15, 61, 205–206, 208, 228
Silurus, 231
   S. glanis, 97, 226
Silverside (see *Menidia menidia notata*)
   waxen (see *Menidia beryllina*)
Sisoridae, 229
Skeletal development, 62–64
Skeletal parts in sound production (see Stridulation)
Snakes, 81, 82
Sonar, 37, 162 (see also Echolocation)
Spadefish, 178, 199 (see also *Chaetodipterus faber*)
Sparidae, 12
Sparisoma, 321
Spatularia, 78
Spawning (see Reproductive behavior)
Spectrograms, 24, 26–27, 29, 262–263, 301, 336–337 (see also Analysis, spectral)
Spectrograph, sound, 23
Spectrum (see Analysis, spectral)
Spheroides maculatus, 194, 198, 205
   S. nephelus, 153
Sphyraena barracuda, 209
Spines, fin, in sound production (see Stridulation)
Spino-occipital nerves (see Nerves, occipital)
Spot, 177–178 (see also *Leiostomus xanthurus*)
Sprat, 204
Springfederapparat (see Elastic spring)
Squeteague, 119, 127, 133–138, 147–148, 178, 199–200 (see also *Cynoscion*)
Squirrelfish, 20–21, 35–36, 48, 51, 203, 220, 297, 342, 348 (see also *Holocentrus*; *Myripristis*)
   longspine, 333 (see also *Holocentrus rufus*)
   reef (see *Holocentrus coruscus*)
Stapes (see *Scaphium*; Weberian ossicles)
Stellifer lanceolatus, 153, 155, 205, 210
Stenotomus chrysops, 194, 196, 343
Stephanolepis hispidus (*Monacanthus hispidus*), 205
Stickleback, 194
   four-spined (see *Apeltes quadracus*)
   two-spined (see *Gasterosteus aculeatus*)
Stress (see Alarm)
Stridulation 25, 118, 157–158, 188, 194, 199–200, 205–206, 215, 335, 345
   fin spines, 12–13, 92–94, 153–155, 157, 197, 226, 337
   skeletal parts, 13, 152, 196–198, 206, 212
   teeth, incisors, 11, 12, 57, 118, 152–153, 155, 157, 196–198, 205
   teeth, pharyngeal, 10–11, 24, 58, 118, 131–132, 149–150, 152, 155–158, 166, 194, 196–197, 205, 209–210
Sturgeon, beluga (see *Huso huso*)
Sunfish, 48 (see also Centrarchidae; *Lepomis*)
   giant (see *Mola mola*)
Surf noise, 176
Surgeonfish, 178
Swellfish, 118
Swim bladder
   auditory function, 228, 264–265 (see also Hearing in fish; Weberian ossicles)
   gas pressure in, 20, 138
   hydrostatic function, 18, 66, 100, 102–104, 119, 205, 229
   respiratory function, 78–88
   sound production by, 4, 24–28, 34, 58, 70–77, 89–91, 95–99, 105–116, 127–139, 157–158, 166, 177, 194–199, 202, 220–221, 226, 264–265, 276, 333 (see also Elastic spring; Muscle, sonic)
   extrinsic musculature, 14–17, 50, 65–70, 74, 107–112, 116–122, 143–146, 152–153, 155, 206–207, 210–211, 213–215, 223–224, 320, 341
   intrinsic musculature, 17–18, 58, 77, 123–127, 153, 206, 210, 222, 311, 335
   gas expulsion, 14, 65–67, 70, 113, 204–205
   mechanisms and acoustics, 18–21, 73–74, 132–141, 147–148, 268–275, 288–293
   resonance, 10–11, 13, 21, 58, 149, 152, 156, 223–224
Swimming sounds, 21–22, 28–29, 203–204, 345 (see also Hydrodynamic sounds)
Synalpheus, 178, 179
Synentognathi, 151, 157
Syngnathidae, 209

Subject Index

Syngnathus fuscus, 195
   S. louisianae, 13, 152
Synodontidae, 206, 235
Synodontis, 98, 101, 226, 229
   S. schal, 205

Tachysuridae (Ariidae), 230
Tautog, 199 (see also *Tautoga onitis*)
Tautoga onitis, 194, 197, 207
Tautoglabrus adspersus, 194, 197
Teeth, 131
   in sound production (see Stridulation)
Temperature
   effects on sonic behavior, 36, 208
   effects on sound velocity, 6
Temperature variations, 314
Territorial behavior, sounds in, 11, 33, 35, 47–48, 50, 296–298, 305, 311–331, 338–340, 342–346
Tetanization, of sonic muscles, 19–20, 50, 220, 270–271
Tetradon, 57
   T. fahaka, 89
Tetraodontidae, 209
Tetraodontiformes, 12
Therapon, 5, 15, 19, 25, 206
   T. jarbua, 343
Tilapia mossambica, 47, 50
Toad, 298
Toadfish, 18–21, 27, 35, 48, 50–51, 58, 123, 139, 141, 162, 177–178, 181–182, 184, 199, 203, 220, 296, 325, 328, 333, 339, 348 (see also *Opsanus*)
Toadfishes (see Batrachoididae)
Torpedo marmorata, 29
Trachinotus glaucus, 321
   T. palometa, 204
   T. biaculeatus, 89
   T. brevirostris, 89, 206
Trichomycteridae, 229
Trichopsis vittatus, 5, 338, 344
Triggerfish, 13–14, 26, 35, 178, 188, 203, 207, 215 (see also *Balistes*; Balistidae; *Melichthys*)
Trigla, 4, 17–18, 76–77, 102, 114, 206, 224
   T. hirundo, 90
   T. lyra, 89–91
Triglidae, 106, 124, 129, 153, 209
Tripus, 60–61, 74, 98, 228, 234, 275 (see also Weberian ossicles)
Trout, 128, 178
   bastard, 184 (see also *Cynoscion nothus*)
Turbulence, 22

Umbrina canariensis, 107
   U. cirrhosa, 89–90, 106, 112–113
   U. ronchus, 107

Velocity, acoustic, 6–7, 283, 293
   particle, 280–294 (see also Displacement, acoustic)
Vertebrae, 60–64 (see also Stridulation)
Video-acoustic installation, 6, 38, 47
Vision in water, 31
Vomer setapinnis, 152, 205

Wallago, 233
   skull of, 230–232, 234, 238–240, 276
Warfare, underwater, 5
Wavelength, relation to frequency, 6
Weakfish, 128 (see also *Labrus squeteague*)
Weather, effects on noise, 170–177
Weberian ossicles, 13, 57, 60–66, 68, 99, 207, 228, 231, 328
World War II, 162, 164, 168–169, 188, 208
Worm, blood (see *Polycirrus eximius*)

Xenopus, 329

Yellow-tail (see *Bairdiella chrysura*)

Zeidae, 209
Zeomorphi, 203
Zeus, 17
   Z. faber, 124, 206

About the Editor

WILLIAM N. TAVOLGA is Professor in the Departments of Biology and Psychology at the City University of New York, and he has been teaching biology at the City College since 1946. He has been Research Associate in the Department of Animal Behavior at the American Museum of Natural History since 1954, and is also currently Senior Research Associate at the Mote Marine Laboratory, Sarasota, Florida. After completing his undergraduate work at the City College, he received his Ph.D. at New York University in 1949, where he worked under Professors Charles M. Breder, Jr., and Roberts Rugh.

He was awarded a Public Health Research Fellowship in 1954–1955 and a Guggenheim Fellowship in 1967–1968. In 1974, he was awarded an Erskine Fellowship by the University of Canterbury, Christchurch, New Zealand, where he gave a series of lectures in animal behavior and animal communication. In addition, he has been principal investigator for grants and contracts from the National Science Foundation and the Office of Naval Research. He is a member of several scientific societies, including the Animal Behavior Society, of which he is a Fellow.

His primary research interests are in animal communication, bioacoustics, and behavior in marine organisms. He has over fifty scientific publications in these areas, as well as in the fields of embryology, parasitology, ecology, and ichthyology. He is author of a textbook in animal behavior and editor of two volumes on marine bioacoustics.

Dr. Tavolga has traveled widely throughout the world, and lectured in most of the areas he has visited. In addition to his scientific pursuits, he has many outside interests, comprising such disparate subjects as enology and amateur radio.

THE LIBRARY
ST. MARY'S COLLEGE OF MARYLAND
ST. MARY'S CITY, MARYLAND 20686